Regular Complex Polytopes

H. S. M. COXETER
LL.D., D.MATH., D.SC., F.R.S., F.R.S.C.
Professor of Mathematics in the University of Toronto

CAMBRIDGE UNIVERSITY PRESS

Published by the Syndics of the Cambridge University Press
Bentley House, 200 Euston Road, London NW1 2DB
American Branch: 32 East 57th Street, New York, N.Y. 10022

Library of Congress Catalogue Card Number: 73-75855

First published 1974

ISBN: 0 521 20125 X

Printed in Great Britain
at the University Printing House, Cambridge
(Brooke Crutchley, University Printer)

Regular Complex Polytopes

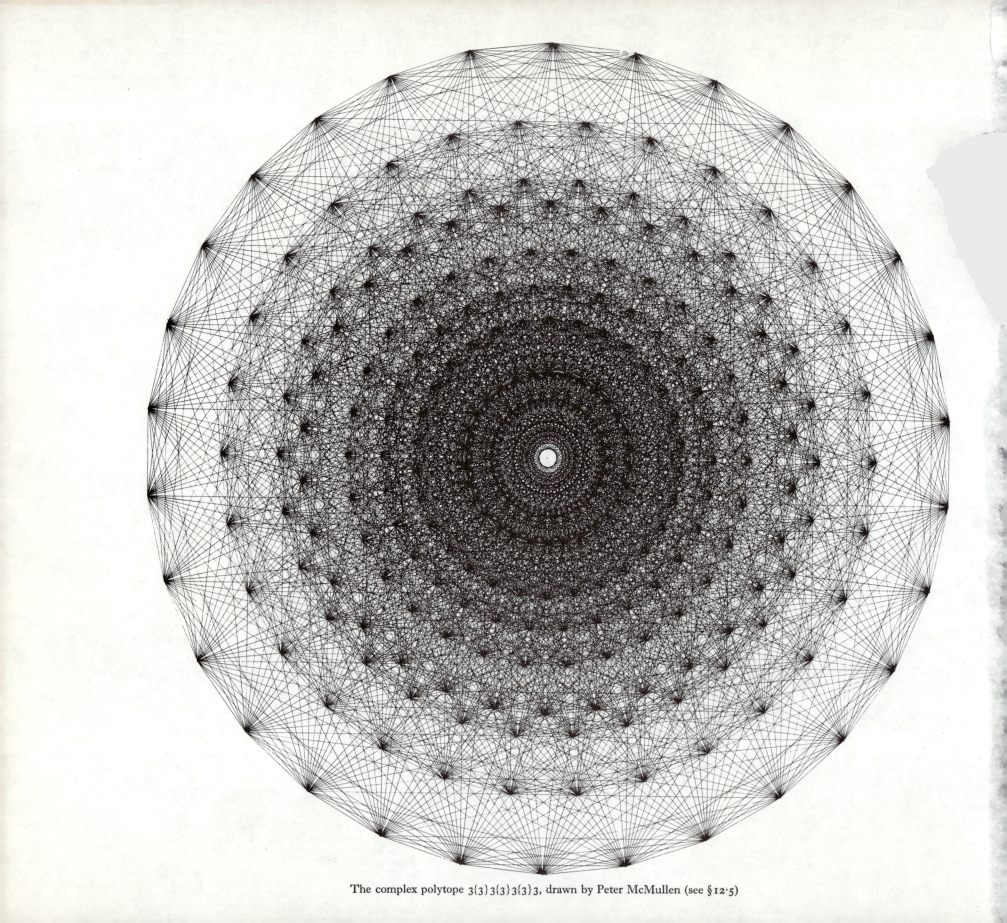

The complex polytope $3\{3\}3\{3\}3\{3\}3$, drawn by Peter McMullen (see §12·5)

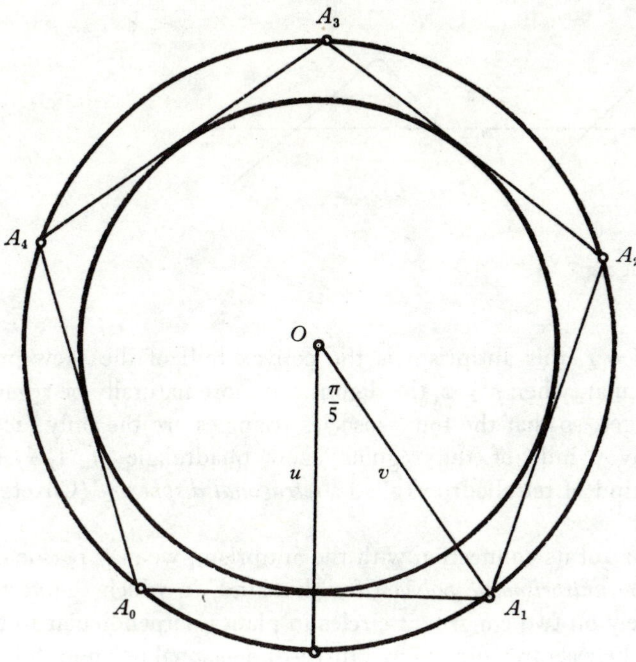

Figure 1·6A: The pentagon {5}

angles $2d\pi/p$ at the centre. For such a polygon, inscribed in the unit circle $|z| = 1$, we may describe the νth vertex as being given by the complex number $\exp(2\nu d\pi i/p)$.

For some purposes it is convenient to include, in the list of polygons, the *digon* {2}, which has two vertices and two coincident edges.

Let d points $P_1, P_2, \ldots, P_d$ be taken alternately on two rays forming an angle π/p at O, in such a way that the triangle $OP_{\nu-1}P_\nu$ has an angle

$$\left(\frac{1}{2}-\frac{\nu}{p}\right)\pi \quad \text{at } P_\nu \quad (\nu = 1, \ldots, d),$$

as in Figure 1·6B (where $d = 5$ and $p = 12$). A practical kaleidoscope can be made by drawing this figure on a horizontal table-top and letting two hinged mirrors stand upright on the two rays. When d overlapping triangles $OP_{\nu-1}P_\nu$ (cut out from a sheet of translucent plastic) have been placed between the mirrors, the whole polygon $\{p/d\}$ will be seen. Its vertices are images of P_d, and its edges are images of the zigzag $P_0P_1\ldots P_d$, which has equal 'angles of incidence and reflection' at $P_1, \ldots, P_{d-1}$.

EXERCISES

1. For a planar polygon $A_0A_1A_2\ldots$, in which both U and V are reflections, the mirror for V is, as we have seen, the bisector of $\angle A_0A_1A_2$. Where is the mirror for U?

2. Obtain expressions for the circumradius and inradius of a $\{p/d\}$ of edge $2l$. What would be a sensible definition for its area?

3. How much of this section remains valid in hyperbolic geometry?

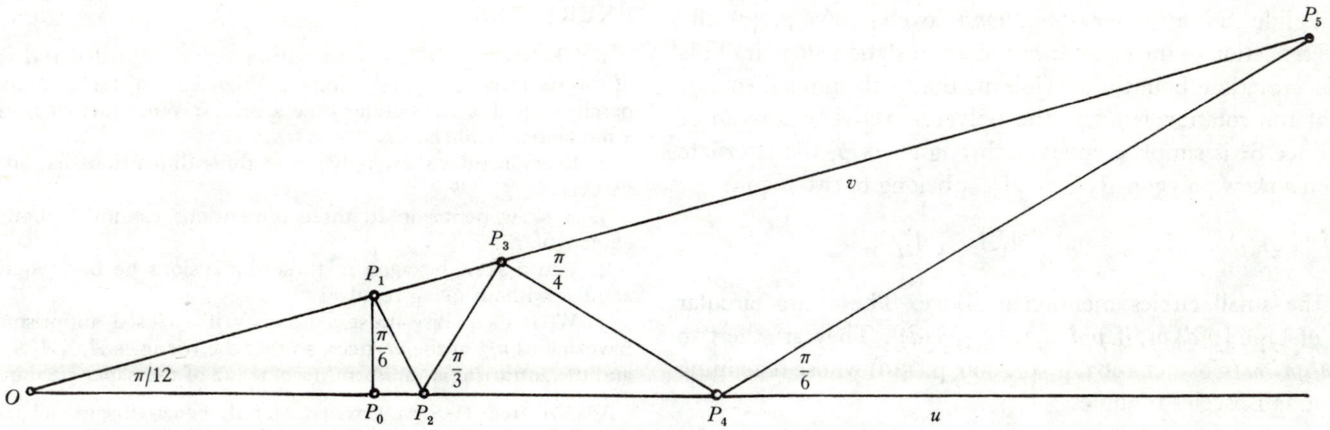

Figure 1·6B: The kaleidoscope for $\mathfrak{D}_{12}$

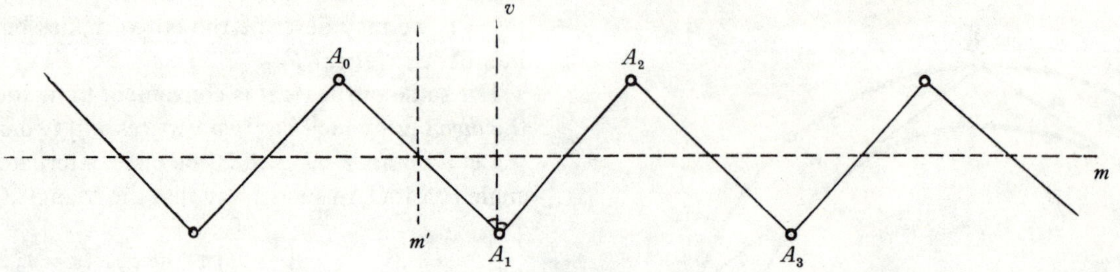

Figure 1·7A: A regular zigzag

1·7 ZIGZAGS AND ANTIPRISMATIC POLYGONS

Continuing the classification of two-dimensional regular polygons, we still have to consider the case when U is a half-turn (namely, the half-turn that interchanges A_0 and A_1) while V is a reflection. Expressing U as the product RR′ of reflections in two perpendicular lines *m* and *m*′, we may choose *m* to be perpendicular to *v* (the mirror for V; see Figure 1·7A). Then

$$S = UV = RR'V = RT,$$

where R is the reflection in *m* while T is a translation along *m*. In other words, S is a *glide* (the kind of isometry that relates successive footprints in a straight walk through snow), and the polygon $A_0A_1A_2\ldots$ is a *regular zigzag*.

When we carry out a similar procedure on the unit sphere, *m* is a great circle (equator) while *m*′ and *v* are two other great circles (meridians) perpendicular to *m*. The polygon is a *spherical zigzag* whose vertices lie alternately on two small circles like the Tropics of Cancer and Capricorn. The 'spherical glide' S is a *rotatory-reflection* (Coxeter 1963, p. 37), the product of the reflection in the equator *m* and a translation along it. This 'translation' is a rotation about the axis joining the north and south poles. If the angle of the rotation is $d\pi/p$, the polygon is a skew $2p$-gon of 'density' *d*. Since S^2 is simply a rotation through $2d\pi/p$, the alternate vertices of such a skew polygon $A_0A_1\ldots A_{2p-1}$ belong to two *p*-gons

$$A_0A_2\ldots A_{2p-2} \quad \text{and} \quad A_1A_3\ldots A_{2p-1},$$

inscribed in the small circles mentioned above. These two circular polygons are of type $\{p/d\}$ or, if $p/d < 2$, $\{p/(p-d)\}$. They are the two bases of an *antiprism* (Coxeter 1963, p. 4; 1969, p. 149) whose remaining faces consist of $2p$ isosceles triangles

$$A_0A_1A_2,\ A_1A_2A_3,\ \ldots,\ A_{2p-2}A_{2p-1}A_0,\ A_{2p-1}A_0A_1.$$

When $d = 1$, this antiprism is the convex hull of the skew polygon. In particular, when $p = 2$, the digons can more naturally be regarded as mere edges, so that the four isosceles triangles are the only faces, and the convex hull of the regular skew quadrangle $A_0A_1A_2A_3$ is a special kind of tetrahedron called a *tetragonal disphenoid* (Coxeter 1963, p. 15).

Because of its connection with the antiprism, we may reasonably use the name *antiprismatic polygon* for this kind, in which the vertices lie alternately on two congruent circles in planes perpendicular to the line joining their centres. For such a three-dimensional polygon, S is a rotatory–reflection; U and V are a half-turn and a reflection. There is just one invariant point, where the axis for the half-turn meets the mirror for the reflection. When *p* is odd and *d* even, as on the right in Figure 1·7B.[†] the edges of the skew $2p$-gon may alternatively be described as the diagonals of the *p* rectangular 'side' faces of a prism whose base is a $\{p/\tfrac{1}{2}d\}$.

EXERCISES

1. The six edges of any tetrahedron may be regarded as diagonals (one each) of the six faces of a parallelepiped. What kind of tetrahedron arises when the parallelepiped is rectangular (like a brick)? What further restriction will yield a tetragonal disphenoid?
2. Every regular skew polygon in three dimensions has an even number of vertices.
3. A skew pentagon in three dimensions cannot be both equilateral and equiangular.
4. Can a skew hexagon in three dimensions be both equilateral and equiangular without being regular?
5. What inequality must p/d satisfy if a closed antiprismatic polygon can have angles $\pi/3$ at the vertices, so that the triangles $A_{\nu-1}A_\nu A_{\nu+1}$ are equilateral and the antiprism is 'uniform' (see p. 422 of the paper cited in the footnote).

† Adapted from H. S. M. Coxeter, M. S. Longuet-Higgins and J. C. P. Miller, Uniform Polyhedra, *Philosophical Transactions of the Royal Society* A **246** (1954), 401–50; see especially Figures 34, 35, 117, 118.

To RIEN
who never abandoned hope that the
book would eventually be completed

By the same author

INTRODUCTION TO GEOMETRY
Wiley, New York

PROJECTIVE GEOMETRY
University of Toronto Press

THE REAL PROJECTIVE PLANE
Cambridge University Press

NON-EUCLIDEAN GEOMETRY
University of Toronto Press

TWELVE GEOMETRIC ESSAYS
Southern Illinois University Press

REGULAR POLYTOPES
Dover, New York

THE FIFTY-NINE ICOSAHEDRA
(with P. Du Val, H. T. Flather, and J. F. Petrie),
University of Toronto Press

GEOMETRY REVISITED
(with S. L. Greitzer), Random House, New York

GENERATORS AND RELATIONS FOR DISCRETE GROUPS
(with W. O. J. Moser), Springer, Berlin

MATHEMATICAL RECREATIONS AND ESSAYS
(with W. W. Rouse Ball), University of Toronto Press

Contents

Preface

This book has occupied much of my time and attention for nearly twenty years. It was inspired by the dissertation of G. C. Shephard, which had the same title.[†] I have made an attempt to construct it like a Bruckner symphony, with crescendos and climaxes, little foretastes of pleasure to come, and abundant cross-references. The geometric, algebraic and group-theoretic aspects of the subject are interwoven like different sections of the orchestra.

Its relationship to my earlier *Regular Polytopes* ('Coxeter 1963' in the Bibliography on page 180) resembles that of *Through the Looking-Glass* to *Alice's Adventures in Wonderland*. The sequel is more profound; it is essentially self-contained, but some of the same characters reappear with recognizable but slightly changed names, and there are many new characters of the same sort, but even more fantastic.

The term *complex polytope* was first used by D. M. Y. Sommerville (1929, p. 98). His 'complex polygon' may have more than two vertices on an edge, or more than two edges at a vertex. He seems to have used the word 'complex' in its colloquial sense without noticing how natural the idea becomes when the coordinates are *complex numbers* $x+yi$ and a Hermitian form is used to define a unitary metric. For instance, there is a generalized octahedron β_n^p whose pn vertices ($p \geqslant 2$, $n \geqslant 1$)[‡] are obtained by permuting the n coordinates $(\epsilon^\nu, 0, 0, \ldots, 0)$, where ϵ is a primitive pth root of unity and $0 \leqslant \nu < p$.

In spite of its bold n-dimensional beginning, Chapter 1 is chiefly concerned with plane and solid *kinematics* of the kind that could interest architects and engineers as well as crystallographers. Chapter 2 introduces some concepts of spherical trigonometry: a sadly neglected subject appearing again in Chapter 3, where the functions

$$\alpha = \tan^2 A, \quad \beta = \tan^2 B, \quad \gamma = \cot^2 b, \quad \delta = \tan^2 c, \quad \epsilon = \cot^2 a$$

of the five 'parts' A, B, b, c, a of a right-angled spherical triangle are arranged as a *frieze pattern*

$$\begin{array}{ccccccccccccccccc}
1 & & 1 & & 1 & & 1 & & 1 & & 1 & & 1 & & 1 & & \\
 & \alpha & & \beta & & \gamma & & \delta & & \epsilon & & \alpha & & \beta & & \ldots & \\
\gamma & & \delta & & \epsilon & & \alpha & & \beta & & \gamma & & \delta & & \epsilon & & \\
 & 1 & & 1 & & 1 & & 1 & & 1 & & 1 & & 1 & & \ldots &
\end{array}$$

[†] Shephard, *Proceedings of the London Mathematical Society* (3), **2** (1952), 82–97.
[‡] Coxeter, *Proceedings of the London Mathematical Society* (2), **41** (1936), 287.

Chapter 3 includes practical instructions for making an *icosahedral kaleidoscope*: a fascinating toy that shows 120 replicas of any small object, the way Sir David Brewster's ordinary kaleidoscope shows 6.

Chapters 4 and 5 provide a new approach to the sixteen regular polytopes in real space of four dimensions. In Chapters 6 and 7, geometric ideas are used for enumerating the finite multiplicative groups of *quaternions*, and these groups are seen to have elegant presentations in terms of two generators and two relations. There is an interesting connection with groups of 2×2 matrices over the field of residues modulo p.

Chapter 8 is an introduction to the analytic geometry of *unitary n*-space, using vectors, Hermitian forms, and inner products. Two-dimensional unitary transformations are investigated more intensively in Chapter 9, with the aid of quaternions, in a manner suggested by Patrick Du Val and D. W. Crowe. A simple criterion (9·42) is found to distinguish reflections from other unitary transformations. This leads to concise presentations for the groups generated by two or three unitary reflections.

The complete list of finite reflection groups in unitary n-space was compiled in 1954 by Shephard and Todd, who found that there are many more of them in the plane than in any higher space. Chapter 10 checks their results (in the two-dimensional case) by a new method: examining all the finite groups of unitary transformations and picking out those that are generated by reflections. In particular, those that are generated by *two* reflections are the symmetry groups of the regular complex polygons. These (including *star* polygons[§]) are enumerated in Chapter 11. Somewhat surprisingly, it is possible to make real drawings of these imaginary figures, and in many cases such a drawing of one complex polygon serves as a *Cayley diagram* for the symmetry group of another.

Chapters 12 and 13 deal with regular polytopes and honeycombs, using definitions suggested by Peter McMullen. There are interesting connections with certain projective configurations such as the 27 lines on the cubic surface. A remarkable presentation (13·83) is found for the simple group of order 25920.

My own chief contribution was, perhaps, the extended Schläfli symbol

$$p_1\{q_1\}p_2 \cdots \{q_{n-2}\}p_{n-1}\{q_{n-1}\}p_n,$$

[§] See p. 92 of Shephard's paper cited in the first footnote

where the p's are integers and the q's are rational numbers (usually 'small' integers). For instance, in this notation β_n^p is

$$2\{3\}2\ldots\{3\}2\{4\}p.$$

Most of the sections end with Exercises, some easy, some challenging. Answers are sketched at the end of the book.

I am grateful to many universities on several continents for giving me opportunities to lecture on various portions of this material between 1966 and 1972.

I offer most cordial thanks to J. H. Conway, D. W. Crowe, P. Du Val, G. C. Shephard and J. A. Todd for help and inspiration, and above all to P. McMullen and J. G. Sunday for reading the whole manuscript and making valuable suggestions. McMullen's M.Sc. thesis (Birmingham, 1966) was closely related to this work. He and B. B. Phadke kindly drew most of the figures. In particular, the frontispiece is McMullen's drawing of the four-dimensional complex polytope

$$3\{3\}3\{3\}3\{3\}3,$$

which has 240 vertices and 2160 '3-edges' (appearing as equilateral triangles).

Du Val kindly allowed me to reproduce the three parts of Figure 2·4A from his monograph of 1964.

Finally, I would express my gratitude to the staff of the Cambridge University Press for their courteous and efficient handling of many problems that inevitably arose.

University of Toronto H. S. M. Coxeter
August 1973

CHAPTER 1

Regular polygons

Θεὸς ἀεὶ γεωμετρεῖ
(God is always doing geometry)
Plato

Since the vertices of an ordinary regular polygon are cyclically permuted by a rotation, it is natural to ask what will happen when the rotation is replaced by an arbitrary isometry. Just as the rotation is the product of two reflections, so the general isometry is the product of two involutory isometries. We still have a dihedral group and a cyclic subgroup. Some of toe regular skew polygons will reappear in Chapters 2 and 4 as Petrie polygons of regular polytopes. The planar polygons will reappear in Chapter 11 as special cases of regular complex polygons.

1·1 ISOMETRIES

In Euclidean space of n dimensions, an *isometry* (or 'congruent transformation') is a point-to-point transformation (of the whole space onto itself) preserving distance. A *reflection* is an isometry that leaves invariant every point on one $(n-1)$-flat or *hyperplane*, the mirror, and interchanges the two half-spaces into which this hyperplane decomposes the whole space. Any isometry is expressible as the product of at most $n+1$ reflections (Coxeter 1963, p. 213). It is *direct* (sense-preserving) or *opposite* (sense-reversing) according as the number of reflections is even or odd. An isometry that leaves at least one point invariant is called an *orthogonal transformation*; it is the product of at most n reflections.

A transformation is said to be *involutory* if it is of period 2, that is, if it is equal to its own inverse without being the identity. An involutory isometry S interchanges each non-invariant point A_0 with its image A_1 and thus leaves invariant the midpoint O of A_0A_1. If it leaves no other point invariant, O is also the midpoint of B_0B_1, where B_0 and B_1 are any other pair of corresponding points. Thus S is the *central inversion* in O, reversing the direction of every line through O. In particular, S reverses each of n mutually perpendicular lines through O, and thus has the same effect as the product of reflections in n mutually perpendicular hyperplanes, each spanned by $n-1$ of these n lines. (This 'central inversion' has no connection with 'inversion' in a circle or sphere.) If, on the other hand, O is not the only invariant point for the involutory isometry S, the set of all invariant points is a flat subspace, say an $(n-r)$-flat. Completely orthogonal to this $(n-r)$-flat is an r-flat (through O) in which S induces an involutory isometry having O for its only invariant point. This isometry is, as we have seen, the product of reflections in r mutually perpendicular $(r-1)$-flats of the r-flat. These mirrors are sections, by the r-flat, of r hyperplanes of the n-space, each containing the invariant $(n-r)$-flat. Hence

(1·11) *Every involutory isometry can be expressed as the product of reflections in r mutually perpendicular hyperplanes, where* $0 < r \leqslant n$.

When $r = 1$, the isometry is simply a reflection; when $r = 2$, a half-turn; when $r = n$, a central inversion.

Some authors use the term 'reflection' for any opposite isometry, and 'rotation' for any direct isometry (Du Val 1964, pp. 8, 35). Others speak of 'reflections' whose mirrors are subspaces other than hyperplanes (Bachmann 1959, p. 1); for instance, they call a central inversion 'reflection in a point'. In the present work it is convenient to use the word *reflection* only for the 'simple reflection' whose mirror is a hyperplane, and *rotation* for the 'simple rotation' which is the product of reflections in two intersecting hyperplanes.

EXERCISES[†]

1. The product of two reflections is a rotation or a translation according as the two mirrors do or do not intersect. The angle of the rotation or distance of the translation is twice the angle or distance between the mirrors.
2. The theory of orthogonal transformations is the same in non-Euclidean geometry as in Euclidean. What complication arises in hyperbolic geometry when we consider the product of two reflections whose mirrors have no common point?

1·2 THE CYCLIC AND DIHEDRAL GROUPS

For any positive integer p, there is a *cyclic group* $\mathfrak{C}_p$, of order p, generated by an element S of period (or 'order') p. This generator may conveniently be represented by a rotation through $2\pi/p$. The elements of the group consist of the identity 1 (sometimes denoted by E) and the rotations $S, S^2, \ldots, S^{p-1}$. The equation

(1·21) $$S^p = 1$$

[†] Whenever an exercise is phrased as a statement, we understand that it is a theorem to be proved. The omission of the words 'show that' or 'prove that' saves space.

is called a *presentation* of the abstract group $\mathfrak{C}_p$ (Coxeter and Moser 1972, p. 1). (In this context, it is tacitly understood that $S^q \neq 1$ for $0 < q < p$.) By a natural extension of these ideas, the free group with one generator (whose presentation is vacuous) is called the *infinite cyclic group* and is denoted by $\mathfrak{C}_\infty$.

Any isometry that leaves a figure invariant as a whole (while possibly permuting parts of the figure) is called a *symmetry operation* (or simply, a 'symmetry'). The various symmetry operations of a given figure evidently form a group; this is called the *symmetry group* of the figure. For instance, $\mathfrak{C}_1$ and $\mathfrak{C}_2$ are the symmetry groups of the letters F and N; $\mathfrak{C}_3$ is the symmetry group of the triquetra, the three-legged symbol for Sicily and the Isle of Man; and $\mathfrak{C}_4$ is the symmetry group of the swastika (Weyl 1952, p. 66).

The rotation S (through an angle $2\pi/p$) is expressible as the product $S = R_1R_2$ of two reflections whose mirrors form an angle π/p (Coxeter 1969, p. 33). Such a pair of mirrors is called a *dihedral kaleidoscope*. The reflections R_1 and R_2 generate a group of order $2p$ in which $\mathfrak{C}_p$ occurs as a subgroup of index 2. This is called the *dihedral group* of order $2p$, and is denoted by $\mathfrak{D}_p$ (Coxeter and Moser 1972, p. 6). $\mathfrak{D}_1$ and $\mathfrak{D}_2$ are the symmetry groups of the letters A and H, $\mathfrak{D}_3$ is the symmetry group of the equilateral triangle, and $\mathfrak{D}_4$ is that of the square.

Since there is only one abstract group of order 2, the groups $\mathfrak{C}_2$ and $\mathfrak{D}_1$ are *isomorphic*,

$$\mathfrak{C}_2 \cong \mathfrak{D}_1,$$

even though one is generated by a half-turn and the other by a reflection. On the other hand, the two groups $\mathfrak{C}_4$ and $\mathfrak{D}_2$, of order 4, are not only geometrically but abstractly distinct, as we can see at once by observing that only the former contains elements of period 4.

In a dihedral kaleidoscope with $p > 2$, an arbitrary point on the mirror for R_2, and the perpendicular drawn from this point to the mirror for R_1, are transformed into the vertices and edges of a regular p-gon $\{p\}$, which has $\mathfrak{D}_p$ for its complete symmetry group. One edge is reversed by the reflection R_1; other edges are reversed by the conjugate reflections

$$R_2R_1R_2, \quad R_1R_2R_1R_2R_1, \ldots.$$

One vertex-angle is reversed by the reflection R_2; other vertex-angles are reversed by the conjugate reflections

$$R_1R_2R_1, \quad R_2R_1R_2R_1R_2, \ldots.$$

If p is even, these two sequences of $\frac{1}{2}p$ reflections are distinct; but if p is odd, the second sequence is merely the first written backwards. In either case, the $2p$ elements of $\mathfrak{D}_p$ consist of p rotations (forming the subgroup $\mathfrak{C}_p$) and p reflections (forming the other coset of this subgroup).

The reflection R_1 transforms the generator $S = R_1R_2$ of $\mathfrak{C}_p$ into the reversed rotation

$$R_1 . R_1R_2 . R_1 = R_2R_1 = S^{-1}.$$

From $\mathfrak{C}_p$, with its obvious presentation (1·21), we can derive a presentation for the abstract group $\mathfrak{D}_p$ by adjoining an element R_1 such that

$$R_1^2 = 1, \quad R_1SR_1 = S^{-1}.$$

Thus $\mathfrak{D}_p$ has the presentation

$$R_1^2 = S^p = (R_1S)^2 = 1$$

or, in terms of $R_2 = R_1S$,

$$(1.22) \qquad R_1^2 = R_2^2 = (R_1R_2)^p = 1,$$

which is clearly equivalent to

$$(1·221) \qquad R_1^2 = R_2^2 = 1, \quad R_1R_2R_1 \ldots = R_2R_1R_2 \ldots$$

with p R's on each side of the last equation. When p is even, this means

$$R_1^2 = R_2^2 = 1, \quad (R_1R_2)^{p/2} = (R_2R_1)^{p/2}.$$

When p is odd, *two* relations suffice:

$$(1·23) \qquad R_1^2 = 1, \quad R_1R_2R_1 \ldots R_2R_1 = R_2R_1R_2 \ldots R_1R_2.$$

For, since $R_1(R_2R_1)^{(p-1)/2} = (R_2R_1)^{(p-1)/2}R_2$, the two generators are conjugate and must have the same period.

Similarly, the *infinite dihedral group* $\mathfrak{D}_\infty$, generated by reflections in two parallel mirrors, has the presentation

$$(1·24) \qquad R_1^2 = R_2^2 = 1.$$

Its subgroup $\mathfrak{C}_\infty$, of index 2, is generated by the product R_1R_2, which is now a *translation*.

EXERCISES

1. What figure is seen when a horizontal right-angled triangle MNO is placed between two vertical mirrors standing on the perpendicular lines OM and ON?
2. The generators of the group $\mathfrak{D}_3$, defined by

$$R_1^2 = 1, \quad R_1R_2R_1 = R_2R_1R_2,$$

are represented by the matrices

$$R_1 = \begin{bmatrix} 1 & 0 \\ 1 & -1 \end{bmatrix}, \quad R_2 = \begin{bmatrix} -1 & 1 \\ 0 & 1 \end{bmatrix}$$

or equally well by

$$R_1 = \begin{bmatrix} 0 & \omega^2 \\ \omega & 0 \end{bmatrix}, \quad R_2 = \begin{bmatrix} 0 & \omega \\ \omega^2 & 0 \end{bmatrix},$$

where $\omega = \exp(2\pi i/3)$.

3. Find generating matrices for $\mathfrak{D}_\infty$.

1·3 THE THEOREM OF LEONARDO DA VINCI

According to Weyl (1952, pp. 64–5, 98) it was Leonardo da Vinci who discovered the following theorem:

(1·31) *The only finite groups of isometries in the real plane are the cyclic and dihedral groups,* $\mathfrak{C}_p$ *and* $\mathfrak{D}_p$ ($p = 1, 2, \ldots$).

To prove this, we observe first (Coxeter 1963, p. 44) that every finite group of isometries leaves invariant at least one point, namely the centroid of all the images of any given point. (These images are merely permuted by each of the isometries.) In other words, every finite group of isometries is a group of orthogonal transformations. Since every orthogonal transformation in the plane is either a reflection or a rotation, and since rotations and reflections multiply like positive and negative numbers (e.g. the product of two reflections is a rotation), the group either consists entirely of rotations or contains equally many rotations and reflections. (The result of multiplying all the rotations and reflections by any one of the reflections is to reproduce all the reflections and rotations, respectively.) Since the only groups having more than one invariant point are $\mathfrak{C}_1$ and $\mathfrak{D}_1$, any rotations that occur all have the same centre, and their angles are multiples of a smallest non-zero angle $2\pi/p$. Thus the only groups consisting entirely of rotations are the cyclic groups $\mathfrak{C}_p$. Finally, if there are equally many rotations and reflections, the group, being derived from $\mathfrak{C}_p$ by adjoining a reflection, is the dihedral group $\mathfrak{D}_p$.

EXERCISE

For what famous buildings is the symmetry group essentially $\mathfrak{D}_4$, $\mathfrak{D}_5$, $\mathfrak{D}_9$?

1·4 THE PRODUCT OF TWO INVOLUTORY ISOMETRIES

It has been known since 1891 that, in real Euclidean n-space, any given isometry S can be expressed as the product of q rotations, r reflections and t translations, mutually commutative, where

$$0 \leqslant q \leqslant \tfrac{1}{2}n, \quad r = 0 \text{ or } 1, \quad t = 0 \text{ or } 1, \quad 2q+r+t \leqslant n$$

(Coxeter 1963, p. 218). Every rotation is associated with a family of parallel planes, each one of which is rotated in itself; the 'axis' of the rotation is an $(n-2)$-flat completely orthogonal to all these planes. In saying that the q rotations are mutually commutative, we mean that the q corresponding families of planes are completely orthogonal to one another: any line in a plane of one family is orthogonal to any line in a plane of any *other* family.

As we observed in §1·2, any rotation or translation can be expressed as the product of two reflections. Let our q rotations be so expressed in the form $Q_\nu Q_\nu'$ ($\nu = 1, \ldots, q$). Let the extra reflection (if $r = 1$) be R, and let the extra translation (if $t = 1$) be expressed as the product TT′ of reflections in two parallel hyperplanes. The commutativity of the reflections associated with different rotations enables us to write

$$S = UV, \quad U^2 = V^2 = 1$$

where, if $t = 0$ (and $r = 0$ or 1), $U = \Pi Q_\nu$ and

$$V = \Pi Q_\nu' \quad \text{or} \quad \Pi Q_\nu'.R,$$

but if $t = 1$ (and again $r = 0$ or 1), $U = \Pi Q_\nu.T$ and

$$V = \Pi Q_\nu'.T' \quad \text{or} \quad \Pi Q_\nu'.RT'.$$

(Here ΠQ_ν means $Q_1 \ldots Q_q$.) If $n > 1$, any reflection can be expressed as the product of another reflection and a rotation; for, if Q and R are reflections in perpendicular mirrors, QR is a *half-turn* (which is a special rotation) and R = Q.QR. This shows that if $n > 1$, we can arrange the decomposition of S in such a way that neither U nor V is the identity. Hence

(1·41) *In real Euclidean space of two or more dimensions, every isometry can be expressed as the product of two involutory isometries.*

EXERCISES

1. Every orthogonal matrix can be expressed as the product of two symmetric matrices.
2. Does Theorem (1·41) remain valid when the geometry is hyperbolic?

1·5 REGULAR POLYGONS IN n DIMENSIONS

The above remarks suggest the following generalization. In real Euclidean n-space, a *polygon* $A_0A_1A_2\ldots$ is a figure consisting of a sequence of points called *vertices* joined in successive pairs by line-segments

$$A_0A_1, \quad A_1A_2, \ldots$$

called *edges* (or 'sides'). In particular, for any isometry S, a non-invariant point A_0 has an *orbit* which is the set of points A_ν into which A_0 is transformed by the powers S^ν, ν running over all the integers; and we may still appropriately call $A_0A_1A_2\ldots$ a *regular* polygon. Since the possible values of ν include the negative integers, the sequence proceeds backwards as well as forwards, and the polygon $A_0A_1A_2\ldots$ should perhaps be more precisely described as

$$\ldots A_{-2}A_{-1}A_0A_1A_2\ldots.$$

It is regular if and only if it is congruent to $\ldots A_{-1}A_0A_1A_2A_3\ldots$.

Since S^ν transforms the point pair $A_\lambda A_\mu$ into a congruent point pair $A_{\lambda+\nu}A_{\mu+\nu}$, the reversed polygon $\ldots A_2A_1A_0A_{-1}A_{-2}\ldots$ is congruent to the original one. More precisely, for each integer, μ, the polygon has a symmetry operation which interchanges all pairs of points whose subscripts add up to μ. Such isometries with $\mu = 1$ or 2 may be identified with the U and V of Theorem (1·41). For the product of the permutations

(1·51) $$U = (A_0A_1)(A_{-1}A_2)(A_{-2}A_3)\ldots$$
$$V = (A_0A_2)(A_{-1}A_3)(A_{-2}A_4)\ldots$$

is

(1·52) $$(\ldots A_{-2}A_{-1}A_0A_1A_2\ldots) = S.$$

Although this result provides an *example* of Theorem (1·41), it is not a *proof*, since for some n-dimensional isometries every orbit lies entirely in a subspace of less than n dimensions. For instance, in four dimensions there is a *Clifford displacement* (Coxeter 1963, p. 217) for which every orbit lies on a circle (and thus in a plane).

In the special case when S is periodic, of period p, so that $S^p = 1$ (but $S^\nu \neq 1$ for $0 < \nu < p$), the polygon has p vertices and p edges, and we call it a *regular p-gon*. If it does not lie in one plane, we call it a regular *skew* p-gon. In denoting it by

$$A_0A_1\ldots A_{p-1},$$

we regard the subscripts as residues modulo p, so that $A_{-1} = A_{p-1}$, $A_0 = A_p$, and so on.

EXERCISES

1. Under what circumstances do the permutations (1·51) and (1·52) uniquely determine the isometries U, V, and UV = S?
2. A polygon $A_0A_1A_2\ldots$ is regular if $A_0A_\lambda = A_1A_{\lambda+1} = A_2A_{\lambda+2} = \ldots$ for each positive integer λ.
3. A pentagon $A_0A_1A_2A_3A_4$ is regular if it is both equilateral and equiangular.
4. A polygon $A_0A_1A_2\ldots$ is regular if its edges are all equal and its angles of each kind are all equal. For instance, a skew polygon in three dimensions has angles of two kinds (analogous to the curvature and torsion of a twisted curve): an ordinary angle $A_{\lambda-1}(A_\lambda)A_{\lambda+1}$ or $\angle A_{\lambda-1}A_\lambda A_{\lambda+1}$, and a dihedral angle $A_{\lambda-1}(A_\lambda A_{\lambda+1})\,A_{\lambda+2}$, which means the angle between the planes $A_{\lambda-1}A_\lambda A_{\lambda+1}$ and $A_\lambda A_{\lambda+1}A_{\lambda+2}$ (Schoute 1902, p. 268).
5. Describe a regular skew pentagon in four dimensions.
6. Express the matrix

$$\begin{bmatrix} \cos\alpha & \sin\alpha & 0 & 0 \\ -\sin\alpha & \cos\alpha & 0 & 0 \\ 0 & 0 & \cos\beta & \sin\beta \\ 0 & 0 & -\sin\beta & \cos\beta \end{bmatrix}$$

as the product of two symmetric matrices. When does it represent a Clifford displacement?

1·6 STRAIGHT AND CIRCULAR POLYGONS

When describing the general regular polygon in terms of its generating isometry S, we naturally take S to be the simplest isometry that serves this purpose: an n-dimensional isometry for an n-dimensional polygon. Thus S = UV, where the involutory isometries U and V are determined by their effect on the vertices, as in (1·51).

If V is a central inversion, the edges A_0A_1 and A_1A_2, going out from A_1 in opposite directions, are segments of one line; thus $n = 1$, and the polygon is the *apeirogon* $\{\infty\}$, whose infinitely many vertices are evenly spaced along the line. S is the translation that takes A_λ to $A_{\lambda+1}$, U is the reflection in the midpoint of A_0A_1, and V is the reflection in A_1. Thus S generates the infinite cyclic group $\mathfrak{C}_\infty$, while U and V generate the infinite dihedral group $\mathfrak{D}_\infty$.

If U is either a reflection or a central inversion, the planes $A_{-1}A_0A_1$ and $A_0A_1A_2$ coincide; thus $n = 2$. Since the case when V is a central inversion has already been covered, the only genuinely two-dimensional polygons are those for which V is a reflection (namely, the reflection in the internal bisector of $\angle A_0A_1A_2$) while U is either a reflection or a half-turn. (In two dimensions, a central inversion is a half-turn!)

In the case when both U and V are reflections (and still $n = 2$), we may assume the mirrors (u and v) to be lines through a point O (as parallel mirrors would take us back to the apeirogon). Thus S is a rotation about O, and O is the common centre of two circles (see Figure 1·6A), one passing through all the vertices while t3e other touches all the edges. We now have a 'circular' polygon, whose *circumcircle* and *incircle* completely determine its shape. If the angle A_0OA_1 is incommensurable with π, the edges $A_{\lambda-1}A_\lambda$ (which are chords of the circumcircle and tangents of the incircle) go round and round for ever, 'blackening' the annulus formed by the circles; and the vertices are dense on the circumcircle. This 'general' circular polygon is not particularly interesting. But if

$$\angle A_0OA_1 = 2d\pi/p,$$

where d and p are coprime integers, we have a *p-gon* of *density d*. This closed polygon is conveniently denoted by $\{p/d\}$.

When $d = 1$, we have the ordinary convex p-gon, $\{p\}$. Its p vertices, joined in suitable non-consecutive pairs, also serve as the vertices of a regular *star* polygon $\{p/d\}$ for any d that is prime to p and less than $\frac{1}{2}p$. (The generator S, of the cyclic group $\mathfrak{C}_p$, is now represented by a rotation through $2d\pi/p$ instead of $2\pi/p$.) In particular, we can have a *pentagram* $\{\frac{5}{2}\}$, an *octagram* $\{\frac{8}{3}\}$, a *decagram* $\{\frac{10}{3}\}$, a *dodecagram* $\{\frac{12}{5}\}$, and so on (Coxeter 1969, p. 37; 1963, p. 93). In other words, there is, for every rational number $p/d > 2$, a regular polygon $\{p/d\}$ whose edges subtend

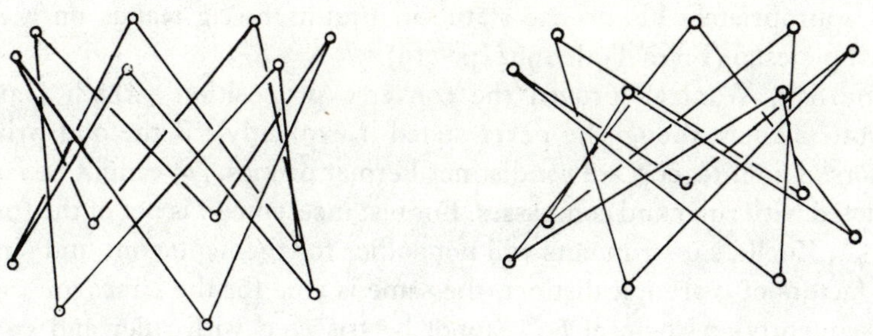

$p = 7, d = 3$ $p = 7, d = 4$

Figure 1·7B: Antiprismatic polygons

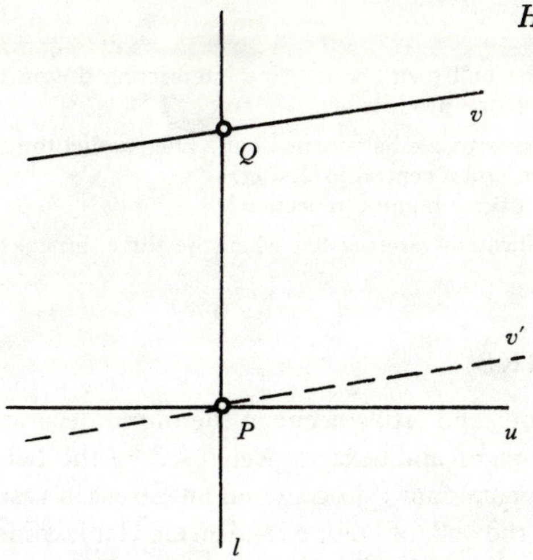

Figure 1·8A: The product of two half-turns

1·8 HELICAL POLYGONS

Since we have already considered the cases when U is a reflection or a central inversion or V is a central inversion, the only remaining three-dimensional regular polygon arises when both U and V are half-turns. In other words, the only remaining three-dimensional isometry is *the product of two half-turns.*

Let u and v be the axes of the two half-turns, and let l be a line cutting both these axes at right angles. If u and v intersect, l is concurrent with them and perpendicular to their plane, and S is a rotation (about l, through twice the angle between u and v). If u and v are parallel, l is coplanar with them, and S is a translation (along l, through twice the distance between u and v).

If u and v are *skew* lines (that is, the lines of two opposite edges of a tetrahedron), l is their unique common intersecting perpendicular, which measures the shortest distance PQ between them, as in Figure 1·8A. Let v' be the line through P parallel to v (that is, perpendicular to l in the plane lv) and let V′ be the half-turn about v'. Then the isometry

$$S = UV = UV'V'V$$

is a *twist*: the product of the rotation UV′ about l and the translation V′V along l (Veblen and Young 1918, p. 320). The angle of the rotation is twice the angle between u and v' (or between the planes lu and lv), and the extent of the translation is twice the distance PQ.

The orbit of any point not on l is inscribed in a circular helix, so we call ... $A_{-1}A_0A_1A_2$ ... a *helical polygon.*

Veblen (*ibid.* p. 324) uses the term *half-twist* for the special case when the skew lines u and v are at right angles; then the angle of the twist is π, the planes $A_{-1}A_0A_1$ and $A_0A_1A_2$ coincide, and the polygon is just a regular zigzag. In every other case the helical polygon is genuinely three-dimensional. Given such a polygon, we can construct the axis l as the common intersecting perpendicular of the lines u and v, where u joins the midpoints of A_0A_1 and $A_{-1}A_2$ while v joins A_1 to the midpoint of A_0A_2. In other words, v is the internal bisector of $\angle A_0A_1A_2$. Of course, U transforms v into the internal bisector of $\angle A_1A_0A_{-1}$, which is the axis of the half-turn

$$V^U = UVU.$$

Thus l may be described more symmetrically as *the common intersecting perpendicular of the internal bisectors of the angles $A_{-1}A_0A_1$ and $A_0A_1A_2$.*

The most intuitively significant difference between a direct isometry and an opposite isometry (see § 1·1) is that only the former can be regarded as a *displacement*, operating continuously in the manner of a motion. Since a translation through zero distance, or a rotation through zero angle, is merely the identity, the term *twist* may be deemed to include the rotation and translation as special cases. In this sense, we have proved that, in Euclidean 3-space,

(1·81) *Every displacement is a twist.*

EXERCISES

1. We have seen that, in Euclidean 3-space, the product of two central inversions, or of half-turns about two parallel lines, or of reflections in two parallel planes, is a translation; the product of half-turns about two intersecting lines, or of reflections in two intersecting planes, is a rotation; the product of half-turns about two skew lines is a twist; and the product of a

reflection with the half-turn about a line intersecting the mirror is a rotatory-reflection. What is the product of:

(i) a reflection with the half-turn about a line parallel to the mirror?
(ii) a half-turn and a central inversion?
(iii) a central inversion and a reflection?

2. What modifications are needed when the three-dimensional geometry is hyperbolic?

1·9 REMARKS

Squares, octagons and {16}'s occur in the mural decorations of ancient Egypt. The pentagon and hexagon were used by the Babylonians about 1500 B.C. The pentagram $\{\frac{5}{2}\}$ occurs on an Etruscan vase of the seventh century B.C., on the walls of Pompeii and on old Gallic coins. The Pythagoreans used it as a symbol of good health, and studied the number τ which expresses the ratio of its edge to that of the pentagon {5} having the same five vertices (τ for τομὴ, 'section'). This 'golden section' number is

$$(1\cdot91)\quad \tau = \tfrac{1}{2}(\sqrt{5}+1) = 2\cos\frac{\pi}{5} = 1+\frac{1}{\tau} = 1+1/1+1/1+1/1+\ldots = 1\cdot6180339887\ldots$$

(Coxeter 1969, Chapter 11). Euclid constructed {3} (I. 1), {4} (IV. 6), {5} (IV. 11), {6} (IV. 15), and {15} (IV. 16). He thus seems to have understood that constructions for $\{p\}$ and $\{q\}$, where p and q are relatively prime, will yield a construction for $\{pq\}$, and that repeated angle bisections (XII. 16) will yield $\{2^k\}$ and $\{2^k3\}$ for any k. Archimedes (287–212 B.C.) used {96} to prove that

$$\frac{223}{71} < \pi < \frac{22}{7}.$$

In 1596, L. van Ceulen used $\{2^{62}\}$ for a closer approximation, which was inscribed on his tombstone in St Peter's Church, Leiden (Ball 1967, p. 344). Meanwhile regular star-polygons were studied systematically by Thomas Bradwardine (1290–1349), who became Archbishop of Canterbury for the last month of his life. Kepler (1571–1630) observed that the edges of polygons $\{7\}, \{\frac{7}{2}\}, \{\frac{7}{3}\}$ of unit circumradius (all having the same 7 vertices) are the positive roots of the equation

$$x^6 - 7x^4 + 14x^2 - 7 = 0.$$

The problem of constructing regular polygons by means of ruler and compasses was revived in 1796, when Gauss proved that $\{p\}$ can be so constructed whenever p is a prime of the form $2^{2^k}+1$, and consequently also whenever p is a product of distinct primes of this form, or the same multiplied by any power of 2. The big surprise was the case of {17}. It was perhaps characteristic of Gauss that he was content to establish the possibility, leaving the details of the construction to others, such as H. W. Richmonh (Ball 1967, p. 95). However, this profound discovery had the effect of diverting young Gauss from philology to mathematics, and appropriately his bronze statue in Braunschweig stands on a 17-gonal pedestal (Fejes Tóth 1964, p. 119).

In 1837, Wantzel[†] proved the converse proposition (which Gauss doubtless knew, though he never stated it explicitly): if the odd prime factors of an integer p are *not* distinct Fermat primes, $\{p\}$ cannot be constructed with ruler and compasses. For instance, since 7 is not of the form $2^{2^k}+1$, Euclid's instruments will not suffice for the heptagon; and since the factors of 9 are not distinct, the same is true for the enneagon (and consequently an angle of 60° cannot be trisected with ruler and compasses).

Since $2^{2^k}+1$ is composite for $5 \leqslant k \leqslant 16$, the greatest odd value of p for which p could possibly be 'contructed' is

$$3.5.17.257.65537 = 2^{32}-1.$$

(Any greater value would exceed the number of particles in the universe.)

Theorem (1·41) appears to be new. The proof has been kindly supplied by Israel Halperin.

The dihedral kaleidoscope was developed by Athanasius Kircher (in 1646) and Sir David Brewster (1819). Its use for exhibiting regular polygons $\{p/d\}$ can be seen in a film, *Dihedral Kaleidoscopes*, which was made for the University of Minnesota in 1965.

Theorem 1·81 was discovered in 1830 by Michel Chasles[‡] (1793–1880), whose investigation of kinematics continued for at least 36 years. Various proofs were devised by him and others. Horace Lamb (1920, p. 20, Ex. 17) mentions one by M. W. Crofton (about 1903). It was he who constructed the axis of the twist $A_0A_1A_2 \to A_1A_2A_3$ as the common intersecting perpendicular of the bisectors of $\angle A_0A_1A_2$ and $\angle A_1A_2A_3$. The extension to n dimensions, which we used in §1·4, was discovered by P. H. Schoute[§] (1846–1913).

† P. L. Wantzel, *Journal de Mathématiques pures et appliquées* **2** (1837), 366–72.
‡ *Bulletin des Sciences Mathématiques du baron de Férussac* **14** (1830), 321–6; reprinted in *Correspondance mathématique et physique de l'observatoire de Bruxelles* **7** (1832), 352–7. See also his paper in *Comptes Rendus* (Paris) **51** (1860), 905–9.
§ Le déplacement le plus général dans l'espace à *n* dimensions. *Annales de l'École Polytechnique de Delft* **7** (1891), 139–58.

CHAPTER 2

Regular polyhedra

Certe in Dei Creatoris mente consistit Deo coaeterna figurarum harum veritas.

(Without doubt the authentic type of these figures exists in the mind of God the Creator and shares His eternity.) Kepler (1611, p. 36)

In this chapter the regular polyhedra are derived from patterns formed by arcs of great circles on a sphere. This approach makes it natural to include, along with the five Platonic solids, the two stellated dodecahedra of Kepler, the two star-cornered 'great' polyhedra of Poinsot and Cayley, the hosohedra of Caravelli, and the dihedra of Klein. In §2·2 we consider a basic definition for the word 'regular' and derive from each regular polyhedron a special polygon (usually skew) called the Petrie polygon, which will be found helpful in §3·4. In §2·3 we discuss the symmetry groups of the regular polyhedra, and in §2·4 we see how these groups can be used to reconstruct the Platonic solids and to construct also the Archimedean solids. In §2·5 we enumerate the Schwarz triangles, more neatly than in *Regular Polytopes* (Coxeter 1963, p. 113).

2·1 SPHERICAL TESSELLATIONS

If the circumcircle of a regular polygon $\{p\}$ is regarded as lying on a sphere, we can replace the edges by arcs of great circles joining the same pairs of vertices, and thus obtain a regular *spherical* polygon, for which we may conveniently use the same symbol $\{p\}$. In particular, the circumcircle may be a great circle, and then all the edges of the spherical polygon are arcs of this same circle. Taking the radius of the sphere as our unit of measurement, we see that in this case the edges are arcs of length $2\pi/p$. The great circle decomposes the spherical surface into two hemispheres, either of which may be called the interior of the polygon. Here p is not necessarily an integer; for any rational number $n/d \geqslant 2$, there is a spherical $\{n/d\}$, of density d.

When regarded as a spherical polygon, the digon $\{2\}$ no longer needs to have coincident edges. There is even a *monogon* $\{1\}$ consisting of a single point and a great circle through it. The spherical *digon* is a 'lune' consisting of two antipodal points joined by two great semicircles (like two meridians joining the north and south poles on a geographical globe). The angle formed by these two semicircles (at either vertex) may take

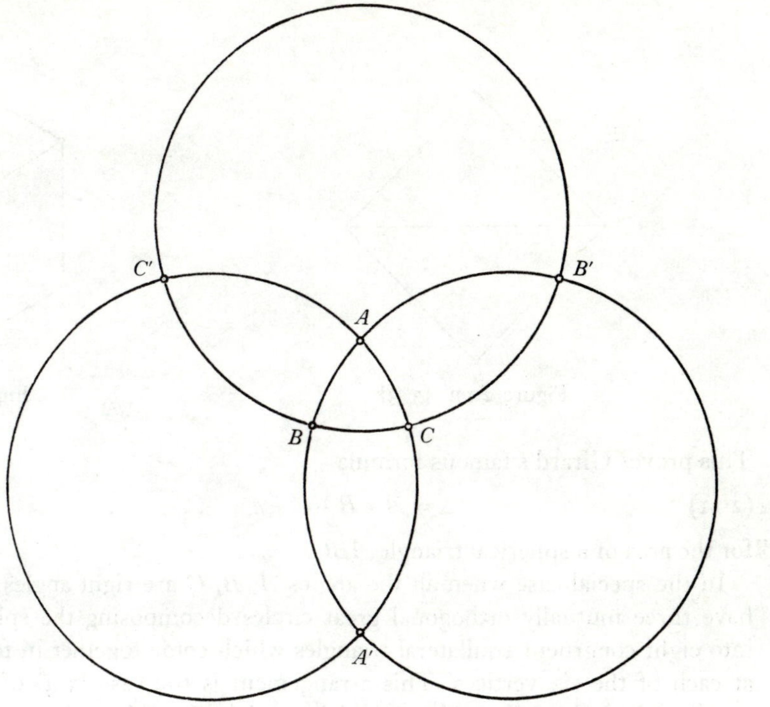

Figure 2·1A: Three great circles

any value from 0 to π. If it is small enough, we can 'stack' several such digons having the same two vertices, to form a larger digon. We see in this manner that the area of any digon is proportional to its angle. Assuming the known value 4π for the area of the whole unit sphere, we deduce that a digon of angle θ has area 2θ.

Figure 2·1A shows, in stereographic projection (Coxeter 1969, p. 93), the three great circles that contain the edges of a spherical triangle ABC. Of course, the two circles through each vertex intersect again at the antipodal point A' or B' or C'. The triangle ABC and its *colunar* triangle $A'BC$ together fill up a digon of angle A, whose area is $2A$. Thus if Δ denotes the area of ABC, that of $A'BC$ is $2A-\Delta$. The three great circles decompose the sphere into four pairs of antipodal triangles:

$$\begin{array}{lll} ABC \text{ and } A'B'C', & \text{each having area } \Delta, \\ A'BC \text{ and } AB'C', & \text{each having area } 2A-\Delta, \\ AB'C \text{ and } A'BC', & \text{each having area } 2B-\Delta, \\ ABC' \text{ and } A'B'C, & \text{each having area } 2C-\Delta. \end{array}$$

Hence
$$\begin{aligned} 4\pi &= 2\Delta+2(2A-\Delta)+2(2B-\Delta)+2(2C-\Delta) \\ &= 4(A+B+C-\Delta). \end{aligned}$$

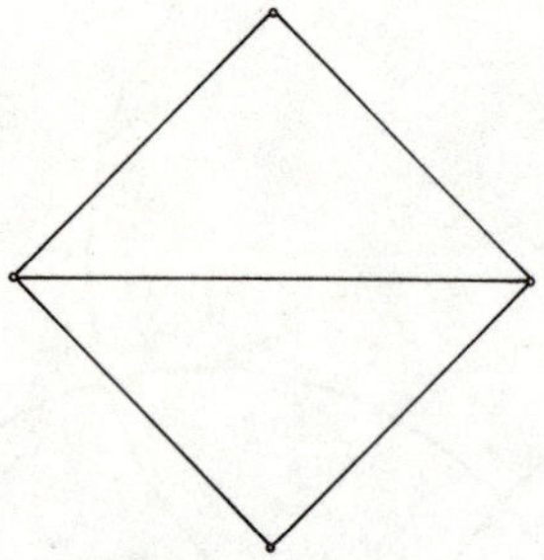

Figure 2.1B: $\{3, 3\}$

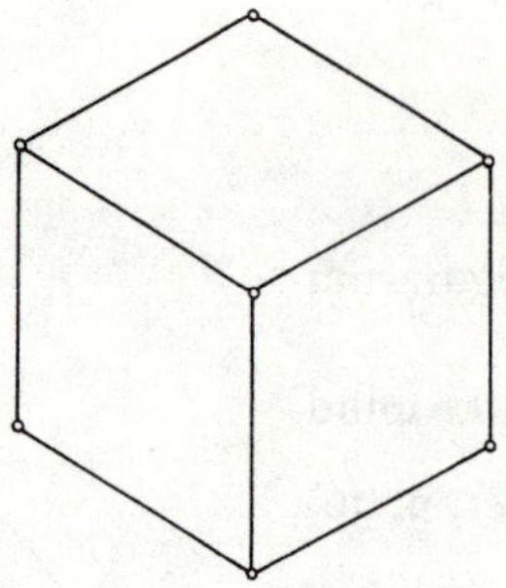

Figure 2.1C: $\{4, 3\}$

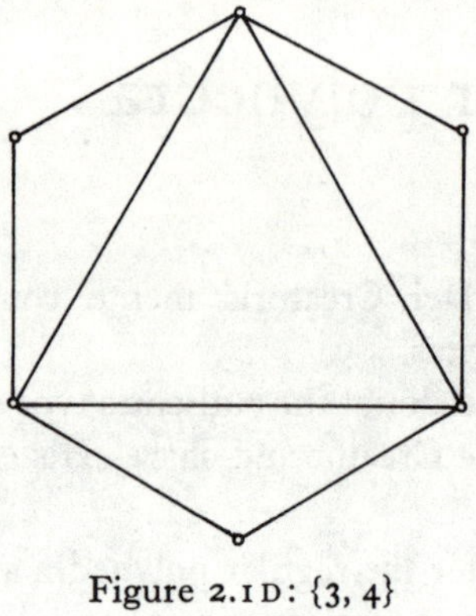

Figure 2.1D: $\{3, 4\}$

This proves Girard's famous formula

(2·11) $$\Delta = A+B+C-\pi$$

for the area of a spherical triangle ABC.

In the special case when all the angles A, B, C are right angles, we have three mutually orthogonal great circles decomposing the sphere into eight congruent equilateral triangles which come together in fours at each of the six vertices. This arrangement is the case $\{3, 4\}$ of the *regular spherical tessellation* $\{p, q\}$, which consists of regular polygons $\{p\}$, of angle $2\pi/q$, fitting together to fill and cover the whole spherical surface (just once if p and q are integers, otherwise several times). The polygons are called *faces*. The centres of the faces are the vertices of the *dual* (or *reciprocal*) tessellation $\{q, p\}$, whose edges cross those of $\{p, q\}$.

The *dihedron* $\{n/d, 2\}$ has n vertices, evenly spaced along a great circle; its n edges are arcs of length $2d\pi/n$; its two faces are coincident spherical $\{n/d\}$'s bounding the two hemispheres. The dual *hosohedron* $\{2, n/d\}$ has two antipodal vertices joined by n great semicircles and n digons of angle $2d\pi/n$. The remaining nine regular spherical tessellations

$$\{3, 3\},\quad \{4, 3\},\quad \{3, 4\},\quad \{5, 3\},\quad \{3, 5\},\quad \{\tfrac{5}{2}, 3\},\quad \{3, \tfrac{5}{2}\},\quad \{\tfrac{5}{2}, 5\}\quad \{5, \tfrac{5}{2}\},$$

are 'blown up' versions of the Platonic and Kepler–Poinsot solids: the tetrahedron, cube (or hexahedron), octahedron, dodecahedron, icosahedron, great stellated dodecahedron, great icosahedron, small stellated dodecahedron, and great dodecahedron (Figures 2·1B–J). The same Schläfli symbol $\{p, q\}$ can be used in either interpretation. The midpoints of the edges that come together at one vertex are the vertices of a polygon called the *vertex figure* (of the tessellation or solid). In the case of a solid, the vertex figure is simply the section by a suitable plane that cuts off one corner. Thus $\{p, q\}$ is a tessellation or solid whose face and vertex figure are $\{p\}$ and $\{q\}$, respectively.

The midpoints of *all* the edges of $\{p, q\}$ are the vertices of a polyhedron $\left\{\begin{matrix}p\\q\end{matrix}\right\}$ whose faces are $\{p\}$'s and $\{q\}$'s, namely one $\{p\}$ inscribed in each face of $\{p, q\}$, and the vertex figure at each vertex. The symbol, with the q below the p, is intended to stress the fact that $\left\{\begin{matrix}p\\q\end{matrix}\right\} \left(=\left\{\begin{matrix}q\\p\end{matrix}\right\}\right)$ can be derived from the reciprocal polyhedron $\{q, p\}$ the same way as from $\{p, q\}$ itself. Clearly, each vertex of this new polyhedron is surrounded by four faces:

(2·12) $$\{p\},\quad \{q\},\quad \{p\},\quad \{q\},$$

in this order. Thus it has twice as many edges as $\{p, q\}$ (or $\{q, p\}$), each belonging to one face of each type. If $p = q$, we have

(2·13) $$\left\{\begin{matrix}p\\p\end{matrix}\right\} = \{p, 4\}.$$

If $p \neq q$, $\left\{\begin{matrix}p\\q\end{matrix}\right\}$ is not quite regular, so let us call it *quasi-regular*, indicating that, although it has faces of two kinds, its edges are all surrounded the same way, namely by one $\{p\}$ and one $\{q\}$. The particular instances are

$$\left\{\begin{matrix}3\\3\end{matrix}\right\},\quad \left\{\begin{matrix}4\\3\end{matrix}\right\},\quad \left\{\begin{matrix}5\\3\end{matrix}\right\},\quad \left\{\begin{matrix}\frac{5}{2}\\3\end{matrix}\right\},\quad \left\{\begin{matrix}\frac{5}{2}\\5\end{matrix}\right\}:$$

the octahedron, cuboctahedron, icosidodecahedron, great icosidodecahedron, and dodecadodecahedron (Coxeter 1963, pp. 18, 101).

The above definition for *vertex figure* can evidently be applied to any *uniform* polyhedron, that is, any polyhedron whose faces are regular while its vertices are all surrounded alike. (Uniform polyhedra include regular and quasi-regular polyhedra, and also 'semi-regular' ones such as the rest of the Archimedean solids.) Moreover, it is desirable to define the vertex figure more precisely as having not only a definite shape but a definite size. This is done either by taking the edge-length of the

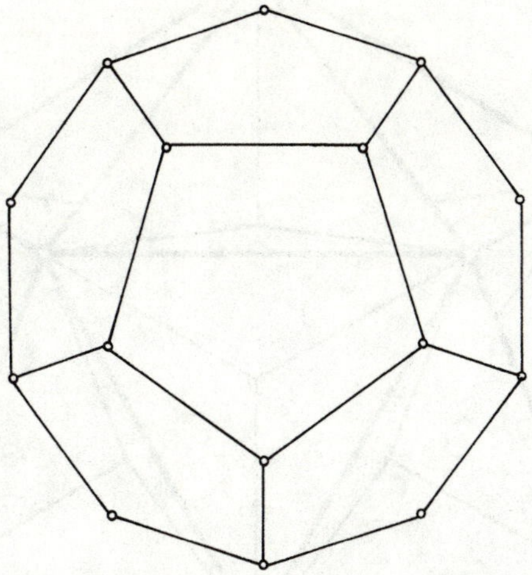

Figure 2.1E: {5, 3}

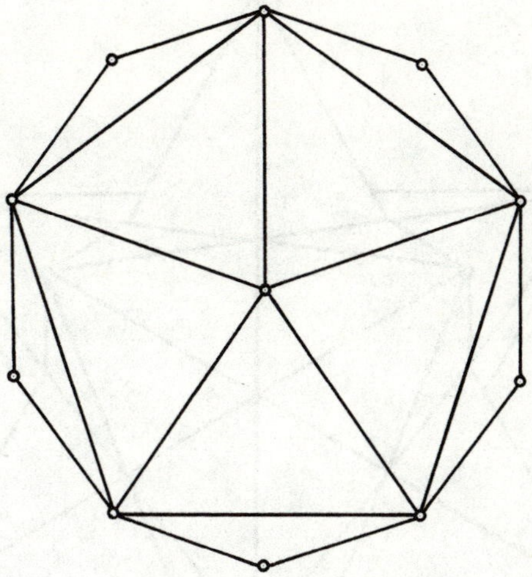

Figure 2.1F: {3, 5}

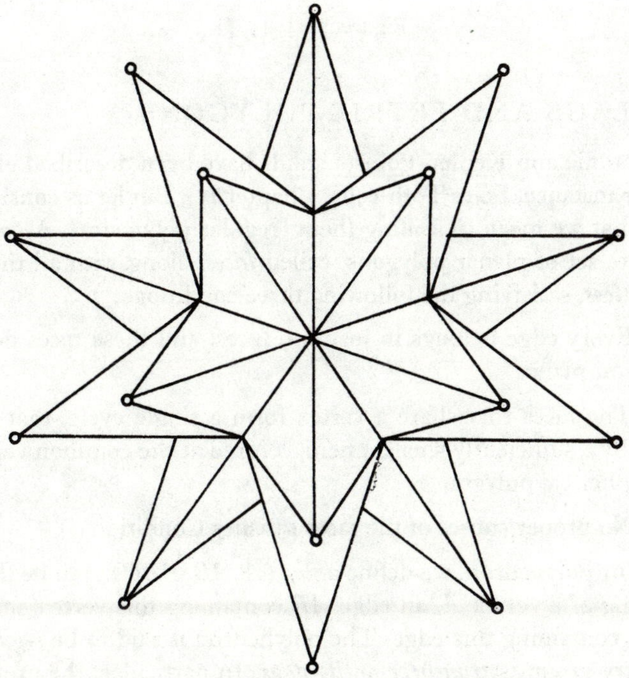

Figure 2.1G: $\{\frac{5}{2}, 3\}$

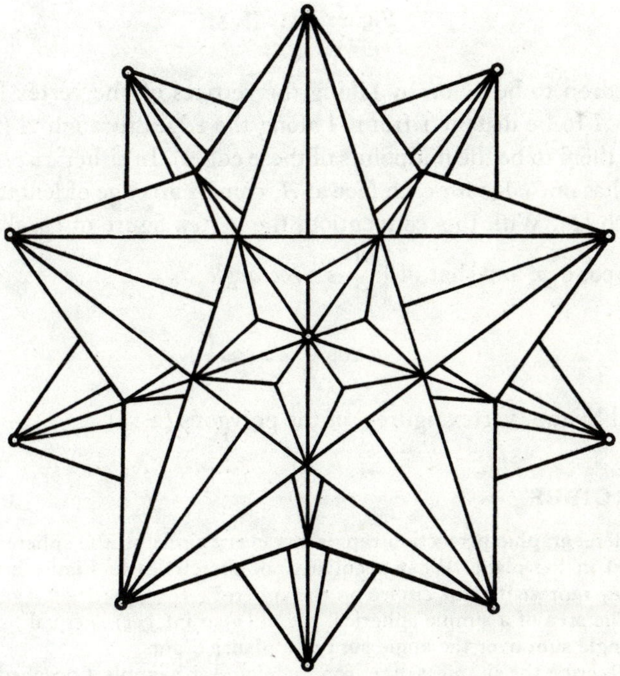

Figure 2.1H: $\{3, \frac{5}{2}\}$

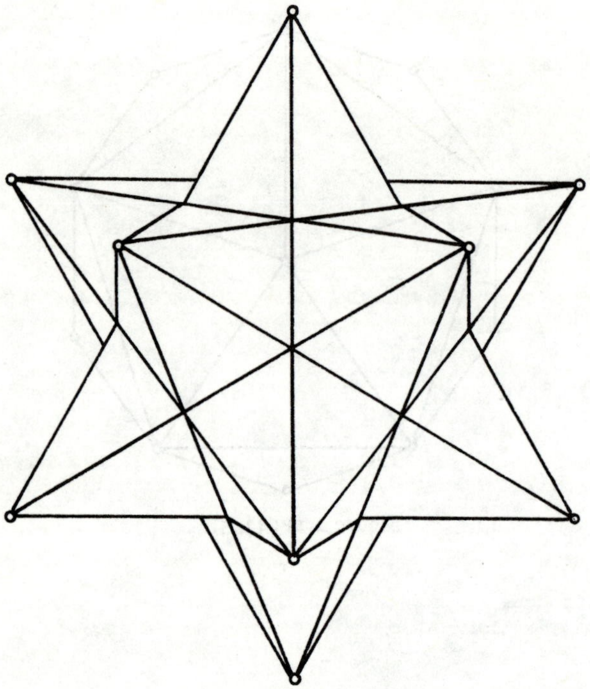

Figure 2.1I: $\{\frac{5}{2}, 5\}$

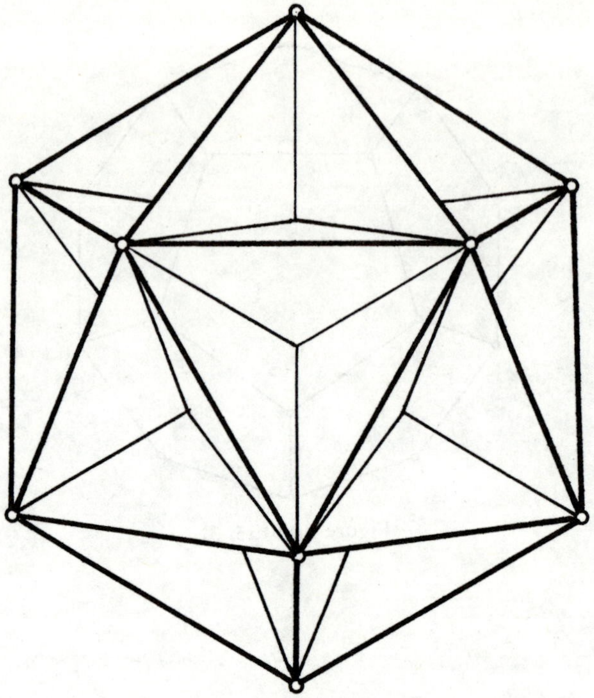

Figure 2.1J: $\{5, \frac{5}{2}\}$

polyhedron to be 2, or by taking the vertices of the vertex figure at a vertex A to be distant 1 from A along the edges through A (instead of taking them to be the midpoints of these edges). In either case, the vertex figure has one edge for each face at A, namely an edge of length $2\cos\pi/p$ for each $\{p\}$. With this convention, the vertex figure of $\{p, q\}$ is a $\{q\}$ of edge $2\cos\pi/p$, and that of $\begin{Bmatrix} p \\ q \end{Bmatrix}$ is a *rectangle*

(2·14) $$2\cos\frac{\pi}{p} \times 2\cos\frac{\pi}{q},$$

formed by the 'vertex figures' of the polygons (2·12).

EXERCISES

1. Stereographic projection represents every circle on the sphere by a circle (or line) in the plane. What peculiarity of the circles in Figure 2·1A ensures that they represent *great* circles on the sphere?
2. The area of a simple spherical p-gon (p an integer) is equal to the excess of its angle sum over the angle sum of a planar p-gon.
3. Describe the simplest dihedron $\{1, 2\}$ and the simplest hosohedron $\{2, 1\}$.
4. If $\{p, q\}$ is a spherical tessellation, $p^{-1}+q^{-1} > \frac{1}{2}$.

2·2 FLAGS AND PETRIE POLYGONS

The Platonic and Kepler–Poinsot solids have been described elsewhere (see, for instance, Fejes Tóth 1964, Chapter IV). But let us consider once more what we mean by calling them 'regular polyhedra'. A *polyhedron* is a finite set of planar polygons, called *faces*, along with all their *edges* and *vertices*, satisfying the following three conditions:

(i) Every edge belongs to just two faces, and these faces do not lie in the same plane.

(ii) The faces that share a vertex form a single cycle, that is, their section by a sufficiently small sphere, centred at the common vertex, is a single spherical polygon.

(iii) No proper subset of the faces satisfies Condition (i).

For any polyhedron, we define a *flag* $(A, AB, ABC\ldots)$ to be the figure consisting of a vertex A, an edge AB containing this vertex, and a face $ABC\ldots$ containing this edge. The polyhedron is said to be *regular* if its symmetry group is *transitive on its flags*. In particular, the group must include symmetries that will transform a given flag into any *adjacent* flag,

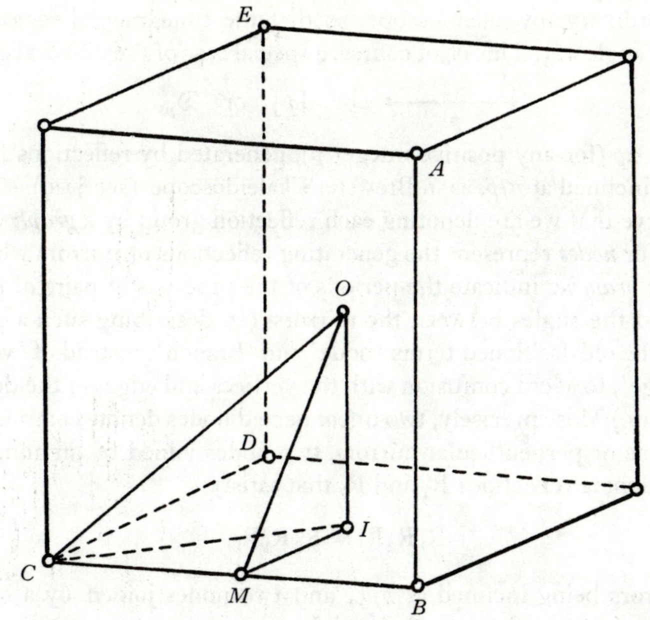

Figure 2·2A: The cube and its characteristic orthoscheme *CMIO*

differing from the given flag in just one of its three elements. Adjacent to the flag (*C*, *BC*, *BCD*...) in figure 2·2A, we find

$$(B, BC, BCD\ldots),\quad (C, CD, BCD\ldots),\quad (C, BC, \ldots ABC).$$

These three flags are derived from (*C*, *BC*, *BCD*...) by reflections in three planes: ρ_1, perpendicularly bisecting the edge *BC*; ρ_2, bisecting the angle *BCD* (perpendicular to the plane *BCD*); and ρ_3, bisecting the dihedral angle formed by the two faces that have *BC* for their common edge. These three *planes of symmetry* of the polyhedron form a trihedral angle or *trihedron* whose edges are the three lines

$$\rho_2 \cap \rho_3 = OC,\quad \rho_3 \cap \rho_1 = OM,\quad \rho_1 \cap \rho_2 = OI,$$

where *M* is the midpoint of the edge *BC*, *I* is the centre of the face *BCD*..., and *O* is the centre of the whole polyhedron. Thus the three planes are

$$(2{\cdot}21)\qquad \rho_1 = OMI,\quad \rho_2 = OCI,\quad \rho_3 = OCM.$$

They reflect one another into the set of all planes of symmetry, most easily visualized by means of the great circles in which they intersect the unit sphere with centre *O* (see Figure 2·4A). The reflections in ρ_1 and ρ_2 generate the symmetry group of the face *BCD*..., which is thus seen to be a regular polygon. The reflection in ρ_3 yields the congruent face ...*ABC*, and so eventually we find that *all the faces are regular and congruent.*

For any regular polyhedron, a *Petrie polygon* is defined as a skew polygon in which every two consecutive edges, but no three, belong to a face. Thus, if ...*ABC* and *BCD*... are two adjacent faces, as in Figure 2·2A, the three edges *AB*, *BC*, *CD* belong to a Petrie polygon. The half-turn about *OM* (which is the product of reflections in ρ_3 and ρ_1) transforms this skew polygon *ABCD*... into *DCBA*..., and the reflection in ρ_2 transforms *DCBA*... into *BCDE*.... The Petrie polygon, being thus shifted one step along itself by a rotatory-reflection, is an *antiprismatic* regular polygon in the sense of §1·7. Since each flag determines two congruent Petrie polygons, the regularity of the polyhedron implies that all the Petrie polygons are congruent, so that we are justified in speaking of *the* Petrie polygon of $\{p, q\}$.

The plane of the rotatory-reflection (that is, the mirror for the component reflection, perpendicular to the axis of the rotation) passes through the midpoints of the edges of the Petrie polygon, thus determining a planar polygon $\{h\}$ that has one edge for each pair of adjacent edges of the Petrie polygon. Since all the edges of this $\{h\}$ are edges of $\begin{Bmatrix} p \\ q \end{Bmatrix}$, and lie in a plane through the centre of $\begin{Bmatrix} p \\ q \end{Bmatrix}$, it is natural to call $\{h\}$ the *equatorial polygon* of $\begin{Bmatrix} p \\ q \end{Bmatrix}$. Since its vertex figure (a line-segment of length $2\cos\pi/h$) is a diagonal of the rectangle (2·14), *h* is given by the simple formula

$$(2{\cdot}22)\qquad \cos^2\frac{\pi}{h} = \cos^2\frac{\pi}{p} + \cos^2\frac{\pi}{q}$$

(Coxeter 1963, p. 19). Thus the particular values of *h* are:

$$4 \text{ for } \begin{Bmatrix} 3 \\ 3 \end{Bmatrix},\qquad 6 \text{ for } \begin{Bmatrix} 4 \\ 3 \end{Bmatrix} \text{ and } \begin{Bmatrix} \frac{5}{2} \\ 5 \end{Bmatrix},\qquad 10 \text{ for } \begin{Bmatrix} 5 \\ 3 \end{Bmatrix},\qquad \tfrac{10}{3} \text{ for } \begin{Bmatrix} \frac{5}{2} \\ 3 \end{Bmatrix}.$$

Projecting the Petrie polygon (of $\{p, q\}$) orthogonally on the plane containing the midpoints of its edges, we obtain a larger $\{h\}$ whose edges have these same midpoints, namely, the vertices of the equatorial polygon of $\begin{Bmatrix} p \\ q \end{Bmatrix}$. Figures 2·1B–J show each of the nine regular polyhedra projected in this manner, so that one specimen of the Petrie polygon appears as an $\{h\}$. In seven of the nine cases, this $\{h\}$ forms the periphery of the drawing. The exceptions are Kepler's two stellated dodecahedra, $\{\frac{5}{2}, 3\}$ and $\{\frac{5}{2}, 5\}$, in which the skew decagram and skew hexagon (respectively) are so wildly non-planar that only small fragments of their edges can be seen.

Analogously, the Euclidean tessellations

$$(2{\cdot}23)\qquad \{4, 4\},\quad \{6, 3\},\quad \{3, 6\}$$

have infinite Petrie polygons which are zigzags (Coxeter 1963, p. 61).

EXERCISES

1. If p and q are integers and $\{p, q\}$ has N_0 vertices, N_1 edges, N_2 faces, then

$$pN_2 = 2N_1 = qN_0 \quad \text{and} \quad N_1^{-1} = p^{-1}+q^{-1}-\tfrac{1}{2}.$$

2. What happens if $p = q = 1$ or 0?
3. Can a convex polyhedron have congruent regular faces without being regular?
4. What kind of polygon is formed by joining the midpoints of the consecutive edges of the Petrie polygon of $\{p, q\}$?
5. When a model of a convex polyhedron is made from an unfolded *net* (Coxeter 1969, p. 151), how many tabs are needed for sticking edges together? (This is George Pólya's version of an important step in von Staudt's proof of Euler's formula; see Coxeter 1963, p. 9.)

2·3 REFLECTION GROUPS AND ROTATION GROUPS

The above definition of a regular polyhedron implies the existence of a number of *planes of symmetry*. Reflections in these planes, or in a suitable subset of them, generate the symmetry group of the polyhedron. More generally, we may consider all the instances of a finite set of planes through one point, so arranged that the whole configuration is symmetrical by reflection in each plane; that is, the various *finite groups generated by reflections*. Although such a group may include rotations and rotatory-reflections as well as simple reflections, we may call it (without serious risk of confusion) a *reflection group*. In each case the planes cut out, from the unit sphere round their common point, a pattern of great circles decomposing the spherical surface into a number of congruent regions whose angles are submultiples of π (Coxeter 1963, p. 78). The typical region is called a *fundamental region* for the group, because every point on the sphere is equivalent (by the action of the group) to some point belonging to this (closed) region, and there is one replica of the region for each element of the group.

The simplest instance is the group

• or [ ] or $\mathfrak{D}_1$,

of order 2, generated by a single reflection, so that the fundamental region is a hemisphere (with no angle). The next simplest is

• • or $[\ ]^2$ or [2] or $\mathfrak{D}_2$,

of order 4, generated by two commutative reflections, i.e. reflections in two perpendicular planes, so that the fundamental region is a digon of angle $\pi/2$. Next comes

 or [3] or $\mathfrak{D}_3$,

of order 6, generated by reflections in two mirrors inclined at $\pi/3$ as in the ordinary toy kaleidoscope, so that the fundamental region is a digon of angle $\pi/3$. This is, of course, a special case of the dihedral group.

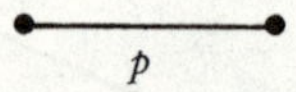 or $[p]$ or $\mathfrak{D}_p$,

of order $2p$ (for any positive integer p), generated by reflections in two mirrors inclined at π/p, as in Brewster's kaleidoscope (see §1·9).

Observe that we are denoting each reflection group by a *graph* whose vertices or *nodes* represent the generating reflections or mirrors while its edges or *branches* indicate the periods of the products of pairs of reflections and the angles between the mirrors. (In describing such a graph, we use the old-fashioned terms 'node' and 'branch', instead of 'vertex' and 'edge', to avoid confusion with the vertices and edges of the derived polyhedra.) More precisely, two unconnected nodes denote commutative reflections or perpendicular mirrors, two nodes joined by an unmarked branch denote reflections R_1 and R_2 that satisfy

$$R_1R_2R_1 = R_2R_1R_2, \tag{2·31}$$

the mirrors being inclined at $\pi/3$; and two nodes joined by a branch marked p denote reflections that satisfy

$$(R_1R_2)^p = 1, \tag{2·32}$$

the mirrors being inclined at π/p.

Regarding these as two vertical mirrors, we can stand them on a horizontal mirror to obtain direct products:

• • • or $[\ ]^3$ or [2, 2],

of order 8, generated by three mutually commutative reflections, so that the fundamental region is an 'octant' (or equilateral right-angled triangle);

 • or $[3]\times[\ ]$ or [3, 2],

or order 12, whose fundamental region has one angle of $\pi/3$ and two right angles; and

 • or $[p]\times[\ ]$ or $[p, 2]$,

of order $4p$.

Reducing the number of right angles to one, we obtain the three true 'polyhedral kaleidoscopes'

 or $[p, 3]$,

where $p = 3$ or 4 or 5 (and, in the first case, the graph is unmarked).

The fundamental region is now a spherical triangle

$$(p\ q\ r)$$

having angles π/p, π/q, π/r. Actually $q = 3$ and $r = 2$, but we retain all three letters for the sake of symmetry, and in order to be able to include the simpler reflection groups

$$[\,],\quad [p],\quad [p, 2],$$

for which the fundamental region reduces to the hemisphere $(1\ 1\ 1)$ (a 'triangle' having three angles π), a digon $(p\ p\ 1)$ (having equal angles π/p at its two antipodal vertices and an extra angle π at a third 'vertex' inserted arbitrarily on either of its two bounding semicircles), or an isosceles triangle $(p\ 2\ 2)$ (whose sides are two semi-meridians and an arc of the equator). There is no such *spherical* triangle $(p\,q\,r)$ with

$$p \geqslant q \geqslant r > 2,$$

because its angle-sum would be too small; for the same reason, the fundamental region cannot be a spherical n-gon with $n > 3$.

Since every finite group of isometries has at least one invariant point (for the same reason as in the two-dimensional case, §1·3), every finite reflection group is generated by mirrors having one common point, and the group may be regarded as acting on the unit sphere with this point for its centre. The above remarks enable us to conclude that every finite reflection group (in Euclidean 3-space, or on a sphere) has a fundamental region $(p\,q\,r)$, where p, q, r are positive integers satisfying

(2·33) $$p^{-1}+q^{-1}+r^{-1} > 1$$

with the restriction that, if one of these integers is 1, the other two must be equal.

Defining a number s by the equation

(2·34) $$s^{-1} = p^{-1}+q^{-1}+r^{-1}-1,$$

we see from Girard's formula (2·11) that the area of $(p\,q\,r)$ is π/s. This clearly holds for $(p\,p\,1)$ as well as for a genuine triangle. Thus

$$s = \tfrac{1}{2}p,\ p,\ 6,\ 12 \quad\text{or}\quad 30,$$

according as the fundamental region is

$$(p\ p\ 1),\quad (p\ 2\ 2),\quad (3\ 3\ 2),\quad (4\ 3\ 2)\quad\text{or}\quad (5\ 3\ 2).$$

When $r = 2$, s is equal to the number of edges of $\{p, q\}$. (See §2·2, Ex. 1.) Since each edge of a Platonic solid belongs to 2 Petrie polygons (skew h gons), there are altogether $2s/h$ Petrie polygons. Every 2 Petrie polygons share one pair of opposite edges; therefore

$$2\binom{2s/h}{2} = s$$

and

(2·35) $$4s = h(h+2).$$

(Compare Coxeter 1970, p. 9, where s is called E and these ideas are developed further.)

Since the elements of the reflection group transform one specimen of the fundamental region into all its replicas, the order of the group is equal to the ratio of the area of the whole sphere to the area of $(p\,q\,r)$, that is,

$$4s.$$

In particular, when $q = 3$ and $r = 2$, $s = 6p/(6-p)$; therefore the order of $[p, 3]$ is

(2·36) $$\frac{24p}{6-p} \quad (p = 2, 3, 4, 5).$$

We saw, in §1·2, that the dihedral group $\mathfrak{D}_p$, generated by two reflections, has a cyclic subgroup $\mathfrak{C}_p$ (of index 2) generated by a rotation which is the product of these two reflections. More generally, every finite reflection group has a subgroup of index 2 which is a *rotation group*, generated by products of pairs of the reflections. Thus the reflection group of order $4s$, whose fundamental region is $(p\,q\,r)$, has a subgroup of order $2s$ generated by rotations through $2\pi/p$, $2\pi/q$, $2\pi/r$ about the vertices P, Q, R of this triangle. These rotations A, B, C are the products of pairs of reflections in the three sides QR, RP, PQ. More precisely, if the vertices are named P, Q, R in clockwise order, the counterclockwise rotations will satisfy the relation ABC = 1 as well as the obvious relations

$$\mathrm{A}^p = \mathrm{B}^q = \mathrm{C}^r = 1.$$

A convenient symbol for this rotation group of order $2s$ is

$$(p,\ q,\ r)$$

(with commas).

Although we have used *three* generators for the sake of symmetry, the relation ABC = 1 shows that any two of A, B, C would suffice. The actual instances (Coxeter 1969, pp. 270–6) are:

the *cyclic* group	$(p, p, 1) \simeq \mathfrak{C}_p$,	of order p,
the *dihedral* group	$(p, 2, 2) \simeq \mathfrak{D}_p$,	of order $2p$,
the *tetrahedral* group	$(3, 3, 2) \simeq \mathfrak{A}_4$,	of order 12,
the *octahedral* group	$(4, 3, 2) \simeq \mathfrak{S}_4$,	of order 24,
the *icosahedral* group	$(5, 3, 2) \simeq \mathfrak{A}_5$,	of order 60.

These are subgroups of index 2 in

$$[p],\quad [p, 2],\quad [3, 3],\quad [4, 3],\quad [5, 3],$$

respectively.

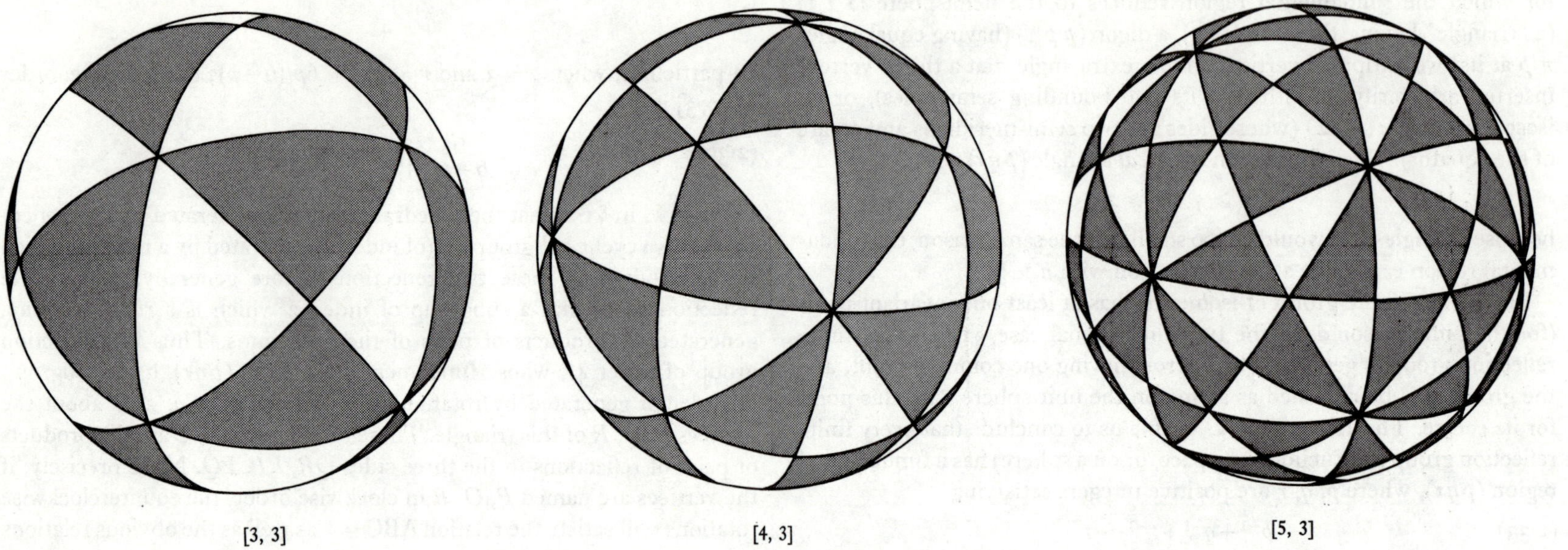

Figure 2·4A: The circles of symmetry for the groups $[p, 3]$

EXERCISES

1. What are the symmetry groups of the two kinds of rhombic dodecahedron (Coxeter 1963, p. 31; 1968, pp. 65–6) whose faces consist of twelve congruent rhombi?
2. Express the generators of $(p, 3, 2)$ as permutations of letters $a, b, c, \ldots$, thus justifying the above symbols $\mathfrak{A}_4, \mathfrak{S}_4, \mathfrak{A}_5$ ($\mathfrak{A}$ for *alternating*, $\mathfrak{S}$ for *symmetric*).
3. Give a simple construction for the product of rotations through given angles about two intersecting lines.
4. If OX and OY are two perpendicular lines in space, any rotation about another line through O can be expressed as a product ABC, where A and C are rotations about OX while B is a rotation about OY. (C. Loewner.)

2·4 WYTHOFF'S CONSTRUCTION

Figure 2·4A shows the patterns of triangles $(p\,q\,2)$ in the cases $p = 3, 4, 5$; $q = 3$. Alternate triangles have been shaded to emphasize the fact that each reflection group (of order $4s$) has a rotation subgroup of index 2, which transforms white triangles into white triangles, and shaded into shaded. Clearly, the $2s/q$ points where the angles are π/q are the vertices (each belonging to q faces) of the regular polyhedron $\{p, q\}$, and the s points where right angles occur are the vertices (each belonging to two p-gons and two q-gons) of the quasi-regular polyhedron $\begin{Bmatrix} p \\ q \end{Bmatrix}$.

In other words, the images of one vertex of the spherical triangle $(p\,q\,2)$ are the vertices of the appropriate polyhedron, for which a con-

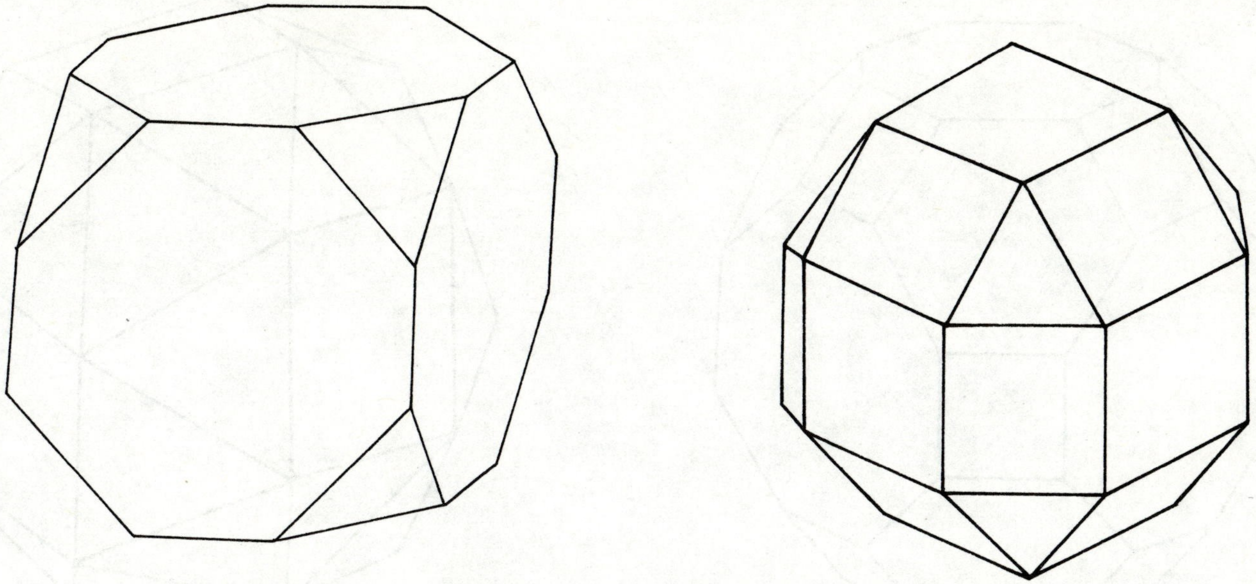

Figure 2·4B. The truncated cube t{4, 3} and the rhombicuboctahedron $r\begin{Bmatrix}4\\3\end{Bmatrix}$

venient symbol is obtained by drawing a ring round one node of the graph

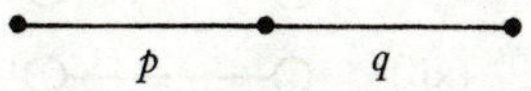

or

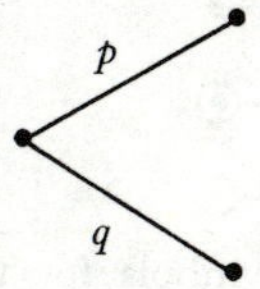

This node is the one representing the mirror on which the chosen point does *not* lie. Thus the symbols

$$\{p, q\} \quad \text{and} \quad \begin{Bmatrix}p\\q\end{Bmatrix}$$

may be regarded as abbreviations for

and

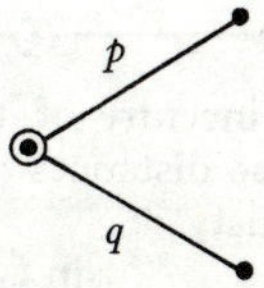

The construction can be extended by considering the images of a point on just one of the mirrors (equidistant from the other two) or on none of the mirrors. The graphical notation is correspondingly extended by allowing more than one node to be adorned with a ring. Again, the ringed nodes represent the mirrors on which the chosen point does not lie. When two nodes are ringed, the point is where the bisector of one angle of the spherical triangle meets the opposite side. (This ensures that the faces of the polyhedron shall be regular.) For instance,

and

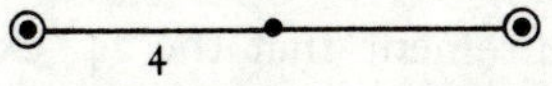

are the *truncated cube* and the *rhombicuboctahedron*, sketched in Figure 2·4B.†

When all three nodes have rings, the chosen point, equidistant from all three mirrors, is the incentre of the spherical triangle ($p\ q\ 2$). This is the point of concurrence of the three angle-bisectors. Its images are the incentres of all the triangles, white and shaded. For instance,

is the *truncated cuboctahedron*. (In this choice of terminology we are following Kepler. N.W. Johnson prefers to call it the 'rhombitruncated cuboctahedron' because an undistorted truncation of the cuboctahedron $\begin{Bmatrix}4\\3\end{Bmatrix}$ would have, among its faces, 12 rectangles instead of 12 squares.) This is sketched in Figure 2·4C, along with the *snub cube* (Kepler's *cubus simus*, which could perhaps have been better named 'snub cuboctahedron'). Its symmetry group is the octahedral group without

† See also the paper by Coxeter, Longuet-Higgins and Miller cited on p. 6.

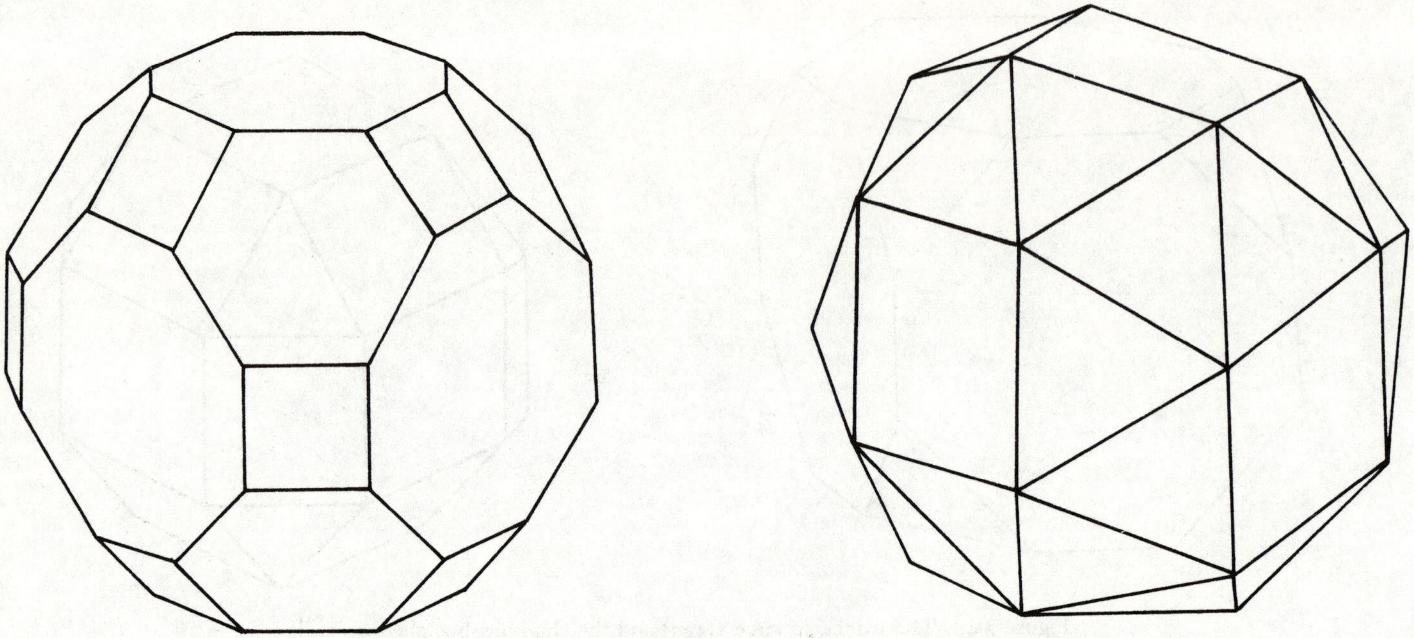

Figure 2·4c. The truncated cuboctahedron t$\begin{Bmatrix}4\\3\end{Bmatrix}$ and the snub cube s$\begin{Bmatrix}4\\3\end{Bmatrix}$

any reflections. The snub cube has one vertex inside each white triangle (4 3 2), not at the incentre but at another point, determined by the requirement that the 24 'extra' triangles must all be equilateral. The appropriate symbol is

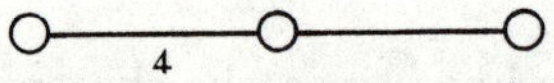

the absence of nodes indicating the absence of mirrors. (This justification was suggested by Mrs Alicia Boole Stott; see Coxeter 1963, pp. 258–9.) The notation extends in an obvious manner (with $s = \infty$) to regular and semi-regular tessellations in the Euclidean plane.

This procedure, whereby each reflection group yields a family of regular or 'semi-regular' polyhedra, is known as *Wythoff's construction.*

EXERCISES

1. Interpret the symbols

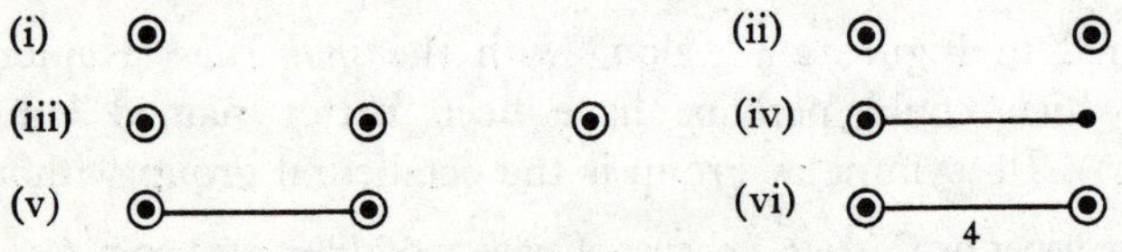

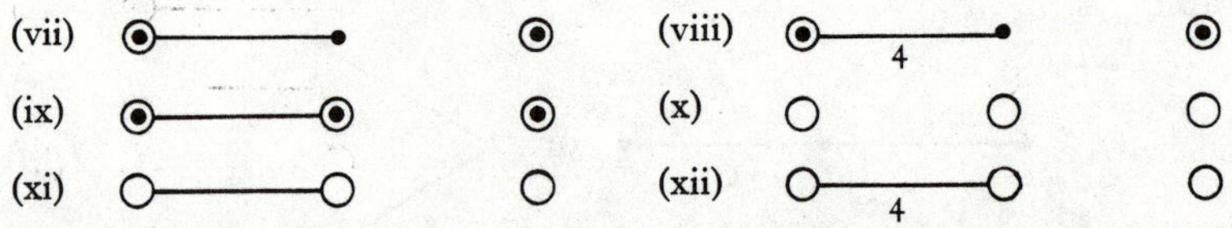

2. Assign such symbols for the rest of the thirteen Archimedean solids, namely the truncated tetrahedron, truncated octahedron, truncated icosahedron, truncated dodecahedron, rhombicosidodecahedron, truncated (or 'rhombi-truncated') icosidodecahedron, and snub dodecahedron (or 'snub icosidodecahedron').

3. Interpret the symbols

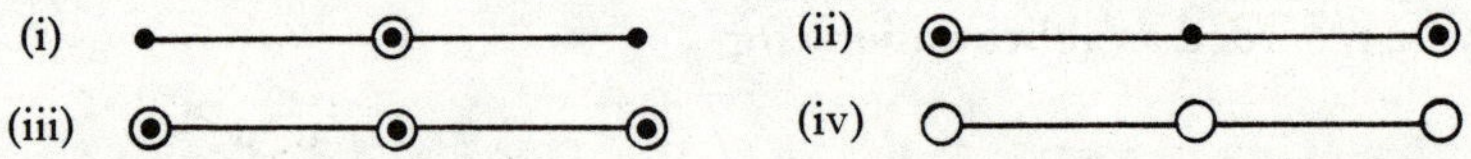

4. The incentre of the spherical triangle ($p\ q\ 2$) may be described as the point whose distances ξ_ν from the three vertices (measured along arcs of great circles) satisfy

$$\sin\xi_1 \sin\frac{\pi}{2p} = \sin\xi_2 \sin\frac{\pi}{2q} = \sin\xi_3 \sin\frac{\pi}{4}.$$

5. The typical vertex of 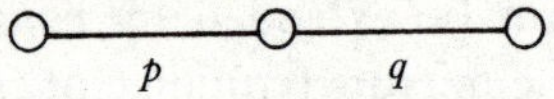 inside a 'white' triangle ($p\ q\ 2$) is the point whose distances ξ_ν from the three vertices satisfy

$$\sin\xi_1 \sin\frac{\pi}{p} = \sin\xi_2 \sin\frac{\pi}{q} = \sin\xi_3 \sin\frac{\pi}{2}.$$

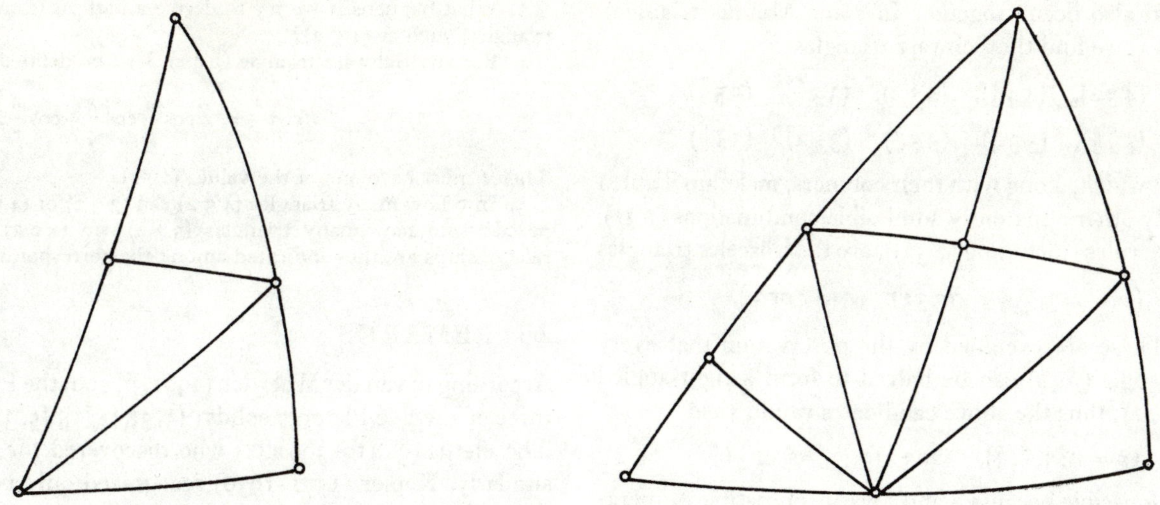

Figure 2·5A: $(5\,\frac{5}{2}\,2)$ and $(3\,\frac{5}{2}\,2)$

2·5 THE SCHWARZ TRIANGLES

We saw, in §2·3, that every finite reflection group (in Euclidean or non-Euclidean 3-space) has for its fundamental region one of the so-called *Möbius triangles*

$$(p\,p\,1),\quad (p\,2\,2),\quad (3\,3\,2),\quad (4\,3\,2),\quad (5\,3\,2).$$

In other words, these are the only spherical triangles which can be reflected in their sides to form a pattern covering the sphere just once. The sides of the triangles are arcs of great circles, and each pattern is symmetrical by reflection in each of the great circles of which it is composed. Although $(p\,q\,r)$ is the smallest triangle formed by these great circles, the pattern includes many other spherical triangles having longer sides and greater angles. These (along with the Möbius triangles themselves) are called *Schwarz triangles*, because they constitute the solution of a problem proposed by H. A. Schwarz (1843–1921) in 1873: to find every possible spherical triangle $(p\,q\,r)$ which will yield, by repeated reflections in its sides, a pattern of such triangles covering the sphere a finite number of times, say d times. Obviously p, q, r are now *rational* numbers. The Möbius triangles are the special cases in which these rational numbers are integers, and consequently $d = 1$. In Schwarz's case we still have the necessary condition (2·33). Since the group generated by reflections in the sides of a Schwarz triangle is equally well generated by reflections in the sides of one of the Möbius triangles, the extension does not yield new groups, merely new ways to generate the familiar reflection groups

$$[p],\quad [p,2],\quad [3,3],\quad [4,3],\quad [5,3].$$

Since the $4s$ elements of the appropriate reflection group transform the Schwarz triangle into $4s$ replicas filling the surface of the sphere d times, and transform the Möbius triangle into the same number of replicas filling the surface once, it follows that the Schwarz triangle is made up of just d replicas of the Möbius triangle; e.g. $(5\,\frac{5}{2}\,2)$ is made up of three $(5\,3\,2)$'s, and $(3\,\frac{5}{2}\,2)$ of seven (see Figure 2·5A). Each angle of the Schwarz triangle is a multiple of one of the angles of the appropriate Möbius triangle.

Referring to Figure 2·1A, we observe that Schwarz triangles may be classified into sets of four (or sometimes fewer) *colunar* triangles

$$(p\,q\,r),\quad (p\,q'\,r'),\quad (p'\,q\,r'),\quad (p'\,q'\,r),$$

where $\quad p^{-1}+p'^{-1} = 1,\quad q^{-1}+q'^{-1} = 1,\quad r^{-1}+r'^{-1} = 1.$

There is evidently a digon $(p\,p\,1)$ for every rational p. Halving this digon, we obtain a doubly right-angled triangle $(p\,2\,2)$. Apart from these two trivial cases, the only possible values for p, q, r are

$$2,\ 3,\ \tfrac{3}{2},\ 4,\ \tfrac{4}{3},\ 5,\ \tfrac{5}{2},\ \tfrac{5}{3},\ \tfrac{5}{4}. \tag{2·51}$$

Moreover, the numerators 4 and 5 cannot occur together (for, if they did, these numbers would also occur together in some Möbius triangle). Inspecting Figure 2·4A, we find the Schwarz triangles

$$(3\,3\,2),\quad (3\,3\,\tfrac{3}{2}),\quad (4\,3\,2),\quad (4\,4\,\tfrac{3}{2}),\quad (5\,3\,2),\quad (3\,3\,\tfrac{5}{2}),\quad (5\,5\,\tfrac{3}{2}),$$
$$(5\,\tfrac{5}{2}\,2),\quad (5\,3\,\tfrac{5}{3}),\quad (\tfrac{5}{2}\,\tfrac{5}{2}\,\tfrac{5}{2}),\quad (5\,3\,\tfrac{3}{2}),\quad (5\,5\,\tfrac{5}{4}),\quad (3\,\tfrac{5}{2}\,2),\quad (3\,\tfrac{5}{2}\,\tfrac{5}{3})$$

(Coxeter 1963, p. 296) which, along with their colunars, make up Table I on page 156. The only other apparently admissible combinations $(p\,q\,r)$, according to the above rules (including (2·33)), are the isosceles triangles

$$(p\,\tfrac{5}{2}\,\tfrac{5}{2})\quad (p = 2, 3, \tfrac{5}{3}, \tfrac{5}{4}),\quad (3\,3\,r)\quad (r = \tfrac{4}{3}\text{ or }\tfrac{5}{3}),$$

and their colunars. These are excluded by the observation that every isosceles Schwarz triangle $(p\,q\,q)$ can be halved to form a right-angled Schwarz triangle $(2p\,q\,2)$; thus the above candidates would yield

$$(2p\,\tfrac{5}{2}\,2)\quad (p = 2, 3, \tfrac{5}{3}, \tfrac{5}{4}),\quad (2r\,3\,2)\quad (r = \tfrac{4}{3}\text{ or }\tfrac{5}{3}).$$

The case $p = 2$ is inadmissible because 4 and 5 are incompatible numerators. The cases $p = 3, \tfrac{5}{3}$ are inadmissible because 6 and $\tfrac{10}{3}$ are not among the allowed values for p, q, r. The case $p = \tfrac{5}{4}$ is inadmissible because $(\tfrac{5}{2}\,\tfrac{5}{2}\,2)$ is the same as $(2\,\tfrac{5}{2}\,\tfrac{5}{2})$, which we have already excluded. Similarly, we cannot have $r = \tfrac{4}{3}$ or $\tfrac{5}{3}$. Hence Table I is complete.

In §2·4 we found that, if p and q are positive integers satisfying

$$p^{-1}+q^{-1} > \tfrac{1}{2},$$

we can apply Wythoff's construction to the right-angled Möbius triangle $(p\,q\,2)$ and obtain a pair of reciprocal Platonic solids (or a pair of dual spherical tessellations)

$$\{p, q\} \quad\text{and}\quad \{q, p\}.$$

When we apply the same procedure to a right-angled Schwarz triangle $(p\,q\,2)$, p and q being rational, we still obtain a pair of reciprocal regular polyhedra (or dual tessellations) $\{p, q\}$ and $\{q, p\}$, but now their density d may be greater than 1. Since the only non-trivial Schwarz triangles with $r = 2$ are

$$(3\,3\,2),\quad (4\,3\,2),\quad (5\,3\,2),\quad (\tfrac{5}{2}\,3\,2),\quad (\tfrac{5}{2}\,5\,2)$$

and their colunars, we verify in this simple manner that the list of regular spherical tessellations in §2·1 is complete.†

EXERCISES

1. Given a Schwarz triangle, how can we tell at a glance whether the relevant group is $[p]$ or $[p, 2]$ or $[3, 3]$ or $[4, 3]$ or $[5, 3]$?
2. The 'density' of the pattern of Schwarz triangles $(p\;q\;r)$ is
$$d = (p^{-1}+q^{-1}+r^{-1}-1)s$$
where s is obtained by applying (2·34) to the corresponding Möbius triangle.

† For the result of applying the same procedure to the general Schwarz triangle, see the monograph *Uniform Polyhedra* cited on page 6, and Wenninger 1971.

3. What happens if we try to derive a pattern from one of the inadmissible triangles, such as $(4\;\tfrac{5}{2}\;2)$?
4. For any Schwarz triangle $(p\;q\;r)$, let t be defined by
$$\cos\frac{\pi}{t} = 2\cos\frac{\pi}{p}\cos\frac{\pi}{q}+\cos\frac{\pi}{r}.$$
Then t must have one of the values (2·51).
5. Into how many triangles $(4\;3\;2)$ can $(3\;3\;2)$ or $(4\;2\;2)$ or $(2\;2\;2)$ be decomposed? Into how many triangles $(5\;3\;2)$ can $(2\;2\;2)$ be decomposed? What relationships are thus indicated among the corresponding groups?

2·6 REMARKS

According to van der Waerden (1961, p. 100) the Pythagoreans knew only three of the five Platonic solids: $\{3, 3\}$, $\{4, 3\}$, $\{5, 3\}$. It was Plato's friend Theaetetus (c. 410–369 B.C.) who discovered the other two. Somewhat similarly, Kepler (1571–1630) recognized only two of the four regular star polyhedra: $\{\tfrac{5}{2}, 3\}$ and $\{\tfrac{5}{2}, 5\}$. Poinsot (1777–1859) rediscovered them and discovered their reciprocals. The principle of duality, which includes the notion that $\{p, q\}$ and $\{q, p\}$ are reciprocal figures, was first fully understood by Gergonne (1771–1859). The regular spherical tessellations (Fejes Tóth 1964, p. 59), which are topologically equivalent to the Platonic solids, were described by Abû'l Wafâ (940–98), and the analogous planar tessellations (2·23) by Kepler. The hosohedron $\{2, p\}$ (in a slightly distorted form) was named by Vito Caravelli (1724–1800), and the dihedron $\{p, 2\}$ by Felix Klein (1849–1925). The symbol $\{p, q\}$, which reveals all these figures as members of one family, was invented by Ludwig Schläfli (1814–95).

The definition of regularity in terms of flags was proposed by Jacques Tits (see also Du Val 1964, p. 63). The Petrie polygon began its useful career in 1925 (see Coxeter 1963, p. 32). Reflection groups were investigated by Brewster (1781–1868, who coined the word *kaleidoscope*), Möbius (1790–1868, who almost anticipated 'Wythoff's construction'), Klein (who derived them from the rotation groups and consequently called them the *extended* groups), and Weyl (1885–1955, who used them in his classification of simple continuous groups; see Bourbaki 1968).

CHAPTER 3

Polyhedral kaleidoscopes

There is something pleasing to a mystic in such a land of mirrors. For a mystic... holds that two worlds are better than one. In the highest sense, indeed, all thought is reflection.

G. K. Chesterton (1912, p. 203)

In §3·1, the angular properties and radii of $\{p, q\}$ are expressed in terms of a kind of *frieze pattern* involving five numbers $\alpha, \beta, \gamma, \delta, \epsilon$ which are functions of p and q. The next section gives detailed information for putting three mirrors together to make an icosahedral kaleidoscope, in which one or two suitably placed line-segments will yield, as images, the complete set of edges of any one of the polyhedra $\{3, 5\}$, $\{5, 3\}$, $\{\frac{5}{2}, 5\}$, $\{5, \frac{5}{2}\}$, $\{\frac{5}{2}, 3\}$. In §3·3, the vertices and edges of certain Archimedean solids are used as Cayley diagrams for rotation groups and reflection groups, so as to establish simple presentations for these groups. In §3·4, each reflection group is shown to be either isomorphic or homomorphic to a rotation group.

3·1 THE CHARACTERISTIC ORTHOSCHEME

The tetrahedron *CMIO* (Figure 2·2A) is called an *orthoscheme* (or 'quadrirectangular tetrahedron') because its faces *MOI*, *COI*, *COM*, *CIM* are four right-angled triangles; in other words, the three edges *CM*, *MI*, *IO* are mutually perpendicular. Figure 3·1A shows an unfolded net which may be cut out from a sheet of thin cardboard, scored along the lines *MO*, *OC*, *CI*, folded downwards along these three edges, and stuck together along the remaining edges, to make a model (for the case $p = 4$, $q = 3$) with the letters ϕ, χ, ψ drawn on the outside. The three planes (2·21), being the mirrors of the polyhedral kaleidoscope, cut out (from the unit sphere round O) the Möbius triangle $(p\,q\,2)$ (Figure 3·1B) whose vertices P, Q, R lie on OI, OC, OM, respectively. Thus the dihedral angles along these edges of the orthoscheme are π/p, π/q, $\pi/2$.

The sides of the Möbius triangle, namely

$$\phi = QR = \angle COM, \quad \psi = RP = \angle MOI, \quad \chi = PQ = \angle IOC,$$

can be found by spherical trigonometry or as follows. Defining two further planes, $\rho_4 = CMI$ (the plane of the face) perpendicular to OI, and $\rho_5 = \rho_0$ (the plane of the vertex figure) perpendicular to OC, we have altogether a *cycle* of five planes ρ_ν such that any two non-adjacent planes in the cycle are perpendicular; that is, if $\theta_{\mu\nu}$ is the non-obtuse dihedral angle formed by ρ_μ and ρ_ν,

$$\theta_{02} = \theta_{13} = \theta_{24} = \theta_{30} = \theta_{41} = \tfrac{1}{2}\pi.$$

It follows that any two of the remaining angles

$$\theta_{12} = \angle CIM = \frac{\pi}{p}, \quad \theta_{23} = \frac{\pi}{q}, \quad \theta_{34} = \angle OMI = \tfrac{1}{2}\pi - \psi,$$

$$\theta_{40} = \angle IOC = \chi, \quad \theta_{01} = \angle OCM = \tfrac{1}{2}\pi - \phi$$

will determine the rest, and any equation relating three of them will remain valid when the five planes (and hence the five angles) are cyclically permuted.

Since
$$\sin\phi = \frac{CM}{CO} = \frac{CI}{CO}\frac{CM}{CI} = \sin\chi \sin\frac{\pi}{p},$$

one such equation is $\cos\theta_{01} = \sin\theta_{40}\sin\theta_{12}$. Applying the permutation, we deduce $\cos\theta_{12} = \sin\theta_{01}\sin\theta_{23}$ and $\cos\theta_{23} = \sin\theta_{12}\sin\theta_{34}$, that is,

$$(3{\cdot}11) \qquad \cos\frac{\pi}{p} = \cos\phi\sin\frac{\pi}{q}, \quad \cos\frac{\pi}{q} = \sin\frac{\pi}{p}\cos\psi.$$

Since
$$\sin\psi = \frac{MI}{MO} = \frac{MC}{MO}\frac{MI}{MC} = \tan\phi\cot\frac{\pi}{p},$$

another such equation is $\cos\theta_{34} = \cot\theta_{01}\cot\theta_{12}$, whence

$$\cos\theta_{40} = \cot\theta_{12}\cot\theta_{23},$$

that is,

$$(3{\cdot}12) \qquad \cos\chi = \cot\frac{\pi}{p}\cot\frac{\pi}{q}.$$

We have thus obtained expressions for ϕ, χ, ψ in terms of p and q (cf. Coxeter 1963, p. 21).

The above arrangement of five planes is essentially the same as the *pentagramma mirificum* of Napier and Gauss (Coxeter 1965, p. 270),

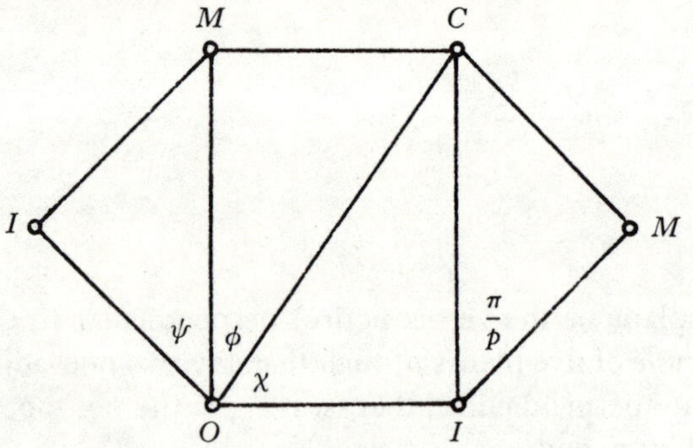

Figure 3·1A: The unfolded orthoscheme

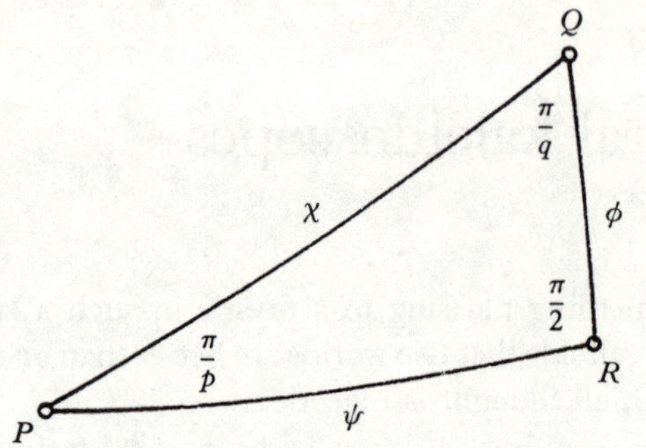

Figure 3·1B: The Möbius triangle

the five angles $\theta_{12}, \ldots, \theta_{01}$ being the five 'parts' of the spherical triangle *PQR*. Gauss expressed the equations in terms of five numbers

$$\alpha = \tan^2\theta_{12}, \quad \beta = \tan^2\theta_{23}, \quad \gamma = \tan^2\theta_{34}, \quad \delta = \tan^2\theta_{40}, \quad \epsilon = \tan^2\theta_{01}$$

and observed that any three of the five relations

$$(3{\cdot}13) \qquad 1+\alpha = \gamma\delta, \quad 1+\beta = \delta\epsilon, \quad 1+\gamma = \epsilon\alpha, \quad 1+\delta = \alpha\beta, \quad 1+\epsilon = \beta\gamma$$

imply the remaining two. In other words, if we make a *frieze pattern*

$$\begin{array}{ccccccccc} & 1 & & 1 & & 1 & & 1 & \\ & & \alpha & & \beta & & \gamma & & \ldots \\ & & & \delta & & \epsilon & & & \ldots \\ & & 1 & & 1 & & 1 & & \end{array}$$

with the rule that every four adjacent numbers

$$\begin{array}{ccc} & b & \\ a & & d \\ & c & \end{array}$$

satisfy $ad-bc = 1$, the pattern is necessarily periodic. (This way of expressing Gauss's equations may seem artificial, but in Chapter 5 we shall find wider frieze patterns helpful in four dimensions.) Thus each regular polyhedron $\{p, q\}$ yields a frieze pattern with

$$\alpha = \tan^2\frac{\pi}{p}, \quad \beta = \tan^2\frac{\pi}{q}, \quad \gamma = \cot^2\psi, \quad \delta = \tan^2\chi, \quad \epsilon = \cot^2\phi.$$

For instance, the cube $\{4, 3\}$ yields

$$\begin{array}{ccccccccccccccc} & 1 & & 1 & & 1 & & 1 & & 1 & & 1 & & \ldots \\ 1 & & 3 & & 1 & & 2 & & 2 & & 1 & & 3 & \\ & 2 & & 2 & & 1 & & 3 & & 1 & & 2 & & \ldots \\ 1 & & 1 & & 1 & & 1 & & 1 & & 1 & & 1 & \end{array}$$

Such patterns for all the regular polyhedra can be derived from the values of $\alpha = \tan^2\pi/p$ and $\beta = \tan^2\pi/q$, as in the following table (which uses the notation of (1·91)):

$\{p, q\}$	α	β	γ	δ	ϵ
$\{3, 3\}$	3	3	$\frac{1}{2}$	8	$\frac{1}{2}$
$\{4, 3\}$	1	3	1	2	2
$\{5, 3\}$	$\sqrt{5}\,\tau^{-3}$	3	τ^2	$4\tau^{-4}$	τ^4
$\{\frac{5}{2}, 3\}$	$\sqrt{5}\,\tau^3$	3	τ^{-2}	$4\tau^4$	τ^{-4}
$\{\frac{5}{2}, 5\}$	$\sqrt{5}\,\tau^3$	$\sqrt{5}\,\tau^{-3}$	τ^2	4	τ^{-2}

The remaining four rows of the table can be supplied immediately by observing that the values of

$$\alpha, \beta, \gamma, \delta, \epsilon \quad \text{for} \quad \{q, p\}$$

are the values of $\quad \beta, \alpha, \epsilon, \delta, \gamma \quad$ for $\quad \{p, q\}$.

Figure 3·1C shows the orthoscheme *CMIO* for $\{p, q\}$ (compare Figures 2·2A and 3·1A) with its six edges expressed as 'monomials' in $\alpha, \beta, \gamma, \delta, \epsilon$. For simplicity we have taken $CM = 1$, so that the edge-length of $\{p, q\}$ is 2. We recognize the edges

$$OC = \sqrt{(\beta\gamma)}, \quad OM = \sqrt{\epsilon}, \quad OI = \sqrt{\frac{\gamma}{\alpha}}$$

as being the radii of three concentric spheres: the *Circumsphere* which passes through all the vertices of $\{p, q\}$, the *Midsphere* (or 'intersphere') which touches all the edges at their midpoints, and the *Insphere* which touches all the face-planes at the face-centres.

In the following discussion of area and volume we shall assume that p and q are integers, so that $\{p, q\}$ is one of the five Platonic solids. In terms of s, where

$$(3{\cdot}14) \qquad s^{-1} = p^{-1}+q^{-1}-\tfrac{1}{2}$$

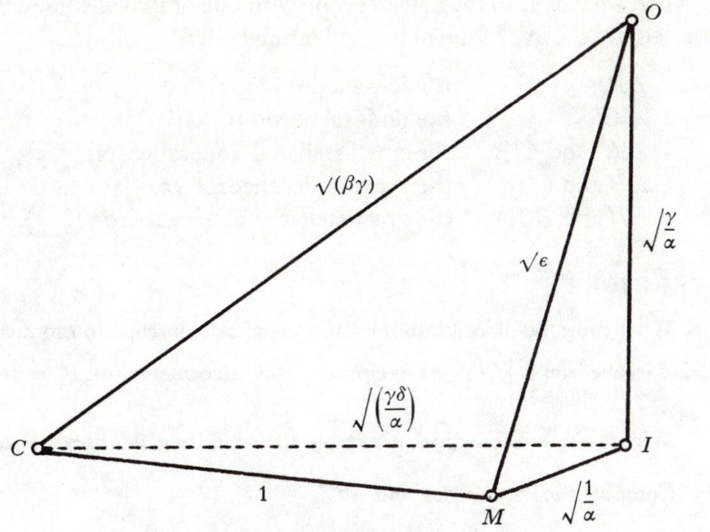

Figure 3.1C: Edge-lengths for the orthoscheme

(see (2·34)), $\{p, q\}$ has s edges, and its surface area S is made up of $4s$ triangles such as CMI. Thus

$$S = 2sCM \times MI = \frac{2s}{\sqrt{\alpha}}$$

or, if we take the edge of $\{p, q\}$ to be $2l$ instead of 2,

$$S = \frac{2s}{\sqrt{\alpha}} l^2.$$

Since the solid $\{p, q\}$ can be dissected into pyramids having height OI and base $\{p\}$, one for each face, its volume is

$$\tfrac{1}{3} OI \times S = \frac{2s\sqrt{\gamma}}{3\alpha} l^3$$

and the 'isoperimetric ratio' (surface cubed over volume squared) is

$$\frac{18s\sqrt{\alpha}}{\gamma}.$$

EXERCISES

1. If $\{p, q\}$ is $\{4, 3\}$, as in Figures 2·2A and 3·1A, the mutually perpendicular edges CM, MI, IO of the orthoscheme $CMIO$ belong to a helical regular polygon generated by a right-handed twist. Draw the analogue of Figure 3·1A for the adjacent orthoscheme $BMIO$, whose twist is left-handed.

2. If CD is the altitude from the hypotenuse AB of a right-angled spherical triangle ABC,

$$\cos^2 CD = \cos^2 A + \cos^2 B.$$

Deduce an alternative derivation of (2·22).

3. Express cosec π/h in terms of γ and ϵ.

4. If reciprocal solids $\{p, q\}$ and $\{q, p\}$ are inscribed in the same sphere, the circumcircles of their faces are congruent (and consequently the two solids, having the same circumsphere, also have the same insphere). If they are Platonic solids with $p > q$, which of them has the greater surface area (and therefore the greater volume)?

5. The sides of the Möbius triangle $(p\ q\ 2)$ lie in the planes

$$\rho_1 = MOI = ROP, \quad \rho_2 = COI = QOP, \quad \rho_3 = COM = QOR.$$

Choosing Cartesian axes such that ρ_1 and ρ_3 are the coordinate planes $x = 0$ and $z = 0$, find an equation for ρ_2 with coefficients involving γ and ϵ.

6. The frieze pattern for the cube (page 22) is the only such pattern having two rows of *positive integers* between its two rows of ones. There is an analogous wider pattern in which the first row of ones is followed by

4 1 2 2 2 1 4 1 2 2 ....

Are there any other frieze patterns having three rows of positive integers between two rows of ones?

7. If a frieze pattern is embellished with borders of 0's outside the present borders of 1's at the top and bottom, it is still true that rotation through $-45°$ yields a set of 2×2 matrices all having determinant 1. Analogously, we might consider a generalized frieze pattern

0		0		0		0		0		0	
	1		1		1		1		1		
α		β		γ		δ		ϵ		...	
	1		1		1		1		1		
0		0		0		0		0		0	

such that rotation through $-45°$ yields a set of 3×3 matrices all having determinant 1. Is this pattern likewise periodic?

3·2 THE ICOSAHEDRAL KALEIDOSCOPE

In the case of $\{4, 3\}$, $\psi + \phi + \chi = \frac{3}{4}\pi$, as we see in Figure 3·1A. Figure 3·2A shows the more interesting case of $\{5, 3\}$, where $\psi + \phi + \chi = \frac{1}{2}\pi$ (Coxeter 1963, p. 74) so that the triangles MOI, COI, COM can be cut out from a square (with one corner truncated along the line CM). In this figure, the orientation has been reversed so as to indicate the specification for a practical *icosahedral kaleidoscope* in which the cardboard is replaced by three front-silvered triangular mirrors, suitably mounted and hinged behind along the edges OC and OM so that the two edges OI can be lifted up and finally brought together to form the third edge of the trihedron. The shapes of the three mirrors are determined by the relative

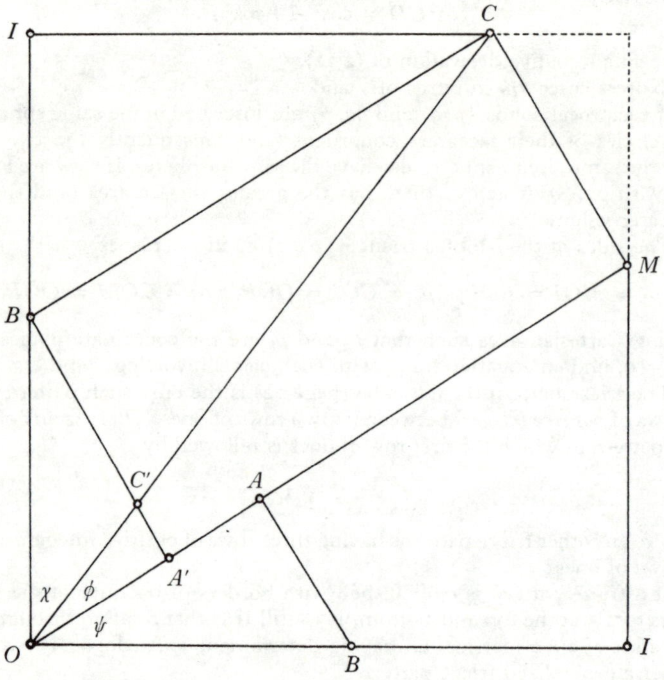

Figure 3·2A: The unfolded kaleidoscope

lengths of the six edges of the orthoscheme $CMIO$. The middle row of the table on page 22 shows that these lengths may be taken to be

$$OC = \sqrt{(\beta\gamma)} = 3^{\frac{1}{2}}\tau \approx 2{\cdot}80, \quad OM = \sqrt{\epsilon} = \tau^2 \approx 2{\cdot}62,$$
$$OI = \sqrt{(\gamma/\alpha)} = 5^{-\frac{1}{4}}\tau^{\frac{5}{2}} \approx 2{\cdot}23,$$
$$MI = 1/\sqrt{\alpha} = 5^{-\frac{1}{4}}\tau^{\frac{3}{2}} \approx 1{\cdot}38, \quad IC = \sqrt{(\gamma\delta/\alpha)} = 2 \times 5^{-\frac{1}{4}}\tau^{\frac{1}{2}} \approx 1{\cdot}72,$$
$$CM = 1.$$

(A convenient unit is the decimetre.)

The remaining line-segments in the figure (namely $A'C'$, $C'B$, AB, all parallel to CM, and BC parallel to OM, so that $A'BCM$ is a rectangle) should be inserted, one by one, in the form of narrow strips of coloured plastic, folded over the edges of the mirrors and stuck behind. Their positions are such that

$$OA = A'B = CM = 1, \quad OB = 5^{\frac{1}{4}}\tau^{-\frac{1}{2}} \approx 1{\cdot}17, \quad OA' = AB = \tau^{-1} \approx 0{\cdot}62,$$
$$OC' = 3^{\frac{1}{2}}\tau^{-2} \approx 0{\cdot}66, \quad BC' = 2\tau^{-2} \approx 0{\cdot}76, \quad BC = A'M = 2.$$

When we look into the kaleidoscope with one or two of these segments inserted, we see five of the nine regular polyhedra, as follows:

AB,	the icosahedron $\{3, 5\}$;
$A'C'$,	the dodecahedron $\{5, 3\}$;
$A'C'$ and $C'B$,	the small stellated dodecahedron $\{\frac{5}{2}, 5\}$;
AB and $C'B$,	the great dodecahedron $\{5, \frac{5}{2}\}$;
AB and BC,	the great stellated dodecahedron $\{\frac{5}{2}, 3\}$.

EXERCISES

1. What could be placed into the icosahedral kaleidoscope to produce (i) the icosidodecahedron $\left\{\begin{matrix}5\\3\end{matrix}\right\}$, (ii) its reciprocal, the triacontahedron (Coxeter 1963, p. 25)?
2. Why is the kaleidoscope unsuitable for exhibiting the great icosahedron $\{3, \frac{5}{2}\}$?
3. Compute the lengths AA' and $A'C'$.

3·3 CAYLEY DIAGRAMS AND PRESENTATIONS

Figure 3·3A shows a fragment of the regular tessellation $\{3, 6\}$ with each of its 'upright' faces marked A or B or C. Every edge belongs to just one of these marked faces, and is regarded as being marked accordingly. To make this clearer, the A edges have been drawn in full lines, the B edges in broken lines, and the C edges in dotted lines. (Ideally, three different colours should be used.) Any path along edges, from one vertex to another (or the same) vertex, may be regarded as representing a *word*; for instance, one path from the lower left vertex to the upper right vertex represents the word

$$ACBA^{-1}B^{-1}C^{-1},$$

while another path between the same two points represents the word

$$A^{-1}B^{-1}C^{-1}ACB,$$

and this last path in the reverse direction represents the inverse word

$$(A^{-1}B^{-1}C^{-1}ACB)^{-1} = B^{-1}C^{-1}A^{-1}CBA.$$

By equating the words represented by different paths between the same two points, we obtain a property of the group generated by rotations A, B, C which are symmetry operations of the whole pattern. For instance, we have

$$ACBA^{-1}B^{-1}C^{-1} = A^{-1}B^{-1}C^{-1}ACB$$

or

$$ACBA^{-1}B^{-1}C^{-1}.B^{-1}C^{-1}A^{-1}CBA = 1.$$

Any word that is equal to the identity, in this sense, is represented by a

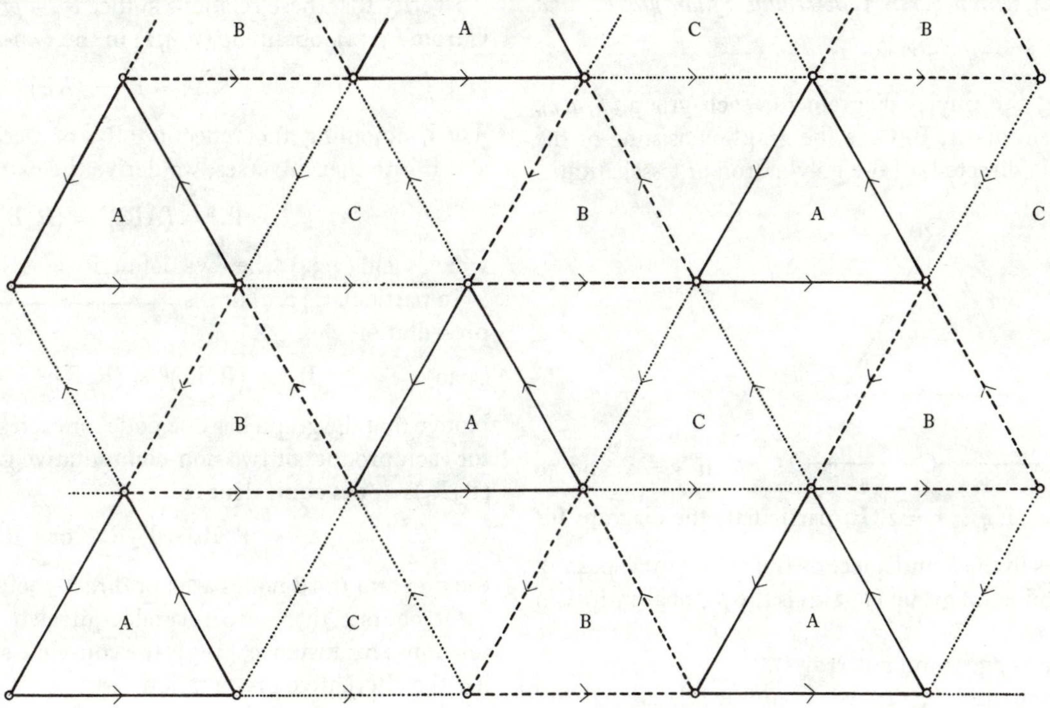

Figure 3·3A: A Cayley diagram for (3, 3, 3)

circuit, that is, a closed path. Since the plane is *simply connected*, any circuit more complicated than a triangle may be regarded as the combination of two simpler circuits juxtaposed. For instance, the rhombus ABC^{-2} can be split up into the two triangles ABC and C^{-3} whose common side cancels out in accordance with the relation $CC^{-1} = 1$. Since every circuit is ultimately a combination of such simple triangles, every true relation satisfied by the rotations A, B, C is an algebraic consequence of the four relations represented by the various types of triangle, namely

$$A^3 = 1, \quad B^3 = 1, \quad C^3 = 1, \quad ABC = 1$$

or, more concisely,

$$(3{\cdot}31) \qquad A^3 = B^3 = C^3 = ABC = 1.$$

This group, being generated by rotations through equal angles $2\pi/3$ about the three vertices of an equilateral triangle of angle $\pi/3$, is (3, 3, 3) in the '(p, q, r)' notation that was introduced at the end of §2·3. Figure 3·3A is a *Cayley diagram* (Coxeter and Moser 1972, p. 21) for (3, 3, 3) in terms of these generators. What we have just done is to use the Cayley diagram to establish the presentation (3·31).

More generally (but still using a *planar* map to form the Cayley diagram), the same ideas can be employed to establish the presentation

$$(3{\cdot}32) \qquad A^p = B^q = C^r = ABC = 1$$

for the group (p, q, r) generated by (counterclockwise) rotations through $2\pi/p$, $2\pi/q$, $2\pi/r$ about the three vertices (in clockwise order) of a triangle PQR having angles π/p at P, π/q at Q, π/r at R. Instead of the triangles marked A, B, C in Figure 3·3A, we now have polygons $\{p\}$, $\{q\}$, $\{r\}$ representing the relations $A^p = 1$, $B^q = 1$, $C^r = 1$; but the 'inverted' triangles (representing $ABC = 1$) remain triangles. If $r = 2$, the digon is conveniently replaced by a single undirected edge. If

$$p^{-1}+q^{-1}+r^{-1} = 1,$$

the plane is still Euclidean, but if the sign = is replaced by < or >, it is hyperbolic or spherical, respectively. Although the sphere, the Euclidean plane and the hyperbolic plane are all simply connected, only the sphere has a *finite area*. We have thus succeeded in proving, by means of a combination of geometry and topology, the following theorem of pure algebra (Coxeter and Moser 1972, p. 55):

(3·33) *The relations* (3·32), *with* $p, q, r > 1$, *determine a finite group if and only if*

$$p^{-1}+q^{-1}+r^{-1} > 1.$$

In the notation of §2·4, the Cayley diagram for each group (p, q, r), in terms of the three generators A, B, C, is the graph consisting of the vertices and edges (suitably directed) of the polyhedron or tessellation

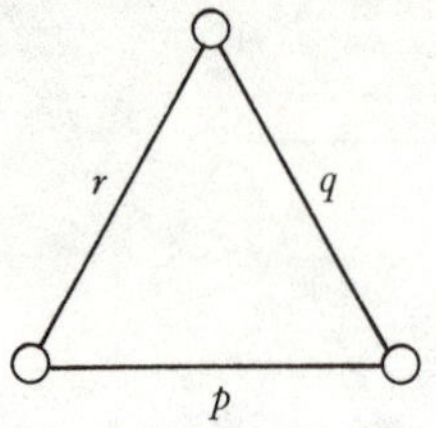

This reduces to 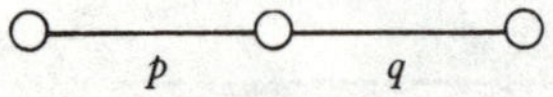

if $r = 2$ and to 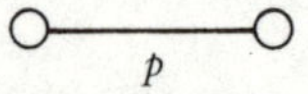

 if $q = r = 2$. In particular, the diagram for (4, 3, 2) is the snub cube, as in the frontispiece of Burnside (1911, p. 423), while the diagram for the dihedral group $(p, 2, 2)$ is the p-gonal antiprism or, if $p = 2$, the tetrahredron  with marks A, B, C on pairs of opposite edges (and no arrows).

Setting $r = 1$ in the presentation (3·32), we obtain the relation $AB = 1$ which forces A and B to have the same period, thus justifying the rule that, if we allow one of p, q, r to have the value 1, the other two must be equal.

We have seen that the reflection group

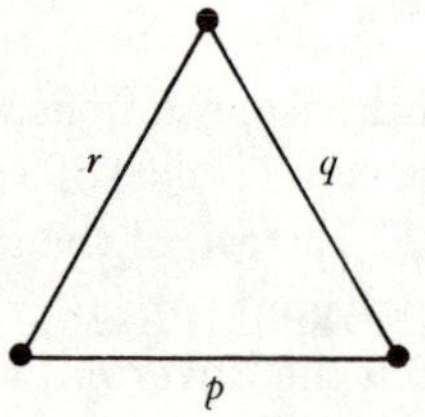

or

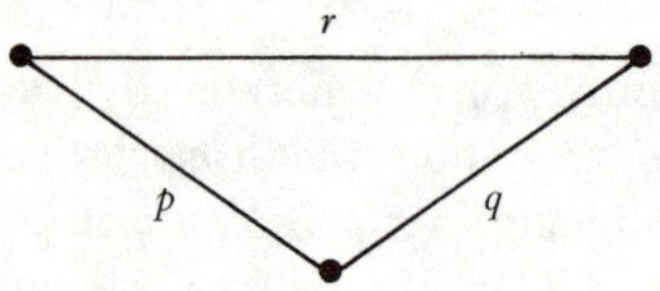

(which reduces to $[p, q]$ when we set $r = 2$) is generated by reflections in the three sides of the triangle $(p\,q\,r)$, and that the corresponding rotation group (p, q, r) is the subgroup of index 2 generated by products of pairs of these reflections. Setting

$$A = R_1R_2, \quad B = R_2R_3, \quad C = R_3R_1$$

in (3·32), we see that the reflections satisfy

(3·34) $$R_\nu^{\,2} = (R_1R_2)^p = (R_2R_3)^q = (R_3R_1)^r = 1.$$

To verify that these relations suffice for a *presentation*, we first eliminate C from (3·32), obtaining (p, q, r) in the two-generator form

(3·35) $$A^p = B^q = (AB)^r = 1.$$

Then, adjoining the reflection R_2, of period 2, which transforms A and B into their inverses, we derive the extra relations

$$R_2^{\,2} = (AR_2)^2 = (R_2B)^2 = 1,$$

which yield (3·34) when we define $R_1 = AR_2$, $R_3 = R_2B$.

In particular, $[p, q]$ or 

(of order $4s$) has the presentation

(3·36) $$R_\nu^{\,2} = (R_1R_2)^p = (R_2R_3)^q = (R_3R_1)^2 = 1.$$

Notice that the graph has one node for each generator R_ν, and one branch for each product of two non-commutative generators. Since the relation $(R_3R_1)^2 = 1$ is equivalent to

$$R_1R_3 = R_3R_1 \quad \text{or} \quad R_1 \rightleftarrows R_3,$$

the first and third nodes are not directly joined.

Of course, there are several equivalent ways to express the same relations; for instance, [4, 3] (the complete symmetry group of the cube) has the alternative presentation

$$R_1^{\,2} = R_2^{\,2} = 1, \quad (R_1R_2)^2 = (R_2R_1)^2, \quad R_2R_3R_2 = R_3R_2R_3, \quad R_1 \rightleftarrows R_3.$$

EXERCISES

1. We have seen that, when the octahedral group (4, 3, 2) is presented in the form
$$A^4 = B^3 = C^2 = ABC = 1,$$
its Cayley diagram is the snub cube. When one of the three generators is eliminated, the corresponding edges of the diagram have to be deleted. After some obvious topological adjustment, the Cayley diagram for the presentation
$$A^4 = B^3 = (AB)^2 = 1$$
may be described as the rhombicuboctahedron. Give analogous descriptions for the presentations of the same group in terms of (i) A and C, (ii) B and C.

2. Describe the Cayley diagram for the group [4, 3] (generated by three reflections).

3·4 FINITE GROUPS GENERATED BY HALF-TURNS

In view of the fact that every rotation group, except $\mathfrak{C}_p$ with p odd, contains at least one half-turn, it is natural to ask which groups contain enough half-turns to generate them. Clearly the single half-turn in $\mathfrak{C}_{2p}$ generates $\mathfrak{C}_2$, and the three half-turns in the tetrahedral group $\mathfrak{A}_4 \simeq (3, 3, 2)$ generate its subgroup $\mathfrak{D}_2 \simeq (2, 2, 2)$ of order 4. It will emerge from the following discussion that all the remaining rotation groups are generated by the half-turns that they contain.

As we saw in (3·35), we can eliminate one of the three generators of (p, q, r) and obtain a presentation involving only three relations. Eliminating A from (3·32), we obtain

$$B^q = C^r = (BC)^p = 1.$$

In particular, the dihedral group $(p, 2, 2)$ is generated by two of its half-turns in the form

$$B^2 = C^2 = (BC)^p = 1.$$

The axes of these half-turns are two intersecting lines forming an angle π/p in (say) a horizontal plane. Let R_0 denote the reflection in this plane, R_1 and R_2 the reflections in vertical planes through the two lines, so that $R_\nu{}^2 = 1$ and

$$B = R_0 R_1 = R_1 R_0, \quad C = R_0 R_2 = R_2 R_0.$$

Then R_1 and R_2 generate the isomorphic dihedral group $[p]$ in the form (1·22).

Another way to exhibit the isomorphism

$$[p] \cong (p, 2, 2)$$

is to use the central inversion I (which is commutative with each R_ν) so that the products

$$U_\nu = IR_\nu \quad (\nu = 1 \text{ or } 2)$$

are half-turns about lines perpendicular to the two vertical mirrors. In terms of these half-turns we have

(3·41) $$U_1{}^2 = U_2{}^2 = (U_1 U_2)^p = 1.$$

Similarly, any reflection group $[p, q]$ that does *not* contain I yields an isomorphic rotation group generated by half-turns

$$U_\nu = IR_\nu \quad (\nu = 1, 2, 3).$$

The presentation

(3·42) $$U_\nu{}^2 = (U_1 U_2)^p = (U_2 U_3)^q = (U_3 U_1)^2 = 1$$

comes from (3·36). Since all the Platonic solids except the tetrahedron are centrally symmetric, the only actual cases of this isomorphism are

$$[p, 2] \cong (2p, 2, 2) \quad (p \text{ odd}) \quad \text{and} \quad [3, 3] \cong (4, 3, 2).$$

In the former case (where $q = 2$), the axes of the half-turns are two horizontal lines inclined at π/p and one vertical line. In the latter (where $p = q = 3$), they are indicated by the points emphasized in Figure 3·4A. These points on the sphere are poles of the great circles that bound the shaded triangle (3 3 2).

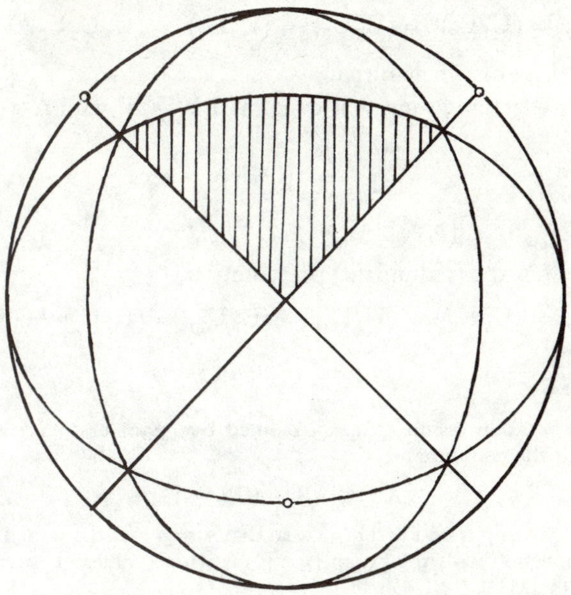

Figure 3·4A: The planes of symmetry for {3, 3}

On the other hand, if $[p, q]$ contains the central inversion, it is the direct product of the $\mathfrak{C}_2$ generated by I and the rotation group $(p, q, 2)$ generated by U_1, U_2, U_3. Since I is an *opposite* isometry, its expression $R_\mu \ldots R_\nu$ (as an element of $[p, q]$) must involve an odd number of R's, and the corresponding word $U_\mu \ldots U_\nu$ is the identity. Hence one way to present $(p, q, 2)$ is by (3·42) with the extra relation $U_\mu \ldots U_\nu = 1$.

In §1·7 we expressed a rotatory-reflection S as the product UV of a half-turn and a reflection. When S is the rotatory-reflection associated with the Petrie polygon for $[p, q]$, as in §2·2, we have $U = R_3 R_1$ and $V = R_2$, so that $S = R_3 R_1 R_2$, which is conjugate to $R_1 R_2 R_3$. Thus $R_1 R_2 R_3$ is a rotatory-reflection involving rotation through $2\pi/h$ (where h is given by (2·22)), and the relations (3·36) must imply

(3·43) $$(R_1 R_2 R_3)^h = 1.$$

When $h/2$ is even, $(R_1 R_2 R_3)^{h/2}$ is a rotation of period 2, that is, a half-turn. This happens in the case of [3, 3] (with $h = 4$).

When $h/2$ is odd, $(R_1 R_2 R_3)^{h/2}$ is a rotatory-reflection of period 2, that is, a central inversion. Thus, for [4, 3] (with $h = 6$) or [5, 3] (with $h = 10$),

(3·44) $$I = (R_1 R_2 R_3)^{h/2} \quad (h = 6 \text{ or } 10),$$

(3·45) $$[p, 3] \cong \mathfrak{C}_2 \times (p, 3, 2) \quad (p = 4 \text{ or } 5),$$

and $(p, 3, 2)$ has the presentation

(3·46) $\quad U_\nu^2 = (U_1 U_2)^p = (U_2 U_3)^3 = (U_3 U_1)^2 = (U_1 U_2 U_3)^{h/2} = 1$

in terms of three of its half-turns.

Another reflection group that contains I is $[p, 2]$ with p even. In this case,

(3·47) $$I = (R_1 R_2)^{p/2} R_3,$$

(3·48) $$[p, 2] \cong \mathfrak{C}_2 \times (p, 2, 2) \quad (p \text{ even}),$$

and $(p, 2, 2)$ has the (redundant) presentation

(3·49) $\quad U_\nu^2 = (U_1 U_2)^p = (U_2 U_3)^2 = (U_3 U_1)^2 = (U_1 U_2)^{p/2} U_3 = 1.$

EXERCISES

1. In the rotation group $(p, q, 2)$, defined by (3·32) or (3·35) with $r = 2$, the period of the commutator

$$A^{-1}B^{-1}AB = (CBA)^{-1}$$

is $\frac{1}{2}h$ (except when p is odd and $q = 2$, so that $h = p$ and the period is h itself).

2. In connection with the group $[p, 2]$, for which values of p can (3·47) be replaced by (3·44)?

3. The number of relations in (3·49) is seven. Reduce this number to five (keeping the same three generators).

3·5 REMARKS

Schläfli used the name *orthoscheme* for an n-dimensional simplex $P_0 P_1 \ldots P_n$ such that the n successive edges $P_0P_1, P_1P_2, \ldots, P_{n-1}P_n$ are mutually perpendicular. Wythoff rediscovered the three-dimensional case and called it a 'double-rectangular' tetrahedron.

Gauss's equations (3·13), which are the basis of the 'frieze pattern', can be seen in his *Werke*, vol. III (Göttingen, 1876), p. 484.

A very accurate icosahedral kaleidoscope (see Figure 3·2A) was made in Minneapolis (by Litton Industries) for use in a film that was never completed because the expected financial support was withdrawn.

Arthur Cayley (1821–95) discovered his diagram for a finite group (§3·3) in 1878. Max Dehn (1878–1952) rediscovered it, with the result that it is sometimes called the *Dehnsche Gruppenbild!* Theorem (3·33) was first proved by G. A. Miller in 1902 in a very complicated way. The proof given here, which has been called 'one of the most remarkable contributions of geometry to algebra', seems to be due to William Threlfall (1888–1949).

The groups generated by half-turns (§3·4) will be found useful in §6·5 and §7·6.

CHAPTER 4

Real four-space and the unitary plane

Spirits have four dimensions.

Henry More (1614–87)

The topics treated in the preceding chapters are here extended from two or three dimensions to four. Since the publication of the second edition of *Regular Polytopes* (Coxeter 1963), several other expositions of the *real* theory have appeared (Fejes Tóth 1964; Du Val 1964; Grünbaum 1967; Bourbaki 1968). In the present chapter, care has been taken to avoid duplicating these earlier treatments while giving a self-contained background for the study of regular *complex* polytopes.

4·1 SPHERICAL HONEYCOMBS

Euclidean 4-space can be explored by means of Cartesian coordinates x_ν ($\nu = 1, 2, 3, 4$), in terms of which a single linear equation

$$b_1x_1 + b_2x_2 + b_3x_3 + b_4x_4 = c$$

represents a *hyperplane* or 3-flat. Two such equations usually represent a *plane* or 2-flat (the intersection of two hyperplanes); but if the equations differ only in the value of c they represent two *parallel* hyperplanes. Three hyperplanes usually meet in a *line* or 1-flat such as

$$\frac{x_1-a_1}{l_1} = \frac{x_2-a_2}{l_2} = \frac{x_3-a_3}{l_3} = \frac{x_4-a_4}{l_4}.$$

Four hyperplanes usually meet in a *point* or 0-flat; e.g. the four hyperplanes $x_\nu = 0$ meet in the *origin* (0, 0, 0, 0). The distance between two points

$$(a_1, a_2, a_3, a_4) \quad \text{and} \quad (b_1, b_2, b_3, b_4)$$

is the square root of $\Sigma(a_\nu - b_\nu)^2$.

The hypersurface $x_1^2+x_2^2+x_3^2+x_4^2 = c^2$ is called a *sphere* (of radius c) or, more precisely, a 3-sphere, to distinguish it from the '2-sphere' which is an ordinary sphere, a '1-sphere' which is a circle, and a '0-sphere' which is a pair of distinct points. The section of the 3-sphere by a plane or hyperplane through its centre (0, 0, 0, 0) is called a *great circle* or *great sphere* respectively. The great circles and great spheres on the unit 3-sphere $\Sigma x_\nu^2 = 1$ are sometimes called the 'lines' and 'planes' of *spherical* 3-space, from which the more familiar *elliptic* 3-space can be derived by identifying all pairs of *antipodal* points

$$(x_1, x_2, x_3, x_4) \quad \text{and} \quad (-x_1, -x_2, -x_3, -x_4).$$

If the circumsphere of a regular polyhedron $\{p, q\}$ is regarded as lying on a 3-sphere, we can replace the edges by arcs of great circles joining the same pairs of vertices, and replace the faces by *spherical* polygons, thus obtaining a regular spherical polyhedron for which the symbol $\{p, q\}$ remains appropriate, although the dihedral angle has increased because of the outward bulging of the faces. If we imagine the polyhedron to grow steadily until its circumsphere becomes a *great* sphere, the dihedral angle will steadily increase from its Euclidean value to π. In this final stage, the spherical polyhedron becomes a spherical tessellation; the great 2-sphere decomposes the 3-sphere into two congruent halves, either of which may be called the 'interior' of the $\{p, q\}$, or we may regard the two halves as the two *cells* $\{p, q\}$ of a very simple 'honeycomb' $\{p, q, 2\}$, analogous to the dihedron $\{p, 2\}$.

More generally, there may be a rational number $r > 2$ such that a spherical $\{p, q\}$ with dihedral angle $2\pi/r$ can be reflected in its face-planes so as to yield the cells of a *regular honeycomb* $\{p, q, r\}$. The most familiar example is the cubic honeycomb $\{4, 3, 4\}$, which is not spherical but Euclidean; the description remains valid, though in this case the number of cells (congruent cubes fitting together to fill the whole Euclidean space) is infinite.

In such a honeycomb $\{p, q, r\}$, the cells that surround an edge are arranged like the edges of a polygon $\{r\}$, and those that surround a vertex are arranged like the faces of a polyhedron $\{q, r\}$. More precisely, the honeycomb has a *vertex figure* $\{q, r\}$, whose faces are the vertex figures of the cells $\{p, q\}$ that come together at one vertex. The *Schläfli symbol* $\{p, q, r\}$ is obtained by 'telescoping' the symbols $\{p, q\}$ and $\{q, r\}$ which describe the cell and vertex figure, respectively.

Each regular honeycomb has a *dual* or *reciprocal* honeycomb, whose vertices, edges, faces and cells correspond to the cells, faces, edges and vertices of the original honeycomb. The relationship is symmetric: in a pair of dual honeycombs, each has a vertex at the centre of a cell of the other, and has an edge joining the centres of two adjacent cells of the other. Since those cells of $\{p, q, r\}$ that surround one vertex have centres which are the vertices of an $\{r, q\}$, reciprocal to the vertex figure $\{q, r\}$, *the dual of* $\{p, q, r\}$ *is* $\{r, q, p\}$. For instance, from the $\{4, 3, 4\}$ whose vertices have integers for their three Cartesian coordinates, we can derive the dual $\{4, 3, 4\}$ by applying the translation that adds $\frac{1}{2}$ to each coordinate.

Dualizing the 'dihedroid' $\{p, q, 2\}$, we obtain a peculiar spherical honeycomb $\{2, q, p\}$ whose two antipodal vertices are joined by great semicircles, digons and hosohedra, corresponding to the vertices, edges and faces of the vertex figure $\{q, p\}$. There is a still more peculiar spherical honeycomb $\{p, 2, q\}$ whose p vertices all lie on one great circle, this circle being the common boundary of a cluster of great hemispheres corresponding to the vertices of a $\{q\}$.

In the more familiar case when p, q, r are all greater than 2, we can replace the edges, which are arcs of great circles, by their chords, so that the cells become ordinary polyhedra $\{p, q\}$ and the whole spherical honeycomb becomes a regular *polytope* (Coxeter 1963, pp. 128–64), for which the Schläfli symbol $\{p, q, r\}$ remains appropriate. When p, q, r are integers satisfying

$$p+2q+r-\frac{4}{p}-\frac{4}{r}<12, \tag{4·11}$$

this is one of the six *convex* regular polytopes

$$\{3, 3, 3\}, \quad \{4, 3, 3\}, \quad \{3, 3, 4\}, \quad \{3, 4, 3\}, \quad \{5, 3, 3\}, \quad \{3, 3, 5\}$$

(Grünbaum 1967, p. 412; Coxeter 1970, p. 25).

Suppose such a convex polytope $\{p, q, r\}$ has N_ν ν-dimensional elements, that is, N_0 vertices, N_1 edges, N_2 faces, and N_3 cells. There is no simple formula for these numbers in terms of p, q, r; but it is quite easy to find their mutual *ratios*. Extending the notation of (3·14), let us write

$$s^{-1}=p^{-1}+q^{-1}-\tfrac{1}{2}, \quad t^{-1}=q^{-1}+r^{-1}-\tfrac{1}{2}, \tag{4·12}$$

so that the cell $\{p, q\}$ has s edges and the vertex figure $\{q, r\}$ has t edges. Since the sN_3 edges of the N_3 cells $\{p, q\}$, and the pN_2 edges of the N_2 faces $\{p\}$, are just all the N_1 edges of $\{p, q, r\}$, each counted r times, we have $sN_3 = pN_2 = rN_1$. Similarly, since each vertex belongs to t faces, each having p vertices, $pN_2 = tN_0$. Thus

$$sN_3 = pN_2 = rN_1 = tN_0. \tag{4·13}$$

EXERCISES

1. Suppose the spherical honeycomb $\{p, q, r\}$ has N_0 vertices, N_1 edges, N_2 faces, N_3 cells. Determine these numbers in the cases

$$\{5, 3, 2\}, \quad \{2, 3, 5\}, \quad \{3, 5, 2\}, \quad \{2, 5, 3\}, \quad \{5, 2, 3\}, \quad \{3, 2, 5\}.$$

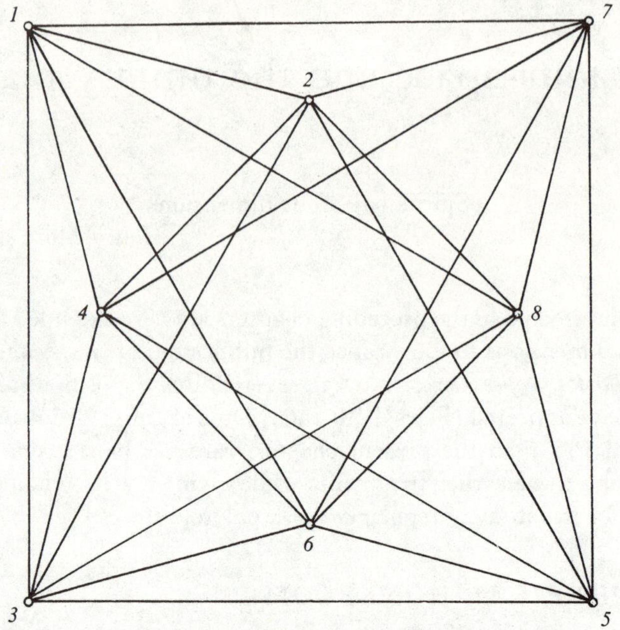

Figure 4·2A: $\{3, 3, 4\}$

2. Every convex regular polytope satisfies the four-dimensional analogue of Euler's formula:

$$N_0-N_1+N_2-N_3=0.$$

4·2 THE CRYSTALLOGRAPHIC REGULAR POLYTOPES

The simplest four-dimensional polytope is the regular *simplex* or *5-cell* $\{3, 3, 3\}$, whose vertices are five mutually equidistant points such as

$$(2, 0, 0, 0),\ (0, 2, 0, 0,),\ (0, 0, 2, 0),\ (0, 0, 0, 2),\ (\tau, \tau, \tau, \tau). \tag{4·21}$$

Any two of these five points form an edge, any three a face, and any four a *cell* (or *facet*). Its complete symmetry group $[3, 3, 3]$ is the symmetric group $\mathfrak{S}_5$, consisting of all the permutations of the 5 vertices. This is generated by the 4 transpositions

$$R_1=(ab), \quad R_2=(bc), \quad R_3=(cd), \quad R_4=(de),$$

which satisfy

$$\begin{aligned} R_\nu{}^2 &= (R_1R_2)^3 = (R_2R_3)^3 = (R_3R_4)^3 \\ &= (R_1R_3)^2 = (R_1R_4)^2 = (R_2R_4)^2 = 1. \end{aligned} \tag{4·22}$$

As a convenient picture we can use a regular pentagon with all its diagonals (Coxeter 1963, p. 120; 1969, p. 398).

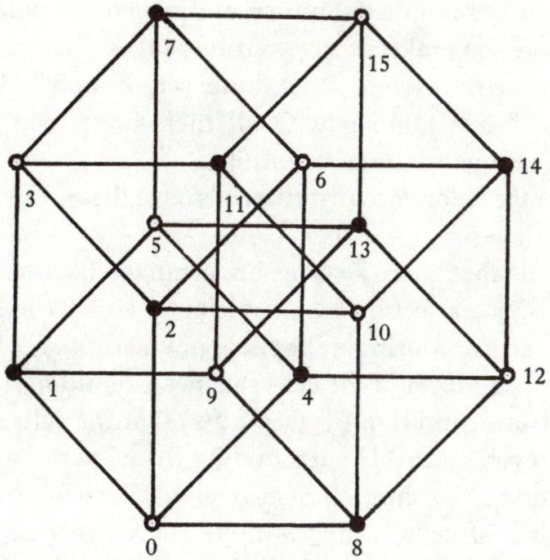

Figure 4·2B: $\{4, 3, 3\}$

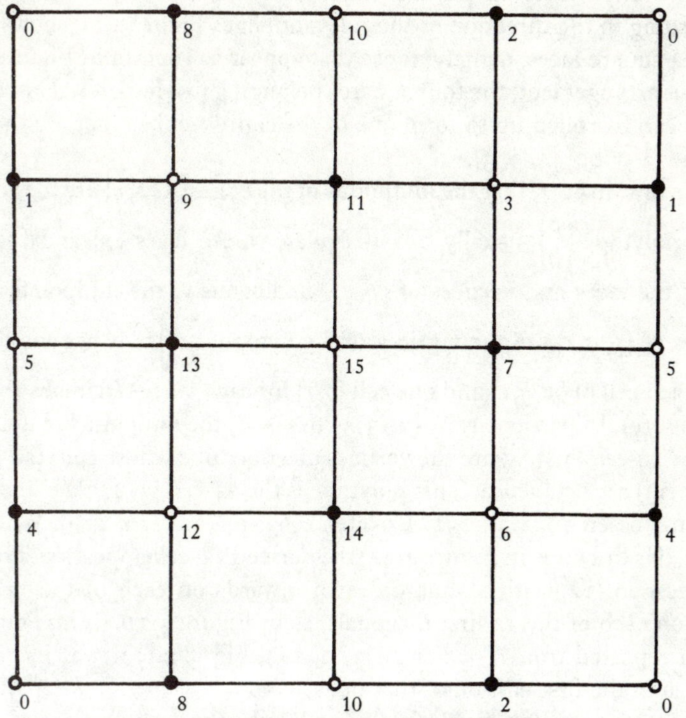

Figure 4·2C: Sixteen of the twenty-four faces

Another regular polytope is the four-dimensional *cross-polytope* or 16-*cell* $\{3, 3, 4\}$, whose eight vertices are equidistant from the origin along the four Cartesian axes:

(4·23) $(\pm 1, 0, 0, 0),\ (0, \pm 1, 0, 0),\ (0, 0, \pm 1, 0),\ (0, 0, 0, \pm 1).$

Any four of these points, one from each pair of opposites, form a cell $\{3, 3\}$. One picture of it consists of a regular octagon with all pairs of vertices joined *except* the pairs of opposites (Coxeter 1963, p. 121). A better one (appearing very modestly in Coxeter 1963, p. 149, Fig. 8·2B) is Figure 4·2A, in which each vertex of either square forms an equilateral triangle with one side of the other.

The reciprocal of $\{3, 3, 4\}$ is the four-dimensional *cube* or *measure polytope* or *tessaract* $\{4, 3, 3\}$, whose 16 vertices may be expressed as

(4·24) $(\pm 1, \pm 1, \pm 1, \pm 1),$

so that its 32 edges are parallel (in sets of eight) to the four Cartesian axes. By fixing any one of the signs, we obtain the eight vertices of a cell $\{4, 3\}$. The common symmetry group $[4, 3, 3]$ of $\{4, 3, 3\}$ and $\{3, 3, 4\}$ is the *wreath product* $\mathfrak{C}_2 \wr \mathfrak{S}_4$, of order $2^4 4! = 384$, which consists of all the permutations of the four coordinates combined with their changes of sign. It is most conveniently generated by three transpositions and one sign-change.

A three-dimensional model of $\{4, 3, 3\}$ can be made by joining corresponding vertices of two ordinary cubes, one inside the other (Hilbert and Cohn-Vossen 1952, p. 150). The simplest two-dimensional projection consists of a regular octagon with a square drawn inwards on each of the eight edges (Coxeter 1963, p. 123). In Figure 4·2B the vertices appear black and white to indicate that alternate vertices belong to two inscribed 16-cells (just as alternate vertices of the ordinary '3-cube' belong to two inscribed tetrahedra; see Coxeter 1969, p. 401). The vertices have been numbered from 0 to 15 by means of the following procedure. Applying to the coordinates (4·24) a dilatation from $(1, 1, 1, 1)$ in the ratio $\frac{1}{2}$, we obtain new coordinates (x_1, x_2, x_3, x_4) in which each of the four x's is either 0 or 1 independently. Then each coordinate symbol is contracted to $x_1 x_2 x_3 x_4$ and regarded as a binary number. Finally, these numbers are converted to the denary scale. For instance, the point $(-1, 1, 1, 1)$ becomes, in turn,

$$(0, 1, 1, 1), \quad 0111, \quad 7.$$

Thus the numbers on pairs of adjacent vertices differ by 1, 2, 4 or 8 according to the direction of the relevant edge. Figure 4·2C shows 16 of the 24 square faces, namely, those that appear as rhombi in Figure 4·2B. In this arrangement, the four squares belonging to a horizontal or vertical strip can be rolled up to form one of the eight 3-cubes that are the cells of the '4-cube' $\{4, 3, 3\}$.

We saw, in §2·1, that the midpoints of the edges of $\{p, q\}$ are the vertices of a polytope $\left\{\begin{matrix}p\\q\end{matrix}\right\}$, usually quasi-regular, whose faces $\{p\}$ and $\{q\}$ arise from the faces and vertices of $\{p, q\}$. Analogously, the midpoints of the edges of $\{p, q, r\}$ are the vertices of a polytope $\left\{\begin{matrix}p\\q, r\end{matrix}\right\}$ that has one cell $\left\{\begin{matrix}p\\q\end{matrix}\right\}$ for each cell of $\{p, q, r\}$ and one cell $\{q, r\}$ for each vertex (namely, the vertex figure). In particular, by (2·13) with $p = 3$, the midpoints of the edges of the 16-*cell* $\{3, 3, 4\}$ are the vertices of a polytope whose cells (all $16+8$ of them) are octahedra. This polytope is the 24-*cell* $\{3, 4, 3\}$ (Hilbert and Cohn-Vossen 1952, p. 152; Coxeter 1963, pp. 145–50, 156). When the 16-cell is drawn as in Figure 4·2A, the derived 24-cell appears as a regular dodecagon $\{12\}$ with a square drawn inwards on each of the 12 edges and on each of the 12 first diagonals, as in Figure 4·2D. (This figure has been repeated from Coxeter 1963, p. 149, Fig. 8·2A; it was incorrectly drawn in the first edition of that book.)

If we take the 8 vertices of the 16-cell to be the permutations of

$$(\pm 2, 0, 0, 0),$$

we see that the midpoints of its 24 edges are the permutations of

$$(\pm 1, \pm 1, 0, 0).$$

The reciprocal of this $\{3, 4, 3\}$ is another $\{3, 4, 3\}$ whose 24 vertices (for a suitable size) are $(\pm 1, \pm 1, \pm 1, \pm 1)$ and the permutations of $(\pm 2, 0, 0, 0)$.

The remaining regular polytopes will be described in §4·6 and §4·7.

EXERCISES

1. Inside a square *1357*, take a point *2* so that *127* is an isosceles triangle with angles 15° at *1* and *7*. What kind of triangle is *235*?

2. Find the side of the smallest square within which eight unit discs can be packed.

3. What kind of arrangement is determined from Figure 4·2C when the number associated with each black point is replaced by its 'opposite' (obtained by subtracting from 15)?

4. The vertices of $\{3, 4, 3\}$ can be distributed among three inscribed $\{3, 3, 4\}$'s. What configuration is obtained when the 4 Cartesian coordinates are reinterpreted as homogeneous coordinates in projective 3-space?

5. Why are $\{3, 3, 3\}$, $\{4, 3, 3\}$, $\{3, 3, 4\}$ and $\{3, 4, 3\}$ called 'crystallographic' polytopes?

6. Are there infinitely many values of n for which $n+1$ mutually equidistant points in Euclidean n-space can have integers for their Cartesian coordinates?

4·3 FLAGS AND ORTHOSCHEMES

A convex polytope (Grünbaum 1967) is most easily defined as the convex hull of a finite set of points. However, our present considerations require something more general, not necessarily convex. Let us say rather that a four-dimensional *polytope* is a finite set of polyhedra, called *cells* (Grünbaum's 'facets'), along with all their faces, edges and vertices, satisfying the following three conditions:

(i) Every face belongs to just two cells, and these cells do not lie in the same hyperplane.

(ii) The cells that share a vertex are arranged like the faces of a polyhedron, that is, their section by a sufficiently small 3-sphere, centred at the common vertex, is a single spherical polyhedron.

(iii) No proper subset of the cells satisfies Condition (i).

It follows from condition (ii) (see § 2·2) that the cells sharing an edge form a single cycle, possibly surrounding the edge more than once.

The definition for a three-dimensional *honeycomb* is almost the same; but the number of cells, which is finite for a spherical honeycomb, is infinite for a Euclidean or hyperbolic honeycomb; also the last phrase in condition (i) has to be deleted, and in condition (ii) we must replace '3-sphere' by '2-sphere', and 'spherical polyhedron' by 'spherical tessellation'.

For any four-dimensional polytope or three-dimensional honeycomb, we define a *flag* $(\Pi_0, \Pi_1, \Pi_2, \Pi_3)$ to be the figure consisting of a vertex Π_0, an edge Π_1 containing this vertex, a face Π_2 containing this edge, and a cell Π_3 containing this face. The polytope or honeycomb is said to be *regular* if its symmetry group is *transitive on its flags* (Du Val 1964, p. 63). In particular, the group must include symmetries which will transform $(\Pi_0, \Pi_1, \Pi_2, \Pi_3)$ into four adjacent flags

$$(\Pi_0', \Pi_1, \Pi_2, \Pi_3), \quad (\Pi_0, \Pi_1', \Pi_2, \Pi_3),$$
$$(\Pi_0, \Pi_1, \Pi_2', \Pi_3), \quad (\Pi_0, \Pi_1, \Pi_2, \Pi_3')$$

in turn. (Since every four-dimensional polytope is associated with a three-dimensional spherical honeycomb, we may conveniently deal first with honeycombs, and make the small changes needed for polytopes afterwards.) These four symmetries are reflections in four planes: ρ_1, perpendicularly bisecting the edge Π_1 (whose ends are Π_0 and Π_0'); ρ_2, perpendicular to the plane of Π_2 and bisecting the angle between Π_1 and

Figure 4·2D: {3, 4, 3}

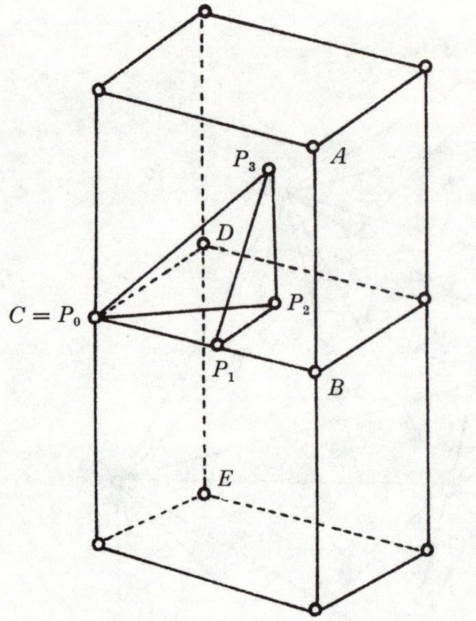

Figure 4·3A: Two cells of {4, 3, 4} and the orthoscheme $P_0P_1P_2P_3$

Π_1'; ρ_3, bisecting the dihedral angle between Π_2 and Π_2'; and ρ_4, which is the plane of Π_2, the common face of Π_3 and Π_3'. Since non-consecutive pairs of these four planes are perpendicular, they form an orthoscheme $P_0P_1P_2P_3$, as in Figure 4·3A (where Π_0 is the vertex P_0 or C, Π_0' is B, Π_1 is the edge BC with midpoint P_1, Π_1' is CD, Π_2 is the face $BCD\ldots$ with centre P_2, Π_2' is $ABC\ldots$, and Π_3 is the cell $ABCD\ldots$ with centre P_3). Thus the four planes (ρ_ν opposite to $P_{\nu-1}$) are

$$(4·31)\quad \rho_1 = P_1P_2P_3,\quad \rho_2 = P_0P_2P_3,\quad \rho_3 = P_0P_1P_3,\quad \rho_4 = P_0P_1P_2.$$

They reflect one another into the set of all planes of symmetry. Let R_ν denote the reflection in ρ_ν. The first three of these four reflections generate the symmetry group of the cell Π_3, which is thus seen to be a regular polyhedron. The reflection R_4 yields the congruent cell Π_3', and so eventually we find that *all the cells are regular and congruent*.

For such a regular honeycomb $\{p, q, r\}$, a *Petrie polygon* is defined as a skew polygon in which every three consecutive edges, but no four, belong to the Petrie polygon of a cell (see §2·2). Thus, if $ABCD\ldots$ and $BCDE\ldots$ are Petrie polygons of two adjacent cells, as in Figure 4·3A, the four edges AB, BC, CD, DE belong to a Petrie polygon of the whole honeycomb. The half-turn $R_1R_3 = R_3R_1$ about P_1P_3 transforms this skew polygon $ABCD\ldots$ into $DCBA\ldots$, while the half-turn $R_2R_4 = R_4R_2$ about P_0P_2 transforms $DCBA\ldots$ into $BCDE\ldots$. Since P_1P_3 and P_0P_2 are two opposite edges of a tetrahedron, they are skew lines, and the product of half-turns about them, namely

$$R_3R_1R_4R_2 = R_3R_4R_1R_2,$$

is a *twist*. The Petrie polygon, being thus shifted one step along itself by a twist, is a *helical* regular polygon in the sense of §1·8.

On sufficiently small spheres drawn round the vertices P_3 and P_0, the planes ρ_ν cut out spherical triangles $(p\,q\,2)$ and $(q\,r\,2)$; therefore the dihedral angles $\theta_{\mu\nu}$ of the orthoscheme $\rho_1\rho_2\rho_3\rho_4$ are

$$(4·32)\quad \theta_{12} = \frac{\pi}{p},\quad \theta_{23} = \frac{\pi}{q},\quad \theta_{34} = \frac{\pi}{r},\quad \theta_{13} = \theta_{14} = \theta_{24} = \frac{\pi}{2},$$

along the edges P_2P_3, P_0P_3, P_0P_1, P_1P_3, P_1P_2, P_0P_2, respectively. The reflections R_ν, which generate the symmetry group $[p, q, r]$ of the honeycomb $\{p, q, r\}$, are conveniently represented by the four nodes of a graph. Adapting the notation of §2·4 to this generalized kaleidoscope having four mirrors, we may regard the symbols

$$[p, q, r],\quad \{p, q, r\},\quad \begin{Bmatrix} p \\ q, r \end{Bmatrix}$$

as abbreviations for

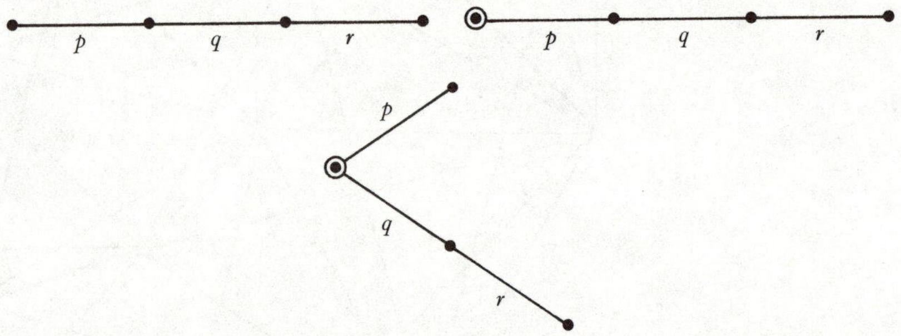

Among the edge-lengths of the orthoscheme $P_0P_1P_2P_3$ (Figure 4·3B), the most important are

$$(4·33)\quad \phi = P_0P_1,\quad \chi = P_0P_3,\quad \psi = P_2P_3,$$

these being analogous to the angles ϕ, χ, ψ of (3·11) and (3·12), which we now denote by $\phi_{p,q}$, $\chi_{p,q}$, $\psi_{p,q}$ because they belong to the cell $\{p, q\}$. Similarly, the corresponding properties of the vertex figure $\{q, r\}$ are denoted by $\phi_{q,r}$, $\chi_{q,r}$, $\psi_{q,r}$.

To express ϕ, χ, ψ in terms of p, q, r, we consider the faces of the orthoscheme and 'solve' these right-angled triangles by the classical formulae of trigonometry: spherical, Euclidean or hyperbolic according to the nature of the honeycomb $\{p, q, r\}$.

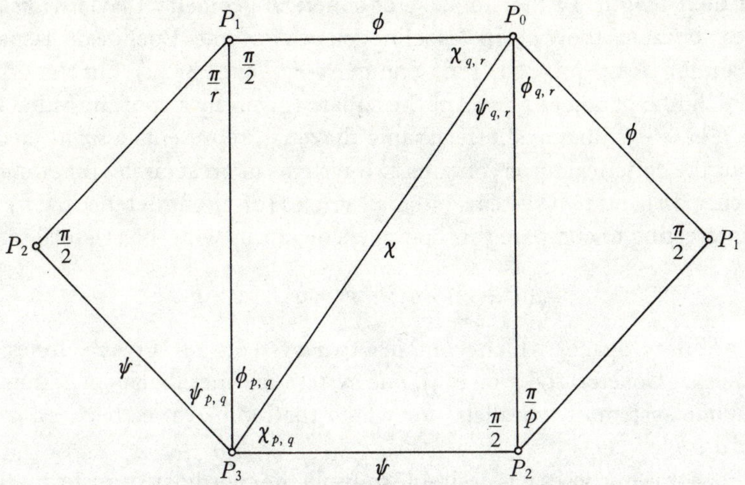

Figure 4·3B: The orthoscheme $P_0P_1P_2P_3$ unfolded

If $P_0P_1P_2$ is a *spherical* triangle, we have $\cos\phi\sin\phi_{q,r} = \cos\pi/p$. But by (3·11) and (2·22),

$$\sin\phi_{p,q}\sin\frac{\pi}{q} = \left(\sin^2\frac{\pi}{q} - \cos^2\frac{\pi}{p}\right)^{\frac{1}{2}} = \sin\frac{\pi}{h_{p,q}},$$

and of course we can replace p and q by q and r. Therefore

$$(4·34)\qquad \cos\phi = \cos\frac{\pi}{p}\sin\frac{\pi}{r}\Big/\sin\frac{\pi}{h_{q,r}},$$

where $h_{q,r} = h_{r,q}$ and

$$h_{3,3} = 4,\quad h_{3,4} = h_{5,\frac{5}{2}} = 6,\quad h_{3,5} = 10,\quad h_{3,\frac{5}{2}} = \tfrac{10}{3}.$$

In particular, for {3, 3, 5},

$$(4·35)\qquad \cos\phi = \sin\frac{\pi}{5}\Big/2\sin\frac{\pi}{10} = \cos\frac{\pi}{10},\quad \phi = \frac{\pi}{10}.$$

If the triangle $P_0P_1P_2$ is hyperbolic instead of spherical, we merely have to replace the cos ϕ in (4·34) by cosh ϕ. This triangle $P_0P_1P_2$ (and hence the whole orthoscheme) is spherical if and only if

$$\cos\frac{\pi}{p}\sin\frac{\pi}{r} < \sin\frac{\pi}{h_{q,r}} = \left(\sin^2\frac{\pi}{r} - \cos^2\frac{\pi}{q}\right)^{\frac{1}{2}},$$

that is, if and only if

$$(4·36)\qquad \cos\frac{\pi}{q} < \sin\frac{\pi}{p}\sin\frac{\pi}{r}$$

(Sommerville 1929, p. 168).

The notation of continued fractions enables us to express (4·34) as

$$\cos^2\phi = \cos^2\frac{\pi}{p}\Big/1 - \cos^2\frac{\pi}{q}\Big/\sin^2\frac{\pi}{r}$$

or, in terms of the abbreviation

$$(4·37)\qquad c_n = \cos^2\frac{\pi}{n},$$

$$(4·38)\qquad \cos^2\phi = c_p/1 - c_q/1 - c_r,\quad \sin^2\phi = 1 - c_p/1 - c_q/1 - c_r.$$

Similarly, from the triangle $P_1P_2P_3$ in Figure 4·3B, or by reciprocation,

$$(4·39)\qquad \cos^2\psi = c_r/1 - c_q/1 - c_p,\quad \sin^2\psi = 1 - c_r/1 - c_q/1 - c_p.$$

EXERCISES

1. Describe the Petrie polygon of the cubic honeycomb {4, 3, 4}, beginning with the points (0, 0, 0), (1, 0, 0), (1, 1, 0), (1, 1, 1). Deduce that the 'cubic' orthoscheme (Figure 3·1C with $\alpha = \gamma = 1$) is a *space-filler*: directly congruent replicas can be packed like bricks to fill the whole Euclidean space. (Michael Goldberg.)

2. For the general orthoscheme in spherical space,

$$\cos\psi = \sin\frac{\pi}{p}\cos\frac{\pi}{r}\Big/\sin\frac{\pi}{h_{p,q}},\quad \cos\chi = \cos\frac{\pi}{p}\cos\frac{\pi}{q}\cos\frac{\pi}{r}\Big/\sin\frac{\pi}{h_{p,q}}\sin\frac{\pi}{h_{q,r}}.$$

4·4 THE SPHERICAL TORUS

The most interesting curved surface in spherical space is the *spherical torus*

$$(4·41)\qquad x_1x_3 = x_2x_4,\quad \Sigma x_\nu^2 = 1.$$

In terms of two angular parameters α and β, such that

$$|\alpha \pm \beta| \leqslant \pi,$$

this surface is given by

$$(4·42)\qquad x_1 = \cos\alpha\cos\beta,\quad x_2 = \cos\alpha\sin\beta,$$
$$x_3 = \sin\alpha\sin\beta,\quad x_4 = \sin\alpha\cos\beta.$$

Since

$$\begin{aligned} &dx_1^2 + dx_2^2 + dx_3^2 + dx_4^2 \\ &\quad = (-x_4d\alpha - x_2d\beta)^2 + (-x_3d\alpha + x_1d\beta)^2 \\ &\qquad\qquad + (x_2d\alpha + x_4d\beta)^2 + (x_1d\alpha - x_3d\beta)^2 \\ &\quad = d\alpha^2 + d\beta^2, \end{aligned}$$

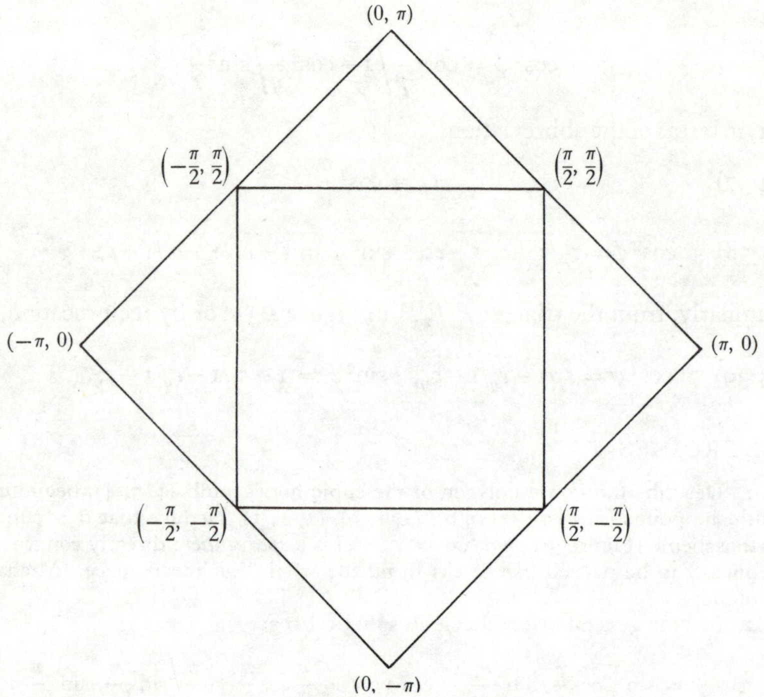

Figure 4·4A: The (α, β) plane

the spherical torus is a ***developable*** surface: we can regard α and β as Cartesian coordinates in a Euclidean plane and thus 'develop' the surface onto the region of the plane enclosed by the four lines

$$\alpha \pm \beta = \pm\pi,$$

which form a square. (See Figure 4·4A.) The periodicity of the parameters compels us to identify each side of the square with the opposite side; for instance, the two points $(\alpha, \pi-\alpha)$ and $(\alpha-\pi, -\alpha)$, on the sides

$$\alpha+\beta = \pm\pi,$$

represent the same point

$$(-\cos^2\alpha, \quad \sin\alpha\cos\alpha, \quad \sin^2\alpha, \quad -\sin\alpha\cos\alpha)$$

on the surface. Thus the 'spherical torus' is (as its name suggests) a torus in the topological sense, even though it is a 'homogeneous' surface (like a plane or a 2-sphere): the neighbourhoods of all its points are exactly alike!

It follows that two lines parallel to the same diagonal of the square (given by two different constant values for α or β) represent two non-intersecting great circles which are equidistant from each other throughout their length. In the language of spherical geometry they are *isocline* lines, because they lie in 'isocline planes' of the Euclidean 4-space (Manning 1914, pp. 123, 186; Sommerville 1929, p. 44). On the other hand, a line parallel to a *side* of the square (given by a constant value for $\alpha+\beta$ or $\alpha-\beta$), being shorter than a diagonal, represents a *small* circle. Thus the spherical torus contains two systems of great circles (of circumference 2π) and two systems of small circles (of circumference $\surd 2\pi$). It is interesting to compare this homogeneous torus with the classical torus

$$(\surd(x^2+y^2)-a)^2+z^2 = b^2 \quad (a > b)$$

in ordinary space, which contains two systems of 'great' circles of radius a (Coxeter 1969, p. 133), one system of 'meridians' of radius b, and one system of 'parallels' for which the radius varies between $a-b$ and $a+b$.

For some purposes it is desirable to apply the coordinate transformation (of period two)

$$(4{\cdot}43) \qquad x_1 \to \frac{x_1+x_3}{\surd 2}, \quad x_2 \to \frac{x_4-x_2}{\surd 2},$$

$$x_3 \to \frac{x_1-x_3}{\surd 2}, \quad x_4 \to \frac{x_4+x_2}{\surd 2},$$

which amounts to a half-turn about the plane

$$\frac{x_1}{x_3} = \surd 2+1 = \frac{x_4}{x_2}.$$

The spherical torus now appears in the form

$$(4{\cdot}44) \qquad {x_1}^2+{x_2}^2 = {x_3}^2+{x_4}^2 = \tfrac{1}{2},$$

with the parametrization

$$(4{\cdot}45) \qquad x_1 = \frac{\cos\xi}{\surd 2}, \quad x_2 = \frac{\sin\xi}{\surd 2}, \quad x_3 = \frac{\cos\eta}{\surd 2}, \quad x_4 = \frac{\sin\eta}{\surd 2},$$

where $\xi = \alpha-\beta, \quad \eta = \alpha+\beta.$

It is sometimes convenient to take the four real Cartesian coordinates in pairs, forming two complex numbers

$$(4{\cdot}46) \qquad u = x_1+x_2 i, \quad v = x_3+x_4 i,$$

so that the distance between points (u, v) and (u', v') is the square root of

$$|u-u'|^2+|v-v'|^2$$

and the inner product of the corresponding vectors is

$$\text{(4·47)} \qquad \operatorname{Re}(u\bar{u}' + v\bar{v}').$$

In terms of these complex coordinates, the spherical torus is simply

$$\text{(4·48)} \qquad |u| = |v| = \sqrt{\tfrac{1}{2}},$$

and its point with parameters ξ and η is given by

$$\text{(4·49)} \qquad u = \sqrt{\tfrac{1}{2}}\exp(\xi i), \quad v = \sqrt{\tfrac{1}{2}}\exp(\eta i).$$

EXERCISES

1. Compute the total area of the spherical torus.
2. Write down equations for the plane that contains
   (i) the great circle on the torus (4·41) or (4·42) given by a particular value for α or β,
   (ii) the small circle on the torus (4·44) or (4·45) given by a particular value for ξ or η.
3. How is the spherical torus related to the rectangular Clifford surface in elliptic space (Coxeter 1965, pp. 134, 143)?

4·5 DOUBLE PRISMS

On the spherical torus (4·49), consider the two systems of p small circles along which ξ or η takes one of the values $0, 2\pi/p, 4\pi/p, \ldots, 2(p-1)\pi/p$. These circles reticulate the surface to form a tessellation of type $\{4, 4\}$ having the p^2 vertices

$$\text{(4·51)} \qquad (\sqrt{\tfrac{1}{2}}\epsilon^{2\mu}, \sqrt{\tfrac{1}{2}}\epsilon^{2\nu}) \quad (\mu, \nu = 0, 1, \ldots, p-1),$$

where $\epsilon = \exp \pi i/p$. The intermediate small circles, along which ξ or η takes one of the values $(2\mu+1)\pi/p$, form the dual tessellation whose p^2 vertices are

$$\text{(4·52)} \qquad (\sqrt{\tfrac{1}{2}}\epsilon^{2\mu+1}, \sqrt{\tfrac{1}{2}}\epsilon^{2\nu+1}).$$

The faces of each tessellation surround the vertices of the other. A typical face of (4·52) has the four vertices

$$\text{(4·53)} \qquad (\sqrt{\tfrac{1}{2}}\epsilon^{\pm 1}, \sqrt{\tfrac{1}{2}}\epsilon^{\pm 1}),$$

whose centroid (in the Euclidean 4-space) is

$$\left(\sqrt{\tfrac{1}{2}}\cos\frac{\pi}{p}, \sqrt{\tfrac{1}{2}}\cos\frac{\pi}{p}\right).$$

Since this point lies midway between each diagonally opposite pair of the vertices, the same four points are the vertices of an ordinary Euclidean square!

There are altogether p^2 such squares and these are the faces of a regular *skew* polyhedron

$$\{4, 4 \mid p\},$$

whose vertex figure is a regular skew quadrangle (Coxeter and Moser 1972, p. 109; Coxeter 1968, pp. 85–7); in other words, they are the square faces of the 'double prism'

$$\{p\} \times \{p\},$$

which is the Cartesian product of two regular p-gons (Coxeter 1963, p. 124). This polytope has p^2 vertices, $2p^2$ edges, p^2 square faces, $2p$ p-gonal faces, and $2p$ cells. The $2p$ cells are uniform p-gonal prisms

$$\{p\} \times \{\,\} \quad \text{or} \quad \{\,\} \times \{p\}$$

(whose side-faces are squares), forming two interlocked rings of p. Each ring of p prisms can be straightened out to form a tall prismatic pillar: p solid p-gonal prisms piled up, base to base, like the stones that form a pillar in a cathedral. In ordinary space the base of the lowest prism is far from the top of the highest; but in 4-space, where rotation takes place about a *plane*, we can bring these two p-gons into coincidence by bending the pillar about the planes of the intermediate bases so as to reconstruct the ring. The whole polytope $\{p\} \times \{p\}$ is obtained by interlocking two such rings so that their bounding polyhedra $\{4, 4 \mid p\}$ are brought into coincidence.

In the special case when $p = 4$, the distinction between p-gons and squares disappears, and we see that $\{4\} \times \{4\}$ is just the four-dimensional cube $\{4, 3, 3\}$. In Figures 4·5A and 4·5B, which resemble Figures 4·2B and 4·2C, the 16 vertices of $\{4, 3, 3\}$ and $\{4, 4 \mid 4\}$ have been marked with two odd digits, which are the exponents of ϵ in the coordinate symbol (4·52). Since $p = 4$, we now have

$$\epsilon = \exp(\pi i/4) = \sqrt{\tfrac{1}{2}}(1+i),$$

so that $\quad \epsilon^3 = \sqrt{\tfrac{1}{2}}(-1+i), \epsilon^5 = \sqrt{\tfrac{1}{2}}(-1-i), \epsilon^7 = \sqrt{\tfrac{1}{2}}(1-i),$

in agreement with the real coordinates

$$(\pm\tfrac{1}{2}, \pm\tfrac{1}{2}, \pm\tfrac{1}{2}, \pm\tfrac{1}{2})$$

for the 16 vertices of $\{4, 3, 3\}$. (See (4·24).)

The peripheral octagon of Figure 4·5A, reappearing as a zigzag in Figure 4·5B, represents a regular skew octagon which is a Petrie polygon of $\{4, 4 \mid 4\}$ (as its various pairs of consecutive edges all belong to different faces). According to the definition in §4.3, the same skew octagon is a Petrie polygon of $\{4, 3, 3\}$, because its various sets of three consecutive edges all belong to different solid cells.

EXERCISE

Find real Cartesian equations for the plane of the square (4·53).

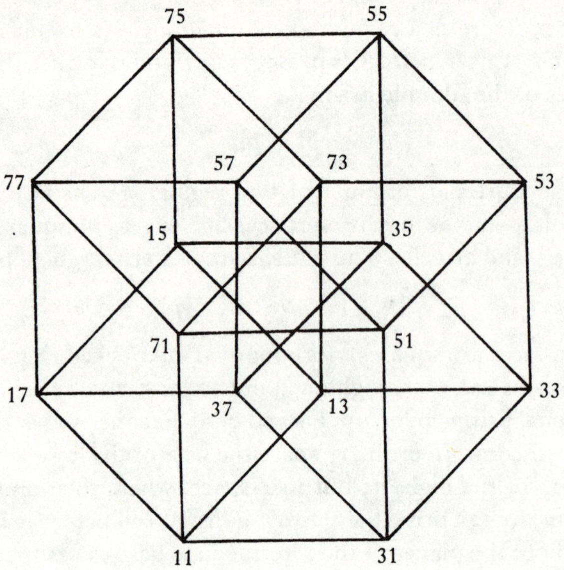

Figure 4·5A: $\{4, 3, 3\}$

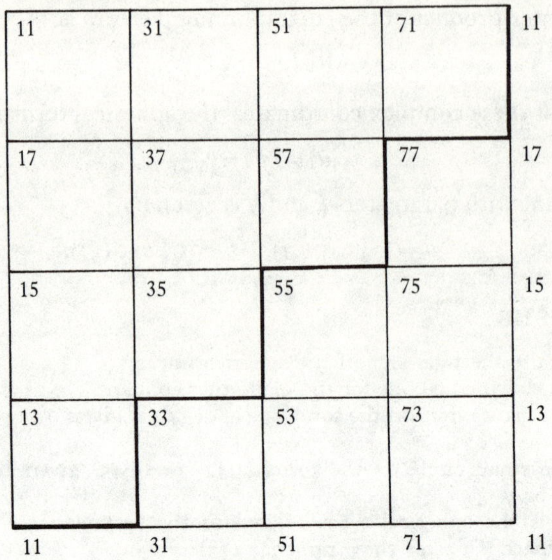

Figure 4·5B: $\{4, 4 \mid 4\}$

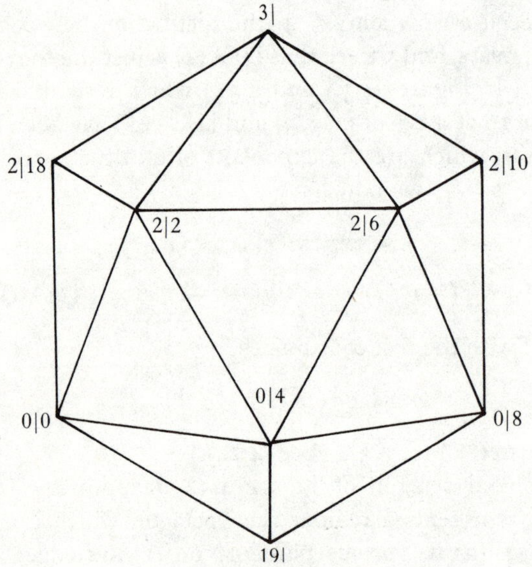

Figure 4·6A: The icosahedral cluster formed by the twenty cells of $\{3, 3, 5\}$ that surround the vertex $1|$

4·6 THE 600-CELL AND THE 120-CELL

We saw in (4·35), that $\phi = \pi/10$ for $\{3, 3, 5\}$. This shows that a regular spherical tetrahedron of dihedral angle $2\pi/5$, being a cell of the spherical honeycomb $\{3, 3, 5\}$, has edge-length $2\phi = \pi/5$. Each vertex of this honeycomb belongs to a cluster of twenty such regular tetrahedra, whose opposite faces form an icosahedron $\{3, 5\}$. (See Figure 4·6A.) Since the icosahedron is centrally symmetric, each edge of the honeycomb belongs to a cycle of ten edges which are arcs of one great circle. (The chords of these arcs are ten edges of the *polytope* $\{3, 3, 5\}$, forming an 'equatorial decagon' as in Coxeter 1963, p. 277.) Let the vertices belonging to these ten edges be denoted by

$$1|,\quad 3|,\quad 5|,\quad \ldots,\quad 19|.$$

(The numbers involved are regarded as residues modulo 20.) The edge joining $1|$ and $3|$ belongs to five triangular faces whose opposite vertices belong to a cycle denoted by

$$2|2,\quad 2|6,\quad 2|10,\quad 2|14,\quad 2|18.$$

Similarly, the edge joining $19|$ and $1|$ belongs to a 'staggered' set of five solid faces whose opposite vertices form the cycle

$$0|0,\quad 0|4,\quad 0|8,\quad 0|12,\quad 0|16.$$

Figure 4·6A shows the vertices adjacent to $1|$. Those adjacent to $3|$ include $2|2$, $2|6$, ... and a new cycle

$$4|0,\quad 4|4,\quad 4|8\quad 4|12,\quad 4|16,$$

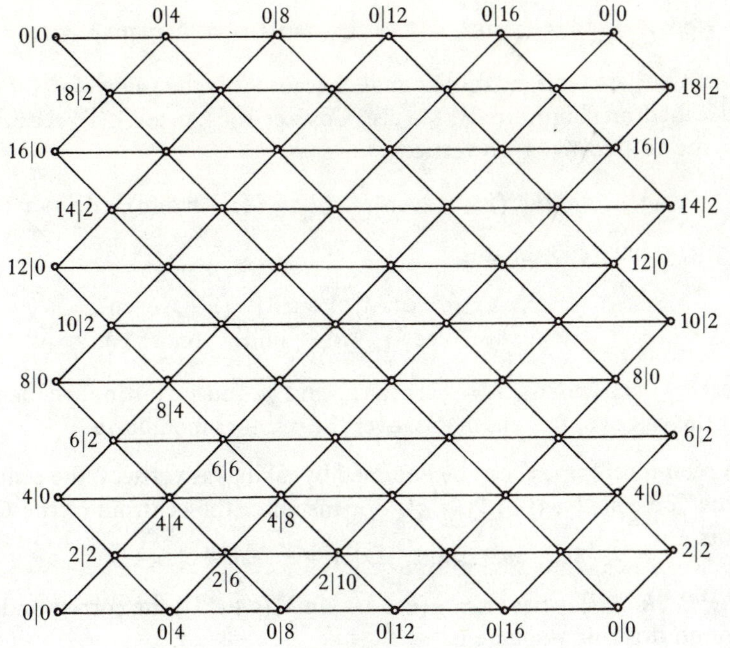

Figure 4·6B: The vertices $2\mu|2\nu$ with $\mu+\nu$ even

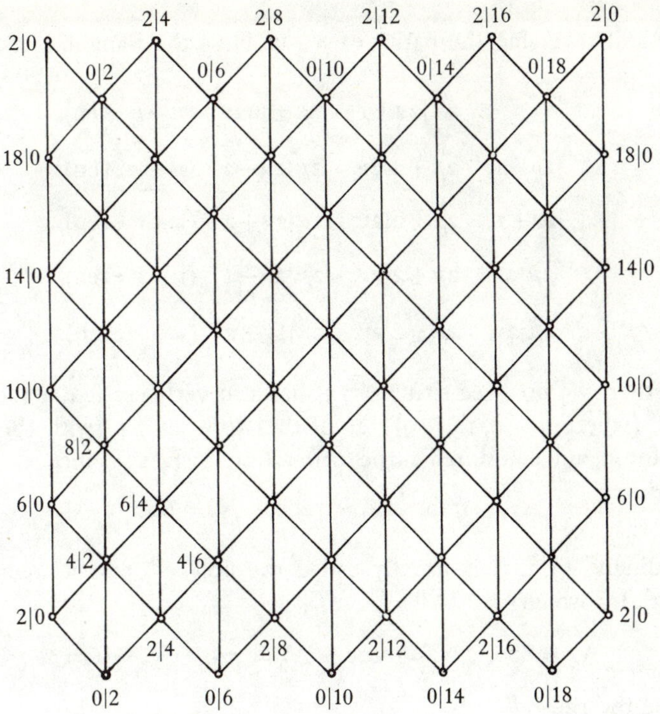

Figure 4·6C: The vertices $2\mu|2\nu$ with $\mu+\nu$ odd

as well as 1| and 5|. Thus the ten 'odd' points are surrounded by overlapping clusters of twenty cells, comprising 150 cells altogether. The hundred 'outer' faces of these cells have vertices $2\mu|2\nu$, where $\mu+\nu$ is even; they form a tessellated torus as in Figure 4·6B.

Each of the 'horizontal' edges belongs to three of the 150 clustered cells, allowing space for two more. For instance, the cycle of five vertices that form triangular faces with 2|2 and 2|6 (see Figure 4·6A) begins with 0|4, 1|, 3|, 4|4. Since we have not yet used symbols $2\mu|2\nu$ with $\mu+\nu$ odd, we naturally let 2|4 denote the remaining vertex in this cycle.

On the other hand, each of the 'oblique' edges in Figure 4·6B belongs to only two of the 150 cells, allowing space for three more. For instance, the vertices 0|4 and 2|6 form cells with 0|8 and 1| (see Figure 4·6A) and with 1| and 2|2. The above naming of 2|4 allows us to describe one of the three remaining cells around this edge as joining it to 2|2 and 2|4. Thus the cycle of five vertices that form faces with 0|4 and 2|6 begins with 0|8, 1|, 2|2, 2|4, and the remaining one is naturally 0|6. The new faces thus introduced form a second tessellated torus, as in Figure 4·6C.

Taking stock of the situation so far, we recognize fifty cells like 3|, 1|, 2|6, 2|2, namely

$$2\mu\pm 1|,\quad 2\mu|2\nu\pm 2\quad (\mu+\nu \text{ odd}),$$

fifty like 1|, 2|6, 2|2, 0|4, namely

$$2\mu-1|,\quad 2\mu|2\nu\pm 2,\quad 2\mu-2|2\nu\quad (\mu+\nu \text{ odd}),$$

fifty like 3|, 2|6, 2|2, 4|4, namely

(4·61) $$2\mu+1|,\quad 2\mu|2\nu\pm 2,\quad 2\mu+2|2\nu\quad (\mu+\nu \text{ odd}),$$

fifty like 2|4, 2|6, 2|2, 0|4, namely

(4·62) $$2\mu|2\nu,\quad 2\mu|2\nu\pm 2,\quad 2\mu-2|2\nu\quad (\mu+\nu \text{ odd})$$

fifty like 2|4, 2|6, 2|2, 4|4, namely

(4·63) $$2\mu|2\nu,\quad 2\mu|2\nu\pm 2,\quad 2\mu+2|2\nu\quad (\mu+\nu \text{ odd}),$$

and a hundred like 0|4, 0|6, 2|4, 2|6, namely

(4·64) $$2\mu|2\nu,\quad 2\mu|2\nu+2,\quad 2\mu+2|2\nu,\quad 2\mu+2|2\nu+2$$

without any restriction on μ and ν (except that they are residues modulo 10). To obtain five tetrahedral cells round *every* edge, we may regard the last hundred as a 'bridge' connecting the five sets of fifty with five new sets, derived by interchanging the left and right sides of the stroke and reversing the parity of $\mu+\nu$. Thus the remaining 250 cells are:

$$|2\nu\pm 1, \quad 2\mu\pm 2|2\nu \quad (\mu+\nu \text{ even}),$$

$$|2\nu-1, \quad 2\mu\pm 2|2\nu, \quad 2\mu|2\nu-2 \quad (\mu+\nu \text{ even}),$$

$$|2\nu+1, \quad 2\mu\pm 2|2\nu, \quad 2\mu|2\nu+2 \quad (\mu+\nu \text{ even}),$$

(4·65) $$2\mu|2\nu, \quad 2\mu\pm 2|2\nu, \quad 2\mu|2\nu-2 \quad (\mu+\nu \text{ even}),$$

(4·66) $$2\mu|2\nu, \quad 2\mu\pm 2|2\nu, \quad 2\mu|2\nu+2 \quad (\mu+\nu \text{ even}).$$

Altogether, we now see that $\{3, 3, 5\}$ has 120 vertices: ten of the form $2\mu+1|$ (where $\mu = 0, 1, \ldots, 9$), ten of the form $|2\nu+1$ (with the same range for ν), and a hundred of the form $2\mu|2\nu$. By (4·13), since $N_0 = 120$, we deduce

$$N_1 = 720, \quad N_2 = 1200, \quad N_3 = 600.$$

Accordingly, $\{3, 3, 5\}$ is usually called the 600-*cell*, and its reciprocal $\{5, 3, 3\}$, for which

$$N_0 = 600, \quad N_1 = 1200, \quad N_2 = 720, \quad N_3 = 120,$$

is called the 120-*cell*.

Using complex coordinates (u, v), as in (4·46), we may conveniently take $2\mu+1|$ to be $(\epsilon^{2\mu+1}, 0)$, where

$$\epsilon = \exp(\pi i/10).$$

Similarly $|2\nu+1$, on the polar great circle, is $(0, \epsilon^{2\nu+1})$, and $2\mu|2\nu$ is

$$(\epsilon^{2\mu}\cos\lambda, \epsilon^{2\nu}\sin\lambda) \quad (\mu+\nu \text{ even}) \quad \text{or} \quad (\epsilon^{2\mu}\sin\lambda, \epsilon^{2\nu}\cos\lambda) \quad (\mu+\nu \text{ odd}),$$

where λ remains to be found.

Since the vertex $2\mu|2\nu$ $(\mu+\nu$ even) is equidistant from its twelve neighbours

$$2\mu\pm 1|, \quad 2\mu|2\nu\pm 4, \quad 2\mu|2\nu\pm 2, \quad 2\mu\pm 2|2\nu\pm 2, \quad 2\mu\pm 2|2\nu,$$

(4·47) yields

$$\operatorname{Re}\epsilon\cos\lambda = \operatorname{Re}(\cos^2\lambda+\epsilon^4\sin^2\lambda) = \operatorname{Re}(1+\epsilon^2)\sin\lambda\cos\lambda = \operatorname{Re}\epsilon^2,$$

that is,

$$\cos\frac{\pi}{10}\cos\lambda = \cos^2\lambda+\cos\frac{2\pi}{5}\sin^2\lambda = \left(1+\cos\frac{\pi}{5}\right)\sin\lambda\cos\lambda = \cos\frac{\pi}{5},$$

or $$\tfrac{1}{2}5^{\frac{1}{4}}\tau^{\frac{1}{2}}\cos\lambda = \cos^2\lambda+\tfrac{1}{2}\tau^{-1}\sin^2\lambda = \tfrac{1}{2}5^{\frac{1}{2}}\tau\sin\lambda\cos\lambda = \tfrac{1}{2}\tau,$$

whence

$$\cos\lambda = 5^{-\frac{1}{4}}\tau^{\frac{1}{2}}, \quad \sin\lambda = 5^{-\frac{1}{4}}\tau^{-\frac{1}{2}}, \quad \tan\lambda = \tau^{-1}, \quad \tan 2\lambda = 2.$$

Thus $\lambda = \frac{1}{2}\arctan 2$, a familiar angle because it is the value of ψ for the dodecahedron (Figure 3·2A; see also Coxeter 1963, p. 293). To recapitulate, the $10+50+50+10$ vertices

$$2\mu+1|, \quad 2\mu|2\nu \quad (\mu+\nu \text{ even}), \quad 2\mu|2\nu \quad (\mu+\nu \text{ odd}), \quad |2\nu+1$$

of $\{3, 3, 5\}$ have coordinates

(4·67) $$(\epsilon^{2\mu+1}, 0), \quad (\epsilon^{2\mu}\cos\lambda, \epsilon^{2\nu}\sin\lambda) \quad (\mu+\nu \text{ even}),$$
$$(\epsilon^{2\mu}\sin\lambda, \epsilon^{2\nu}\cos\lambda) \quad (\mu+\nu \text{ odd}), \quad (0, \epsilon^{2\nu+1}),$$

where $\epsilon = \exp(\pi i/10)$, $\lambda = \frac{1}{2}\arctan 2$, and μ and ν run independently over the values $0, 1, \ldots, 9$, that is, over the residues modulo 10.

A reciprocal $\{5, 3, 3\}$ can be obtained by taking, as vertices, the centres of the 600 cells $\{3, 3\}$ of $\{3, 3, 5\}$. For instance, the centroid of the four points

$$2\mu|2\nu, \quad 2\mu+2|2\nu, \quad 2\mu|2\nu+2, \quad 2\mu+2|2\nu+2$$

is $(k\epsilon^{2\mu+1}, k\epsilon^{2\nu+1})$, where $k = \frac{1}{4}(\cos\lambda+\sin\lambda)(\epsilon+\epsilon^{-1})$; the corresponding point on the unit 3-sphere is

$$(\sqrt{\tfrac{1}{2}}\,\epsilon^{2\mu+1}, \quad \sqrt{\tfrac{1}{2}}\,\epsilon^{2\nu+1}).$$

Letting μ and ν range over their proper values, we see that 100 of the 600 vertices of $\{5, 3, 3\}$ lie on the spherical torus (4·48) and belong to a $\{4, 4 \mid 10\}$.

The 'diagonal' points, for which $\mu = \nu$, form a decagon inscribed in a great circle, just like the decagon $2\mu+1|$ $(\mu = 0, 1, \ldots, 9)$. In fact, the decagon

$$(\sqrt{\tfrac{1}{2}}\,\epsilon^{2\mu+1}, \quad \sqrt{\tfrac{1}{2}}\,\epsilon^{2\mu+1}) \quad (\mu = 0, 1, \ldots, 9)$$

can be derived from the decagon $(\epsilon^{2\mu+1}, 0)$ by applying the transformation

(4·68) $$u' = \sqrt{\tfrac{1}{2}}(u+v), \quad v' = \sqrt{\tfrac{1}{2}}(u-v),$$

which is an isometry (somewhat resembling (4·43)) since

$$\tfrac{1}{2}(u+v)(\bar{u}+\bar{v}) = u\bar{u}+v\bar{v}.$$

The same isometry transforms each of the 120 vertices of $\{3, 3, 5\}$ into one of the vertices of $\{5, 3, 3\}$; for instance, it transforms the point $0|0$, which is

$$(5^{-\frac{1}{4}}\tau^{\frac{1}{2}}, 5^{-\frac{1}{4}}\tau^{-\frac{1}{2}}) \quad \text{into} \quad (2^{-\frac{1}{2}}5^{-\frac{1}{4}}\tau^{\frac{3}{2}}, 2^{-\frac{1}{2}}5^{-\frac{1}{4}}\tau^{-\frac{3}{2}}),$$

which is the centre of the (spherical) tetrahedron whose four vertices are $\pm 1|$ and $0|\pm 2$, that is, $1|$, $19|$, $0|2$, $0|18$. In this manner we obtain

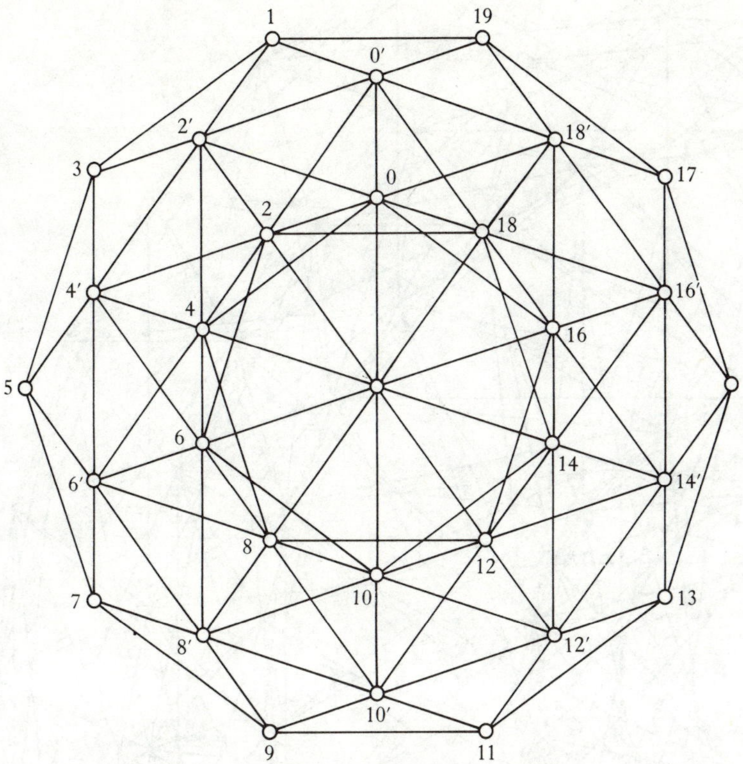

Figure 4·6D: {3, 3, 5}

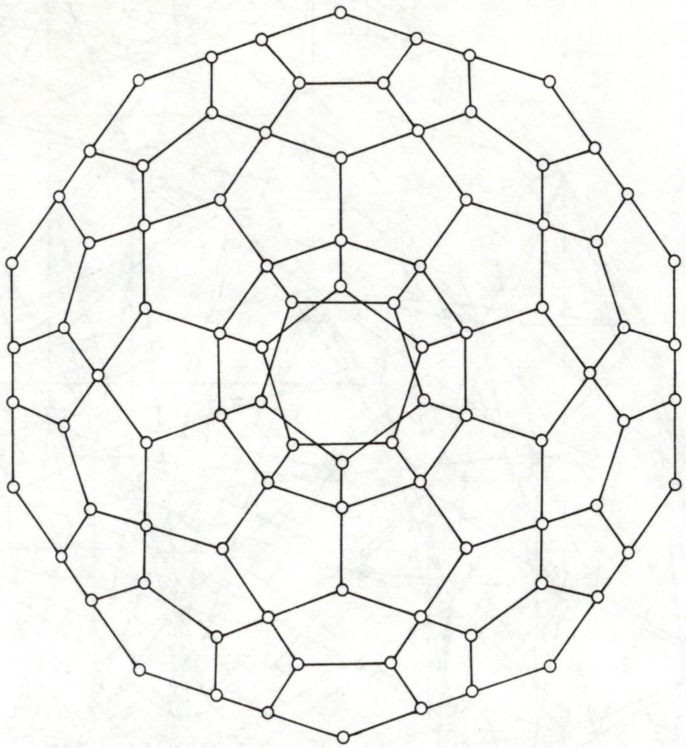

Figure 4·6E: {5, 3, 3}

the vertices of one of the ten {3, 3, 5}'s that can be inscribed in {5, 3, 3} (Coxeter 1963, p. 270). The extra transformations

$$u' = \epsilon^{2\mu}u, \quad v' = v \quad (\mu = 1, 2, 3, 4)$$

serve to complete a set of five {3, 3, 5}'s that together just use up the 600 vertices of {5, 3, 3}. The remaining five of the ten are obtained by using

$$u' = \sqrt{\tfrac{1}{2}}(u-v), \quad v' = \sqrt{\tfrac{1}{2}}(u+v)$$

instead of (4·68).

Figure 4·6D shows the orthogonal projection of {3, 3, 5} on the plane $u = 0$. (This plane serves as an Argand plane for $v = x_3 + x_4 i$.) The unmarked point in the middle is the common image of all the ten vertices $1|, 3|, \ldots, 19|$. For each ν, the vertex $2\mu|2\nu$ has an image marked 2ν or $2\nu'$ according as $\mu+\nu$ is even or odd; for instance, 0′ is the common image of the five points $2|0, 6|0, 10|0, 14|0, 18|0$. Finally, the peripheral points 1, 3, ... are the images of $|1, |3, \ldots$. Notice how strikingly this two-dimensional projection of {3, 3, 5} resembles one view of Donchian's three-dimensional projection (Coxeter 1963, p. 256, Plate VII, no. 1).

Figure 4·6E shows the corresponding two-dimensional projection of the reciprocal polytope {5, 3, 3}. The obvious dodecahedron in the middle is the common image of the ten cells whose centres are

$$2\mu+1| \quad (\mu = 0, \ldots, 9).$$

Each face of this central dodecahedron belongs also to another which appears in a slightly distorted 'edge first' projection. Beyond these cells we see some others in the same kind of projection but more drastically foreshortened. Finally, the ten cells whose centres are $|2\nu+1$ $(\nu = 0, \ldots, 9)$ appear as mere line-segments, forming the peripheral decagon. (For Donchian's analogous three-dimensional projection, see Coxeter 1969, p. 404, Plate III. For a different approach, see Coxeter 1970, §9.)

EXERCISES

1. In the 600-cell (4·67), which vertices are joined by edges to the vertex $4|2$?
2. What convex polytope has for its vertices the hundred points $2\mu|2\nu$?

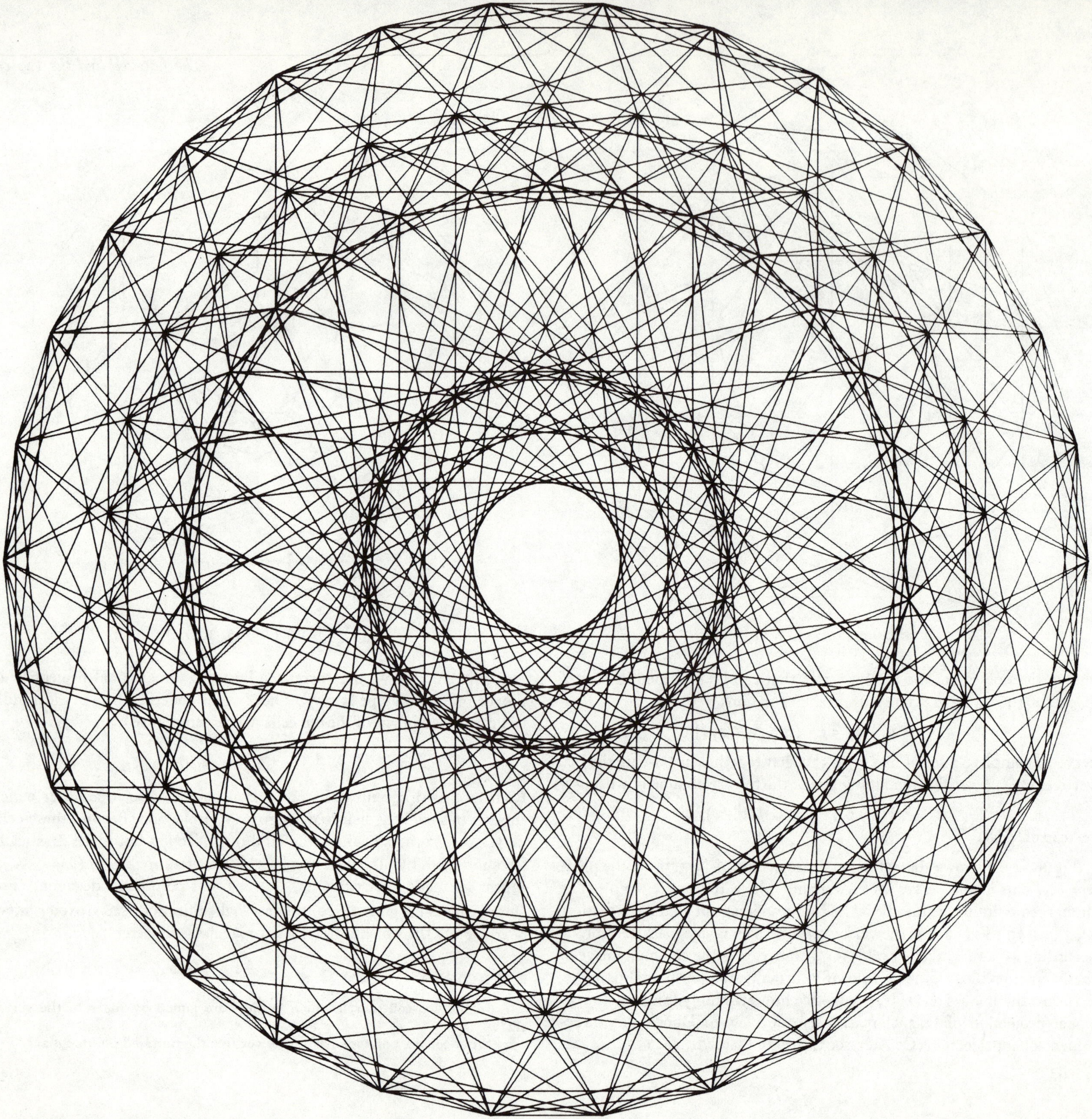

Figure 4·7A: $\{3, 3, 5\}$, $\{3, 5, \frac{5}{2}\}$, $\{5, \frac{5}{2}, 5\}$, $\{5, 3, \frac{5}{2}\}$

Figure 4·7B: $\{\frac{5}{2}, 5, 3\}$, $\{5, \frac{5}{2}, 3\}$

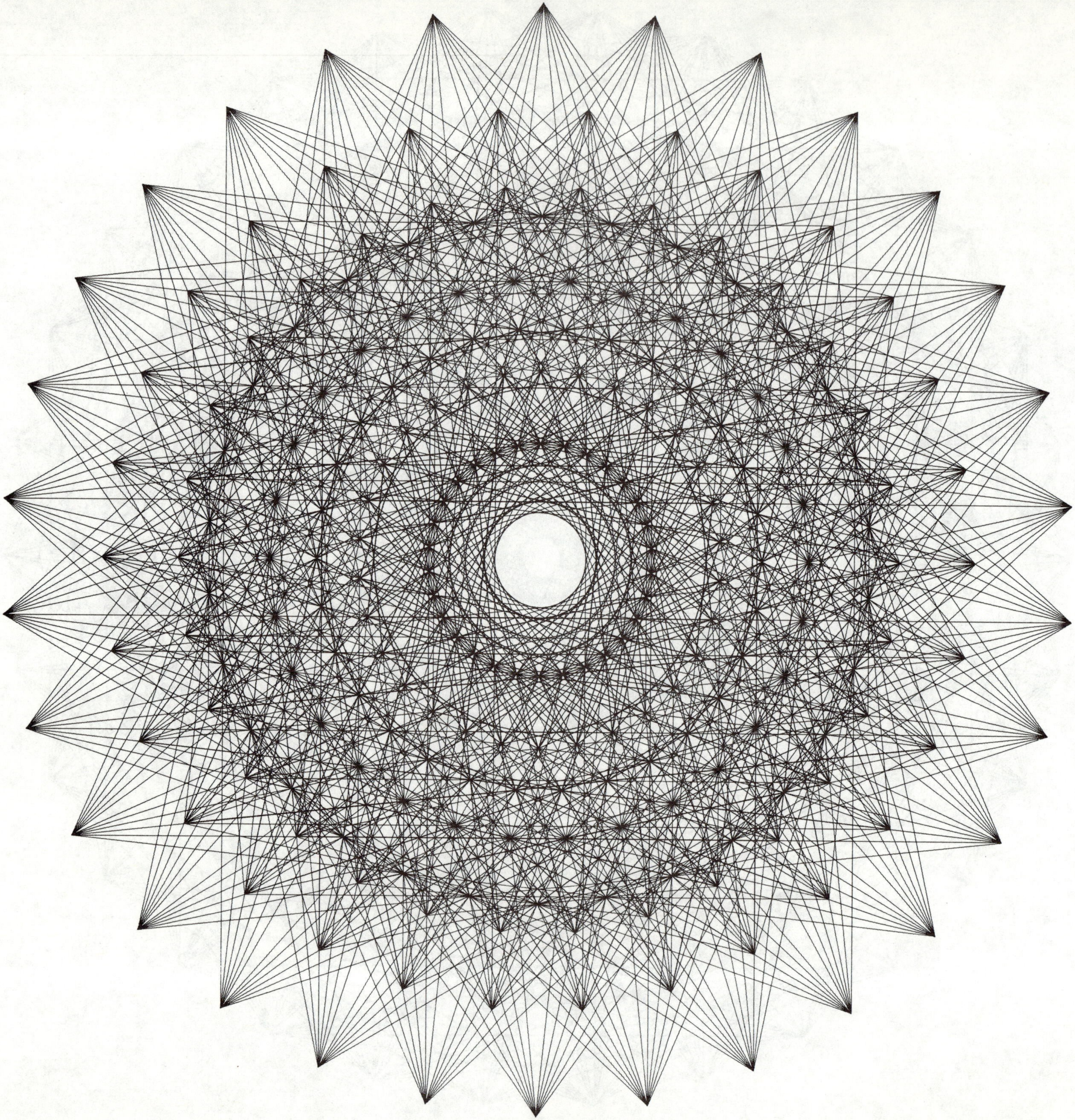

Figure 4·7C: $\{\frac{5}{2}, 3, 5\}$, $\{\frac{5}{2}, 5, \frac{5}{2}\}$, $\{3, \frac{5}{2}, 5\}$, $\{3, 3, \frac{5}{2}\}$

Figure 4·7D: The great grand stellated 120-cell $\{\frac{5}{2}, 3, 3\}$

3. The spherical torus (4·48), midway between the great circles $|v| = 0$ and $|u| = 0$, intersects 300 cells of $\{3, 3, 5\}$ (which are also cells of the polytope described in Ex. 2, above). What pattern is formed by the sections of the corresponding 300 spherical tetrahedra?

4. Which vertices of the 600-cell (4·67) are equidistant from the two opposite vertices $5|$ and $15|$? What shape is the section of the polytope by the hyperplane that contains these thirty points? How does it appear in Figure 4·6D?

4·7 THE TEN STAR POLYTOPES

Looking again at Figure 2·1B, we see that Cayley's names for the Kepler–Poinsot polyhedra can be justified and generalized by declaring that, in certain cases, a polytope Π yields a *stellated* Π when all its edges are dilated, a *great* Π when all its faces are dilated, and a *grand* Π when all its cells are dilated. The ratio of the dilatation is occasionally negative (so that the magnified element is inverted) and of course the dilatation must be such that the bits fit together properly. (This procedure was suggested by J. H. Conway.) Thus the pentagram $\{\frac{5}{2}\}$ is the *stellated* $\{5\}$; the (small) stellated dodecahedron $\{\frac{5}{2}, 5\}$ is the *stellated* $\{5, 3\}$; $\{5, \frac{5}{2}\}$ is the *great* $\{5, 3\}$ (with ratio $-\tau^3$); the great stellated dodecahedron $\{\frac{5}{2}, 3\}$ is the *great* $\{\frac{5}{2}, 5\}$ (it is also the *stellated* $\{5, \frac{5}{2}\}$); and $\{3, \frac{5}{2}\}$ is the *great* $\{3, 5\}$ (with ratio $-\tau^4$).

The four-dimensional analogues of the pentagram and the Kepler–Poinsot polyhedra are the ten star polytopes of Schläfli and Hess (see Coxeter 1963, pp. 264–7). Nine of them may be placed so as to have the same vertices as $\{3, 3, 5\}$. Three of these nine then have also the same *edges* as $\{3, 3, 5\}$ (see Figure 4·7A or the frontispiece of Coxeter 1963), two others both have the same (longer) edges (Figure 4·7B) and the remaining four all have the same (still longer) edges (Figure 4·7C). Finally, the tenth star polytope (Figure 4·7D) has the same vertices as $\{5, 3, 3\}$ (see Coxeter 1969, p. 403).

These twelve 'pentagonal' polytopes are related as follows: $\{\frac{5}{2}, 5, 3\}$ is the *stellated* $\{5, 3, 3\}$; $\{5, \frac{5}{2}, 5\}$ is the *great* $\{5, 3, 3\}$; $\{\frac{5}{2}, 3, 5\}$ is the *great* $\{\frac{5}{2}, 5, 3\}$ and also the *stellated* $\{5, \frac{5}{2}, 5\}$; $\{5, 3, \frac{5}{2}\}$ is the *grand* $\{5, 3, 3\}$; $\{\frac{5}{2}, 5, \frac{5}{2}\}$ is the *grand* $\{\frac{5}{2}, 5, 3\}$ and also the *stellated* $\{5, 3, \frac{5}{2}\}$; $\{5, \frac{5}{2}, 3\}$ is the *great* $\{5, 3, \frac{5}{2}\}$ and also the *grand* $\{5, \frac{5}{2}, 5\}$; $\{\frac{5}{2}, 3, 3\}$ is the *stellated* $\{5, \frac{5}{2}, 3\}$, also the *great* $\{\frac{5}{2}, 5, \frac{5}{2}\}$ and the *grand* $\{\frac{5}{2}, 3, 5\}$. The reciprocal of $\{\frac{5}{2}, 5, 3\}$ is $\{3, 5, \frac{5}{2}\}$, whose cells are icosahedra, and this yields $\{3, \frac{5}{2}, 5\}$, the *great* $\{3, 5, \frac{5}{2}\}$, which is the reciprocal of $\{5, \frac{5}{2}, 3\}$. Finally, the reciprocal of $\{\frac{5}{2}, 3, 3\}$ is $\{3, 3, \frac{5}{2}\}$, the *grand* $\{3, 3, 5\}$.

Conway summarizes these relationships diagrammatically in figure 4·7E, where *stellation* is indicated by a vector going NNE, *greatening* by one going ENE, and *aggrandisement* by one going ESE. The row of numbers at the bottom indicates *density* (Coxeter 1963, p. 263).

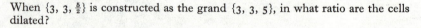

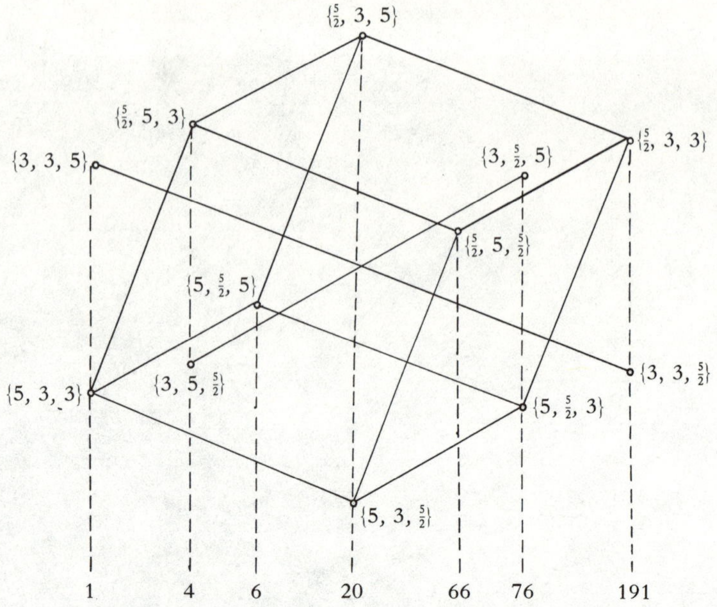

Figure 4·7E: The twelve pentagonal polytopes

EXERCISE

When $\{3, 3, \frac{5}{2}\}$ is constructed as the grand $\{3, 3, 5\}$, in what ratio are the cells dilated?

4·8 A FAMILY OF REGULAR COMPLEX POLYGONS

We saw, in §4·5, that the plane faces of the double prism $\{p\} \times \{p\}$ consist of p^2 squares and $2p$ p-gons. (When $p = 4$, this distinction can still be made, actually in three ways. One way is plainly visible in Figure 4·5A, where the 16 squares are foreshortened into rhombi while the 8 p-gons appear as undistorted squares.) Unlike the p^2 squares, which form $\{4, 4 \mid p\}$, the $2p$ p-gons do not form a polyhedral surface: their $2p^2$ edges are all distinct, and each of the p^2 vertices belongs to only 2 of the p-gons.

Recalling that each vertex of any polygon belongs to just 2 of its edges, we naturally ask whether there is any sense in which our $2p$ precariously connected p-gons might be regarded as the 'edges' of some kind of generalized polygon: a *regular complex polygon.* A clue is provided by the complex coordinates u and v, in terms of which the vertices of each p-gon satisfy a linear equation

$$v = \sqrt{\tfrac{1}{2}}\,\epsilon^{2\nu+1} \quad \text{or} \quad u = \sqrt{\tfrac{1}{2}}\,\epsilon^{2\mu+1}$$

and thus lie on a line of the *unitary plane.* Let us call this line an *edge* of the complex polygon, even though it contains p vertices where p may exceed 2. We shall see later (in chapter 11) that an appropriate 'Schläfli symbol' is

$$p\{4\}2,$$

where the p indicates that there are p vertices on each edge while the 2 indicates that there are 2 edges through each vertex. For a real polygon $\{q\}$, both these numbers are 2; thus the symbol $\{q\}$ may be regarded as an abbreviation for $2\{q\}2$, in agreement with the fact that, when $p = 2$, $p\{4\}2$ reduces to an ordinary square. Let us call such a generalized edge a *p-edge*.

Omitting the nine squares from $\{3\}\times\{3\}$ (Figure 11·5A on page 108), we are left with six triangles representing the six 3-edges of $3\{4\}2$. Similarly, the eight 4-edges of $4\{4\}2$ are represented by the undistorted squares in Figure 4·5A. The analogous drawing for any $p\{4\}2$ consists of a peripheral $\{2p\}$ with a $\{p\}$ drawn inwards on each edge. When p is even,† each edge of the $\{2p\}$ belongs to a $\{p\}$ whose opposite edge belongs to a concentric star-polygon

$$\left\{\frac{2p}{p-1}\right\}.$$

When p is odd, p of the p^2 vertices are projected together into the centre of the $\{2p\}$; but this awkwardness can be remedied by using, instead of the peripheral $\{2p\}$, an equilateral (but not equiangular) $2p$-gon whose vertices lie alternately on two concentric circles.

Similarly, among the 32 triangular faces of $\{3, 3, 4\}$, the eight that appear as equilateral triangles

124, 235, 346, 457, 568, 671, 782, 813

in Figure 4·2A, may be regarded as the eight 3-edges of a regular complex polygon $3\{3\}3$ (Coxeter 1968, pp. 121–5; 1969, p. 237, Ex. 3). The above numbering of the points reveals this configuration 8_3 as a self-inscribed octagon (in the complex plane), or as a pair of mutually inscribed quadrangles (Hilbert and Cohn–Vossen 1952, p. 101).

Again, among the 72 diametral squares of the 24 octahedral cells of $\{3, 4, 3\}$, the 24 that appear as squares in the dodecagonal projection (Figure 4·2D) may be regarded as the 24 4-edges of a regular complex polygon $4\{3\}4$. The simplest coordinates for its 24 vertices are

$$(\epsilon^\mu, 0), \quad (0, \epsilon^\nu) \quad (\mu \text{ and } \nu \text{ even})$$

and

$$(\epsilon^\mu/\sqrt{2}, \quad \epsilon^\nu/\sqrt{2}) \quad (\mu \text{ and } \nu \text{ odd}),$$

where $\epsilon = \exp(\frac{1}{4}\pi i) = \sqrt{\frac{1}{2}}(1+i)$. One 4-edge, having the equation $u+v = 1$, contains the 4 vertices $(1, 0)$, $(0, 1)$, $(\frac{1}{2}(1 \pm i), \frac{1}{2}(1 \mp i))$.

† For $p = 6$, see N. W. Johnson, A geometric model for the generalized symmetric group, *Canadian Mathematical Bulletin* **3** (1960), 133–42; especially p. 135.

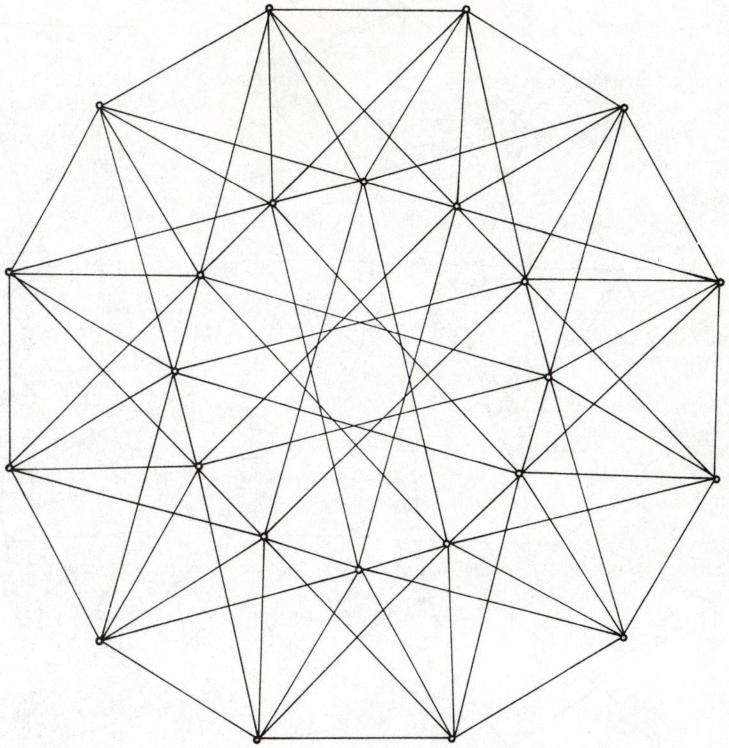

Figure 4·8A: $3\{4\}3$

Figure 4·8A shows the same projection of $\{3, 4, 3\}$ with 24 of the edges omitted. The 72 that remain form 24 triangles (among the 96 faces of the polytope) which appear as undistorted equilateral triangles in the figure. These may be regarded as the twenty-four 3-edges of a regular complex polygon $3\{4\}3$.

Similarly, after omitting 360 or 120 of the 720 edges that appear in Figure 4·7A, we find that the remaining 360 (Figure 4·8B) or 600 (Figure 4·8C) form 120 undistorted triangles (faces of $\{3, 3, 5\}$ or $\{3, 5, \frac{5}{2}\}$) or pentagons (faces of $\{5, \frac{5}{2}, 5\}$ or $\{5, 3, \frac{5}{2}\}$). These triangles and pentagons may be regarded as the 3-edges of $3\{5\}3$ and the 5-edges of $5\{3\}5$, respectively. The 120 vertices of the latter have the coordinates (4·67). One of the 120 5-edges of $5\{3\}5$ has the equation $v = \cos\lambda$ and contains the 5 vertices

$$(\epsilon^{2\mu}\sin\lambda, \cos\lambda) \quad (\epsilon = \exp(\pi i/10);\ \mu = 1, 3, 5, 7, 9;\ \lambda = \tfrac{1}{2}\operatorname{arc}\tan 2).$$

In the same spirit we may omit half the 720 edges that appear in Figure 4·7C, and find that the remaining 360 (Figure 4·8D) form 120

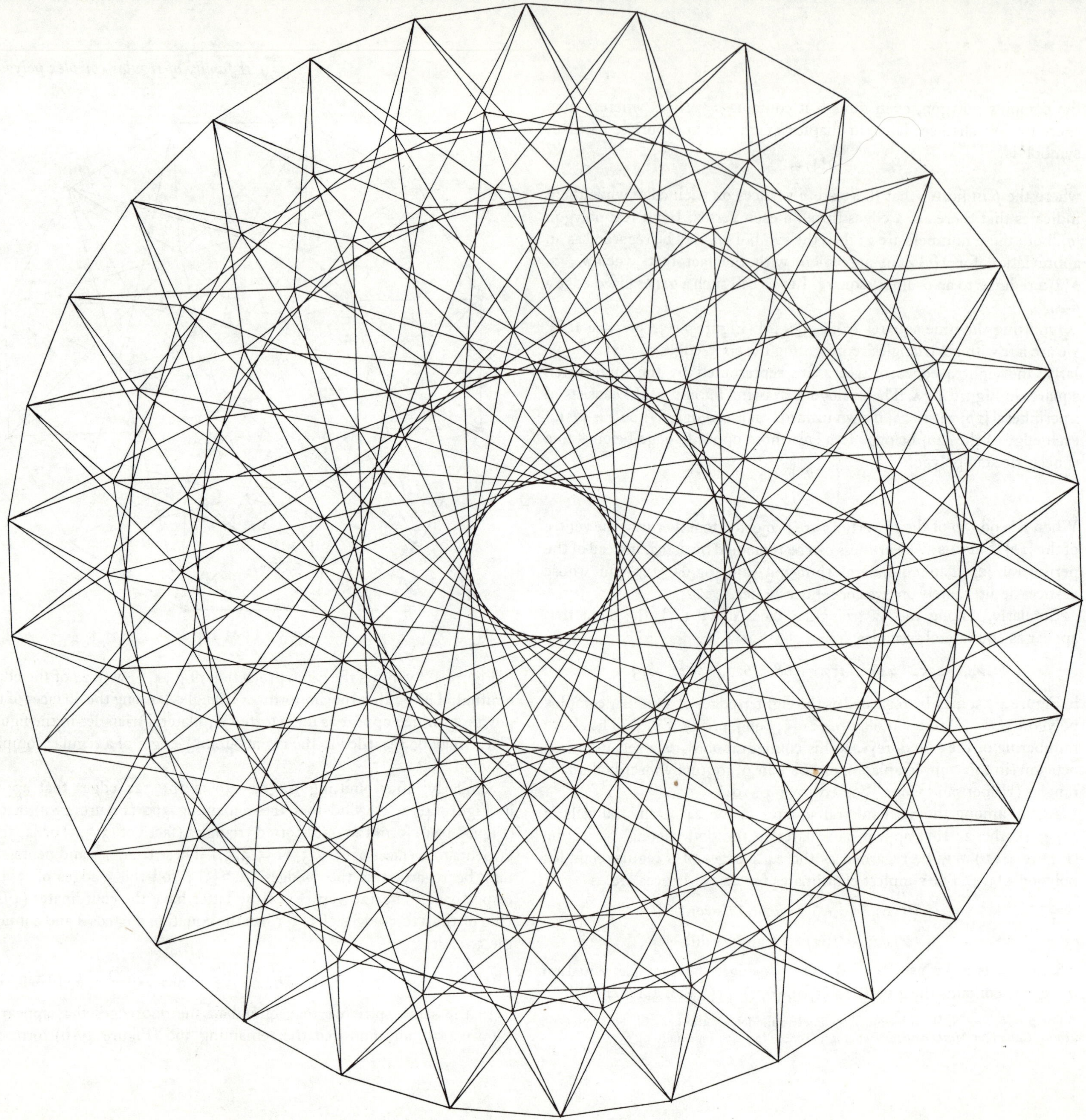

Figure 4·8B: 3{5}3

Figure 4·8c: $5\{3\}5$

Figure 4·8D: $3\{\frac{5}{2}\}3$

Figure 4·8E: $5\{\frac{5}{2}\}5$

undistorted triangles (faces of $\{3, \frac{5}{2}, 5\}$ or $\{3, 3, \frac{5}{2}\}$) which may be regarded as the 120 3-edges of a regular complex star polygon $3\{\frac{5}{2}\}3$. Finally, we may omit half the 1200 edges that appear in Figure 4·7B, and find that the remaining 600 (Figure 4·8E) form 120 undistorted pentagons (faces of $\{5, \frac{5}{2}, 3\}$) which may be regarded as the 120 5-edges of a regular complex star polygon $5\{\frac{5}{2}\}5$.

At the present stage we have been content to introduce the concept of a regular complex polygon by a few examples, leaving the precise definition till Chapter 11. In the symbol $p_1\{q\}p_2$, p_1 is the number of vertices on an edge, p_2 is the number of edges at a vertex, but q is more subtle. For the moment, let us pretend that it indicates the *omitted* polygons $\{q\}$.

EXERCISES

What figures are formed by:

1. The 24 edges of $\{3, 4, 3\}$ that were omitted from Figure 4·2D to make Figure 4·8A?
2. The 120 edges of $\{3, 3, 5\}$ that were omitted from Figure 4·7A to make Figure 4·8C?

4·9 REMARKS

It seems to have been Helmut Emde† who first exploited the idea of considering, instead of the regular polytopes in Euclidean 4-space, the corresponding honeycombs in spherical 3-space. Although his bibliography includes Schläfli, his classification is unnecessarily complicated because he refused to make use of the Schläfli symbol. As his drawings and stereoscopic photographs are strikingly beautiful, it seems worth while to identify them as follows:

		$\{n, 2, m\}$	on pages 47–50,
		$\{2, n, 2\}$	on page 51,
$\{2, 4, 3\}$	and	$\{2, 5, 3\}$	on page 53,
		$\{2, 3, 4\}$	on page 54,
		$\{2, 3, 5\}$	on page 55,
$\{3, 3, 2\}$	and	$\{3, 3, 3\}$	on page 56,
$\{3, 3, 4\}$	and	$\{3, 3, 5\}$	on page 57,
		$\{3, 4, 2\}$	on page 58,
$\{3, 4, 3\}$	and	$\{3, 5, 2\}$	on page 59,
		$\{4, 3, 2\}$	on page 60,
$\{4, 3, 3\}$	and	$\{5, 3, 2\}$	on page 61,
		$\{5, 3, 3\}$	on pages 62–6.

† Homogene Polytope, *Bayerische Akademie der Wissenschaften (Math.-Naturwiss. Klasse) Abhandlungen (Neue Folge)* **89** (1958), 1–67.

The idea of exploring spherical or elliptic space by means of two complex coordinates can be traced back to Edouard Goursat.‡

The elegant construction for the spherical honeycomb $\{3, 3, 5\}$ in §4·6 was described in a letter from Raphael M. Robinson, who is well known for his work on factorization of large numbers and for his proof of van der Waerden's conjecture that the vertices of the snub cube are the centres of 24 circles packed as closely as possible on a sphere§. Robinson described his work of November 1968 as follows:

> First I computed the edge of a regular tetrahedron with dihedral angle 72°, and found it to be 36°. Twenty tetrahedra of this size combine to form a regular icosahedron. If one takes a great circle, divides it into ten equal arcs, and constructs an icosahedron around each division point, then each two consecutive icosahedra have five tetrahedra in common. Thus there are 150 tetrahedra in all, with 60 vertices. If a congruent figure is constructed around the polar great circle, then we will have all the 120 vertices and half of the 600 tetrahedra expected in the $\{3, 3, 5\}$. We must see how to place the two figures relative to each other. The 50 outer vertices of each lie on a torus, and indeed form the alternate lattice points if we divide the torus into tenths in both directions. Place the two figures so that the vertices of one are opposite the empty lattice points of the other. (The long way around one torus corresponds to the short way around the other.) Each outer vertex of one figure is then equally distant from 4 vertices of the other. I actually computed the distance, and found it to be 36° .... It is then clear that we can fill in the space between the two figures with 300 additional regular tetrahedra: 100 with faces on the first figure and one vertex on the second, 100 with faces on the second figure and one vertex on the first, and 100 with one edge on each figure ....
>
> The centres of the last 100 tetrahedra lie on the torus equidistant from the two original polar great circles. They form the lattice points when the torus is divided into tenths both ways. The diagonals are great circles which are divided into tenths, so that the alternate centres are at the distance 36° .... This suggested to me that we could probably inscribe the $\{3, 3, 5\}$ in the $\{5, 3, 3\}$. After some further study I found that this is true. However, ... only 20 of the 100 points on the central torus appear as vertices of any one $\{3, 3, 5\}$. These 20 points lie 10 each on two polar great circles lying on the torus. There are two ways to divide the 100 points into five sets of 20 points, and these correspond to

‡ Sur les substitutions orthogonales et les divisions régulières de l'espace, *Annales Scientifiques de l'École Normale Supérieure* (3), **6** (1889), 9–102.

§ Fejes Tóth (1964, p. 232); R. M. Robinson, Arrangement of 24 points on a sphere, *Mathematische Annalen* **144** (1961), 17–48.

two different sets of five {3, 3, 5}'s inscribed in the {5, 3, 3} dual to the original {3, 3, 5}.

The coordinates (4·67) had already been obtained by G. C. Shephard (in his Cambridge Ph.D. thesis of 1951, where he described the complex polygon now known as 5{3}5). They were rediscovered by Conway, who drew Figures 4·6D and 4·6E. The spectacular figures 4·7D, 4·8B, C were drawn with immense labour by Peter McMullen. George Olshevsky, Jr programmed an electronic 'incremental plotter' to draw Figures 4·7A, B, C and 4·8D, E.

CHAPTER 5

Frieze patterns

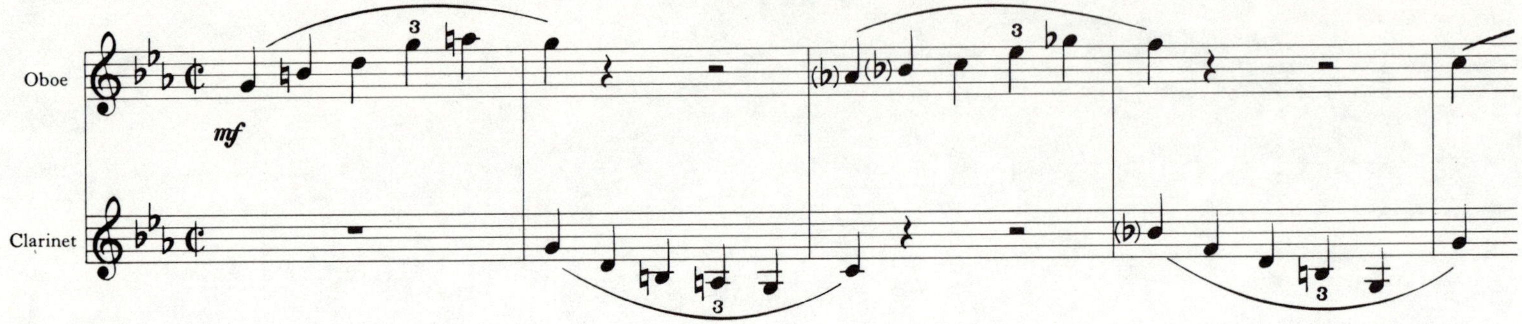

Anton Bruckner
(Fourth Symphony, bars 131–4)

$$\begin{array}{ccccccccccccccccc}
 & 1 & 1 & 1 & 1 & 1 & 1 & 1 & 1 & 1 & 1 & 1 & 1 & 1 & 1 & 1 & \\
 & 1 & 2 & 4 & 1 & 2 & 2 & 3 & 1 & 2 & 4 & 1 & 2 & 2 & 3 & 1 & \\
\dots & 1 & 7 & 3 & 1 & 3 & 5 & 2 & 1 & 7 & 3 & 1 & 3 & 5 & 2 & 1 & \\
 & 1 & 3 & 5 & 2 & 1 & 7 & 3 & 1 & 3 & 5 & 2 & 1 & 7 & 3 & 1 & \dots \\
 & 1 & 2 & 2 & 3 & 1 & 2 & 4 & 1 & 2 & 2 & 3 & 1 & 2 & 4 & 1 & \\
 & 1 & 1 & 1 & 1 & 1 & 1 & 1 & 1 & 1 & 1 & 1 & 1 & 1 & 1 & 1 &
\end{array}$$

When lecturing on frieze patterns, I sometimes begin by writing out a particular example on the blackboard, with horizontal rows of ones at the top and bottom and a zigzag of ones on the left. After filling in the numbers quietly for a few minutes, I am usually interrupted by some member of the audience who has guessed the local rule of formation, the 'unimodular rule' $ad-bc = 1$ of §3·1. I then proceed more rapidly, till the periodicity of the pattern has become apparent. The next twenty minutes may be spent explaining how this global property follows from the local rule. Then someone will probably ask whether the rule can be modified so as to allow the two borders to contain numbers greater than 1. The present chapter is a description of such modified frieze patterns. It appears that, although these first and last rows need not consist entirely of ones, each must be a delayed replica of the other, like the occurrences of the theme in a fugue.

We recall that an n-dimensional regular polytope is decomposed by its hyperplanes of symmetry into a number of congruent simplexes forming a implicial subdivision of the polytope. Such a simplex is an orthoscheme $O_0 O_1 \dots O_n$, where O_0 is a vertex and O_s is the centre of a s-dimensional element, so that, if $s<t<u$, O_sO_t is perpendicular to O_tO_u. In §5·4 we shall see how the angular properties of the orthoscheme can be read off from a diagram consisting of $n+1$ collinear points

$$X_0, X_1, \dots, X_n \quad \text{such that} \quad X_sX_t = (O_sO_t)^2,$$

semicircles with diameters X_sX_t, and a vertical line through each point X_s. An arbitrary inversion yields a new diagram formed by $n+2$ points on a circle, and arcs through pairs of them orthogonal to this circle. If $X_sX_tX_uX_v$ is a convex quadrangle, Ptolemy's theorem tells us that

$$X_sX_t.X_uX_v - X_sX_u.X_tX_v + X_sX_v.X_tX_u = 0,$$

suggesting the investigation of numbers (s, t) (functions of integers s and t) that satisfy

$$(s,t)(u,v)+(s,u)(v,t)+(s,v)(t,u) = 0$$

(Coxeter 1963, p. 160). This functional equation provides the rule for constructing our 'modified frieze patterns'.

5·1 SOME EXAMPLES

In §3·1 we considered frieze patterns of order 5. (The *order* is one more than the number of rows.) Here are some of order 6:

(5·11)

$$\begin{array}{cccccccccccccccccc}
& 1 && 1 && 1 && 1 && 1 && 1 && 1 && 1 && \cdots \\
4 && 1 && 2 && 2 && 2 && 1 && 4 && 1 && 2 & \\
& 3 && 1 && 3 && 3 && 1 && 3 && 3 && 1 && \cdots \\
2 && 2 && 1 && 4 && 1 && 2 && 2 && 2 && 1 & \\
& 1 && 1 && 1 && 1 && 1 && 1 && 1 && 1 && \cdots
\end{array}$$

(5·12)

$$\begin{array}{cccccccccccccccccc}
& 1 && 1 && 1 && 1 && 1 && 5 && 1 && 1 && \cdots \\
4 && 2 && 2 && 2 && 2 && 4 && 4 && 2 && 2 & \\
& 3 && 3 && 3 && 3 && 3 && 3 && 3 && 3 && \cdots \\
2 && 2 && 4 && 4 && 2 && 2 && 2 && 2 && 4 & \\
& 1 && 1 && 5 && 1 && 1 && 1 && 1 && 1 && \cdots
\end{array}$$

(5·13)

$$\begin{array}{cccccccccccccccccc}
& \tau^3 && 1 && 1 && 1 && 1 && \tau^{-3} && \tau^3 && 1 && \cdots \\
4 && 2\tau && 2 && 2 && 2 && 2\tau^{-1} && 4 && 2\tau && 2 & \\
& 3 && \sqrt{5} && 3 && 3 && \sqrt{5} && 3 && 3 && \sqrt{5} && \cdots \\
2 && 2 && 2\tau^{-1} && 4 && 2\tau && 2 && 2 && 2 && 2\tau^{-1} & \\
& 1 && 1 && \tau^{-3} && \tau^3 && 1 && 1 && 1 && 1 && \cdots
\end{array}$$

(5·14)

$$\begin{array}{cccccccccccccccccc}
& \tau && \tau^{-1} && \tau && \tau^{-1} && \tau && \tau^{-1} && \tau && \tau^{-1} && \cdots \\
2 && 2 && 2 && 2 && 2 && 2 && 2 && 2 && 2 & \\
& \sqrt{5} && \sqrt{5} && \sqrt{5} && \sqrt{5} && \sqrt{5} && \sqrt{5} && \sqrt{5} && \sqrt{5} && \cdots \\
2 && 2 && 2 && 2 && 2 && 2 && 2 && 2 && 2 & \\
& \tau^{-1} && \tau && \tau^{-1} && \tau && \tau^{-1} && \tau && \tau^{-1} && \tau && \cdots
\end{array}$$

(5·15)

$$\begin{array}{cccccccccccccccccc}
& \tau && \tau^{-1} && 1 && \tau && \tau^{-1} && 1 && \tau && \tau^{-1} && \cdots \\
2\tau && 2 && 2\tau^{-1} && 2\tau && 2 && 2\tau^{-1} && 2\tau && 2 && 2\tau^{-1} & \\
& \sqrt{5}\tau && \sqrt{5}\tau^{-1} && 3 && \sqrt{5}\tau && \sqrt{5}\tau^{-1} && 3 && \sqrt{5}\tau && \sqrt{5}\tau^{-1} && \cdots \\
2\tau && 2 && 2\tau^{-1} && 2\tau && 2 && 2\tau^{-1} && 2\tau && 2 && 2\tau^{-1} & \\
& \tau && \tau^{-1} && 1 && \tau && \tau^{-1} && 1 && \tau && \tau^{-1} && \cdots
\end{array}$$

The first of these five patterns is precisely analogous to those in §3·1. In the rest, the top and bottom rows include numbers different from 1, and consequently the 'unimodular' rule

$$ad - bc = 1$$

has to be replaced by the slightly more complicated rule

$$ad - bc = ps,$$

where p and s are the numbers determined in the top row by the diagonals through c (containing a and d, respectively). For instance, a pattern of order 7 might include

$$\begin{array}{cccccccccc}
& p && \cdot && \cdot && s && t \\
&& \cdot && b && \cdot && \cdot & \\
& \cdot && a && d && \cdot && \cdot \\
&& \cdot && c && f && \cdot & \\
& \cdot && \cdot && e && \cdot && \cdot \\
s && t && \cdot && \cdot && p &
\end{array}$$

To make the bottom row serve just as well as the top, we assume that the same numbers p and s reappear in the bottom row on the diagonals through b (containing d and a respectively), so that p plays the same role for

$$\begin{array}{ccc} & d & \\ c && f \\ & e & \end{array} \quad \text{as for} \quad \begin{array}{ccc} & b & \\ a && d. \\ & c & \end{array}$$

These rules remain valid when a and d are located in the top row or the bottom, provided we insert a row of zeros above the top (so that $b = 0$ when a and d are at the top) and another row of zeros below the bottom (so that $c = 0$ when a and d are at the bottom).

More precisely, we define a *frieze pattern of order n* to be an arrangement of numbers (s, t) in staggered rows

$$\begin{array}{ccccccccccccc}
(0,0) && (1,1) && (2,2) && (3,3) && (4,4) &&&& \\
& (0,1) && (1,2) && (2,3) && (3,4) && \cdots &&& \\
(-1,1) && (0,2) && (1,3) && (2,4) && (3,5) &&&& \\
& (-1,2) && (0,3) && (1,4) && (2,5) &&&&& \\
\cdot & \cdot & \cdot & \cdot & \cdot & \cdot & \cdot & \cdot & \cdot & \cdot &&& \\
&&& (-1,n-2) && (0,n-1) && (1,n) && (2,n+1) &&& \\
&&&& (-1,n-1) && (0,n) && (1,n+1) &&&& \cdots
\end{array}$$

(Coxeter 1963, p. 160), such that

(5·16) $\qquad (s,s) = (s,s+n) = 0, \quad (s,s+1) = (s+1,s+n) > 0,$

(5·17) $\qquad (s-1,t)(s,t+1)-(s,t)(s-1,t+1) = (s-1,s)(t,t+1).$

In speaking of 'staggered rows', we mean that $(s-1,t)$ and $(s,t+1)$ are consecutive entries in one row while $(s-1,t+1)$ comes between them in the next row.

The equation $(s,s+1) = (s+1,s+n)$ indicates a kind of periodicity that was obscured in §3·1, where each expression was equal to 1. But it still seems remarkable (at first sight) that these rules imply the complete periodicity

$$(s,t) = (s+n,t+n).$$

EXERCISES

1. Draw sufficient portions of the most general frieze patterns of orders 2, 3, 4. How can the last of these be interpreted as a line in projective 3-space?

2. If the top row of a frieze pattern of order 4, 5, or 6 consists entirely of 1's while the next row likewise consists of one number continually repeated, what is this number?

3. The patterns in Ex. 2 are said to be 'of period 1'. Is there a pattern of order 7 and period 1?

4. In (5·13), the diagonal that goes from τ^3 to τ^{-3} is an arithmetical progression.

5·2 PROOF OF THE PERIODICITY

We observe first that the whole pattern, for $s \leqslant t \leqslant s+n$, is determined by the entries in two consecutive diagonals, such as

(5·21) $\qquad f_t = (-1,t), \quad g_t = (0,t) \quad (0 \leqslant t < n).$

In fact, $\quad (1,2) = \dfrac{f_1 g_2 - f_2 g_1}{f_0}, \qquad (2,3) = \dfrac{f_2 g_3 - f_3 g_2}{f_0},$

$$(1,3) = \frac{(1,2)g_3 + (2,3)g_1}{g_2}, \quad (1,4) = \frac{(1,3)g_4 + (3,4)g_1}{g_3},$$

and so on. By (5·16),

$$f_{-1} = g_0 = f_{n-1} = g_n = 0, \quad f_0 = g_{n-1}.$$

Next we find that, for $s \leqslant t \leqslant s+n$,

(5·22) $\qquad (s,t) = (f_s g_t - f_t g_s)/f_0.$

In fact, this definition of (s,t) (for *all* integers s and t) implies

(5·23) $\qquad (s,t)+(t,s) = 0,$

(5·24) $\qquad (s,t)(u,v)+(s,u)(v,t)+(s,v)(t,u) = 0.$

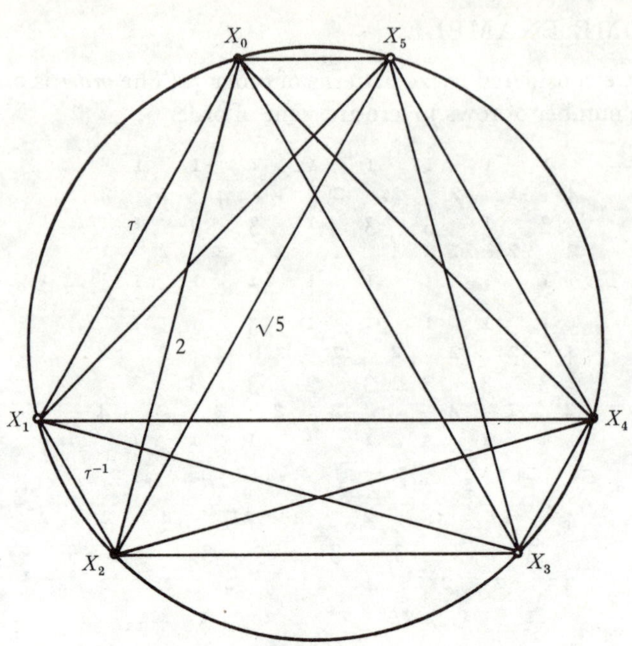

Figure 5·3A: The diagram for (5·14)

We can derive (5·17) by setting $u = t+1$, $v = s-1$ in (5·24) and then using (5·23). Another special case of (5·24) is

$$(s,t)(s+1,s+n)+(s,s+1)(s+n,t)+(s,s+n)(t,s+1) = 0.$$

Since $(s+1,s+n) = (s,s+1) \neq 0$ and $(s,s+n) = 0$, it follows that

(5·25) $\qquad (s,t)+(s+n,t) = 0$

and

(5·26) $\qquad (s,t) = (t,s+n) = (s+n,t+n).$

Thus the pattern is symmetrical by a *glide*: the product of a translation and a reflection.†

This kind of periodicity is well illustrated in (5·13) (where $n = 6$). The translation and reflection occur separately in (5·15), where the period is consequently 3 instead of 6. In (5·11) and (5·12) and (5·14), there are reflections in vertical lines; in (5·14) these are so plentiful that the period is only 2.

† This proof of the periodicity is an improved version of one given in my Frieze patterns, *Acta Arithmetica* **18** (1971), 297–310. See also my Cyclic sequences and frieze patterns, *Vinculum* **8** (Melbourne, Australia, 1971), 4–7, and Martin Gardner's account in *Scientific American* **226** (April 1972), 101.

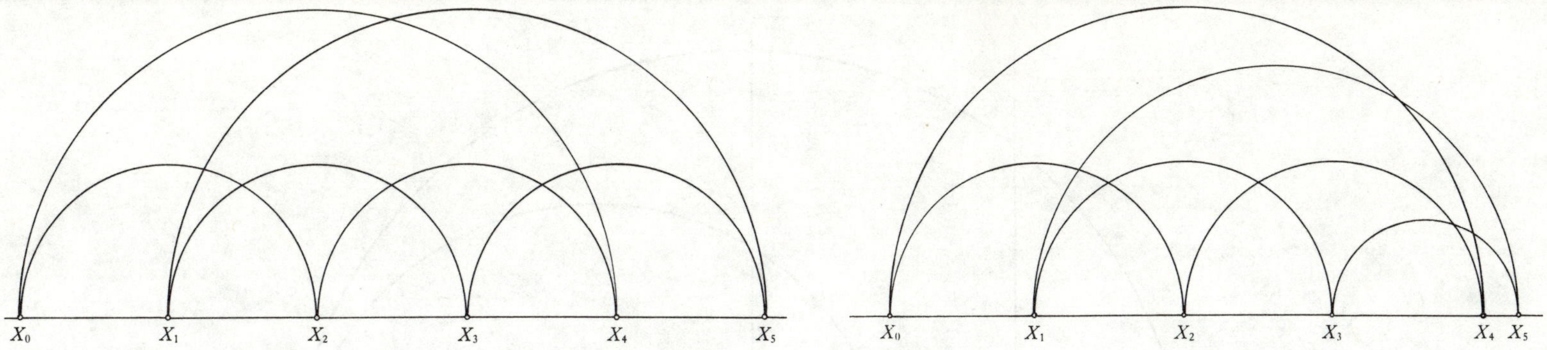

Figure 5·3B: The diagrams for (5·12) and (5·13)

EXERCISES

1. Deduce (5·23) from (5·24).
2. How can the frieze pattern be extended, above and below, to make a complete wallpaper design, still satisfying (5·24)?
3. How is f_{s+n} related to f_s, and g_{s+n} to g_s?
4. Evaluate the determinants

$$\begin{vmatrix} (0,0) & (0,1) & (0,2) \\ (1,0) & (1,1) & (1,2) \\ (2,0) & (2,1) & (2,2) \end{vmatrix} \quad \text{and} \quad \begin{vmatrix} (0,0) & \dots & (0,3) \\ & \dots \quad \dots & \\ (3,0) & \dots & (3,3) \end{vmatrix}.$$

5·3 PTOLEMAIC PATTERNS

Any n distinct points on a circle, named $X_0, X_1, \dots, X_{n-1}$ in their natural order, determine a frieze pattern in which

$$(s,t) = X_s X_t \quad (0 \leqslant s < t < n).$$

We naturally call this a *Ptolemaic* frieze pattern, because the essential relation (5·24) is obtained by applying Ptolemy's theorem (Coxeter and Greitzer 1967, p. 42) to the quadrangle $X_s X_t X_u X_v$. For instance, (5·14) is Ptolemaic, as we see in Figure 5·3A. So too are (5·12) and (5·13), if we allow the points to be on a line (as in Figure 5·3B) instead of a circle. In the case of (5·12),

$$(0,1) = (1,2) = (2,3) = (3,4) = (4,5) = 1,$$
$$(5,6) = (0,5) = X_0X_5 = 5, \quad (-1,1) = (1,5) = X_1X_5 = 4,$$

and so on; in the case of (5·13),

$$(0,1) = (1,2) = (2,3) = (3,4) = 1, \quad (4,5) = \tau^{-3},$$
$$(-1,0) = (0,5) = X_0X_5 = 4 + \tau^{-3} = \tau^3,$$
$$(-1,1) = X_1X_5 = 3 + \tau^{-3} = 2\tau,$$

and so on.

EXERCISES

1. Consider again the pattern of order 3 (in §5·1, Ex. 1) involving positive numbers a, b, c. What condition must these three numbers satisfy if the pattern is Ptolemaic?
2. What generalization of §5·1, Exx. 2 and 3 is suggested by the theory of Ptolemaic patterns?

5·4 REAL POLYTOPES IN FOUR DIMENSIONS

In §4·3 we regarded $\{p, q, r\}$, satisfying (4·36), as a three-dimensional spherical honeycomb. Instead, let us regard it as a four-dimensional polytope and consider its characteristic orthoscheme $O_0O_1O_2O_3O_4$, analogous to the tetrahedron $CMIO$ of Figure 2·2A. Now O_0 is a vertex, O_1 is the midpoint of one of the edges meeting at this vertex, O_2 is the centre of one of the faces $\{p\}$ meeting at this edge, O_3 is the centre of one of the cells $\{p, q\}$ meeting at this face, and O_4 is the centre of the whole polytope $\{p, q, r\}$. The four-dimensional orthoscheme formed by these five points is a simplex whose successive edges O_0O_1, O_1O_2, O_2O_3, O_3O_4 are mutually perpendicular. Its five bounding hyperplanes

$$O_1O_2O_3O_4, \quad O_0O_2O_3O_4, \quad O_0O_1O_3O_4, \quad O_0O_1O_2O_4, \quad O_0O_1O_2O_3$$

contain the four great spheres (4·31) and the cell $\{p, q\}$, respectively. In other words, the 'hypersolid angle' at the vertex O_4 of the four-dimensional orthoscheme determines, on the unit 3-sphere with centre O_4, the three-dimensional spherical orthoscheme described in Figure 4·3B (with P_ν at unit distance from O_4 along the edge O_4O_ν, for $\nu = 0, 1, 2, 3$). Therefore, among the ten dihedral angles of the four-dimensional orthoscheme, all are right angles except those opposite to the edges $O_0O_1, O_1O_2, O_2O_3, O_3O_4$. These, being the angles between adjacent pairs of the five hyperplanes in the above order, are

$$\frac{\pi}{p}, \quad \frac{\pi}{q}, \quad \frac{\pi}{r}, \quad \tfrac{1}{2}\pi - \psi.$$

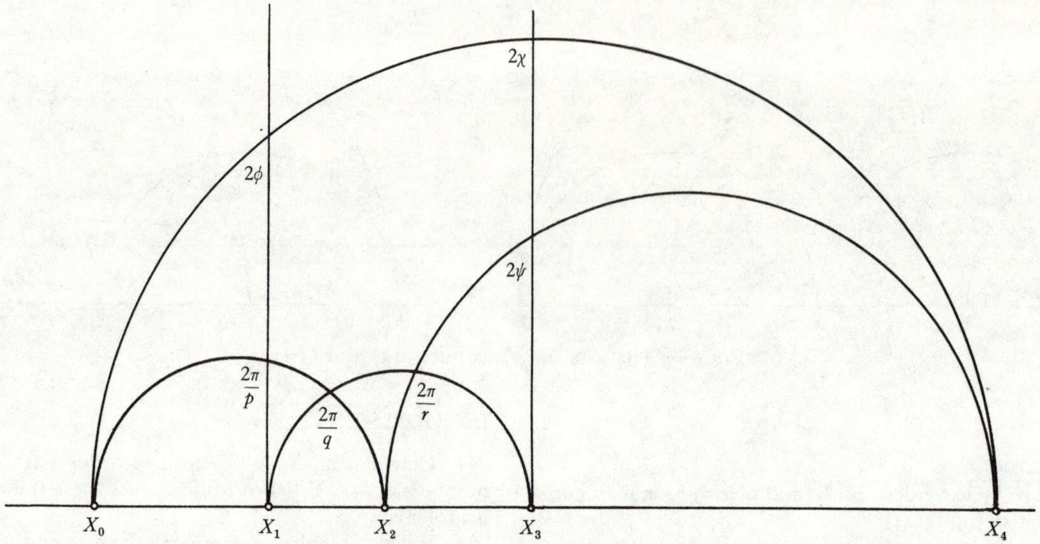

Figure 5·4A: A diagram with $X_5 = X_{-1}$ at infinity

(The last is half the dihedral angle of the whole polytope $\{p, q, r\}$.) Instead of (4·33), we now have

$$(5·41) \qquad \phi = \angle O_0 O_4 O_1, \quad \chi = \angle O_0 O_4 O_3, \quad \psi = \angle O_2 O_4 O_3.$$

The five hyperplanes are easily seen to belong to a cycle of six, such that all except cyclically adjacent pairs are perpendicular. The sixth, being perpendicular to the line $O_4 O_0$, may be described as the hyperplane that contains the vertex figure $\{q, r\}$ of $\{p, q, r\}$. The angle that it makes with $O_0 O_1 O_2 O_3$ (perpendicular to $O_3 O_4$) is

$$\angle O_0 O_4 O_3 = \chi.$$

Similarly, the angle that it makes with $O_1 O_2 O_3 O_4$ (perpendicular to $O_0 O_1$) is

$$\angle O_4 O_0 O_1 = \tfrac{1}{2}\pi - \angle O_0 O_4 O_1 = \tfrac{1}{2}\pi - \phi.$$

Thus the six angles

$$(5·42) \qquad \frac{\pi}{p}, \quad \frac{\pi}{q}, \quad \frac{\pi}{r}, \quad \tfrac{1}{2}\pi - \psi, \quad \chi, \quad \tfrac{1}{2}\pi - \phi$$

form a cycle (like the five considered in §3·1) such that any equation relating four of them will remain valid when the six hyperplanes (and hence the six angles) are cyclically permuted.

Recalling that the triangle $O_s O_t O_u$ with $s < t < u$ is right-angled at O_t, let us place five points X_0, X_1, X_2, X_3, X_4 in natural order along a line in such a way that

$$X_0 X_1 = (O_0 O_1)^2, \quad X_1 X_2 = (O_1 O_2)^2, \quad X_2 X_3 = (O_2 O_3)^2, \quad X_3 X_4 = (O_3 O_4)^2.$$

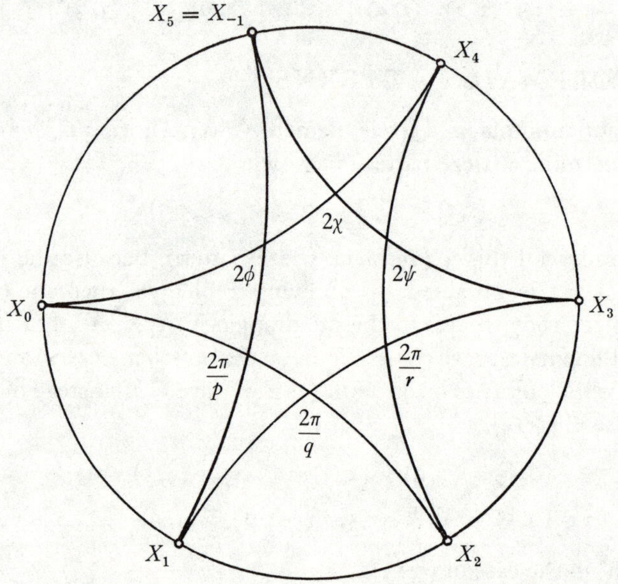

Figure 5·4B: The principal angles in $\{p, q, r\}$
(2ϕ is the angle subtended at the centre by an edge)

Then (by Pythagoras) we will have

(5·43) $$X_sX_t = (O_sO_t)^2$$

in the remaining cases too. In Figure 5·4A we see semicircles on diameters X_0X_4, X_0X_2, X_1X_3, X_2X_4, and lines through X_1 and X_3 perpendicular to the line of the X's. Since the point marked $2\pi/p$ forms, with X_1X_2, a triangle similar to $O_0O_1O_2$ (Coxeter 1965, p. 276), the angle between the vertical line at X_1 and the semicircular arc X_0X_2 is indeed twice the angle $O_0O_2O_1$ which is π/p. For analogous reasons it can be seen that the angles 2ϕ, 2χ, 2ψ have been correctly marked (although we cannot yet be sure about $2\pi/q$ and $2\pi/r$).

By inversion in a suitable circle, Figure 5·4A becomes Figure 5·4B, in which it is clear that four of the angles of the curvilinear hexagon in the middle are the first four of the cycle

$$\pi-2\psi,\quad 2\chi,\quad \pi-2\phi,\quad \frac{2\pi}{p},\quad \frac{2\pi}{q},\quad \frac{2\pi}{r}.$$

Since each member of this cycle is the same function of the preceding three, the angles $2\pi/q$ and $2\pi/r$ are in fact correctly marked in both figures.

The vertical lines in Figure 5·4A may be regarded as joining X_1 and X_3 to the point at infinity X_{-1}. By expressing the ratio X_1X_2/X_0X_2 as a cross-ratio involving X_{-1}, we see that

$$\cos^2\frac{\pi}{p} = \left(\frac{O_1O_2}{O_0O_2}\right)^2 = \frac{X_1X_2}{X_0X_2} = \{X_{-1}X_2, X_0X_1\}.$$

Since cross-ratios are inversive invariants (Coxeter and Greitzer 1967, p. 112), this expression for $\cos^2\pi/p$ as a cross-ratio remains valid in Figure 5·4B. Similarly

$$\cos^2\frac{\pi}{q} = \{X_0X_3, X_1X_2\},\quad \cos^2\frac{\pi}{r} = \{X_1X_4, X_2X_3\},$$
$$\sin^2\psi = \{X_2X_{-1}, X_3X_4\},\quad \cos^2\chi = \{X_3X_0, X_4X_{-1}\},$$
$$\sin^2\phi = \{X_4X_1, X_{-1}X_0\}.$$

By deleting the arcs X_0X_4, X_2X_4, $X_{-1}X_3$ and inserting $X_{-1}X_2$ and X_0X_3 (Coxeter 1965, p. 277; 1970, Fig. 2) we obtain the analogous diagram for $\{p, q\}$, with the conclusion that the properties ψ, χ, ϕ for this polyhedron are given by

$$\sin^2\psi_{p,q} = \{X_1X_{-1}, X_2X_3\},\quad \cos^2\chi_{p,q} = \{X_2X_0, X_3X_{-1}\},$$
$$\sin^2\phi_{p,q} = \{X_3X_1, X_{-1}X_0\}.$$

Adding 1 to each subscript, we deduce

$$\sin^2\psi_{q,r} = \{X_2X_0, X_3X_4\},\quad \cos^2\chi_{q,r} = \{X_3X_1, X_4X_0\},$$
$$\sin^2\phi_{q,r} = \{X_4X_2, X_0X_1\}.$$

Such expressions for sines and cosines immediately yield equally simple expressions for cosines and sines. For instance, since

$$\cos^2\frac{\pi}{p} = \{X_{-1}X_2, X_0X_1\},\quad \sin^2\frac{\pi}{p} = \{X_{-1}X_0, X_2X_1\}.$$

Although we have chosen to express the cross-ratios in terms of distances X_sX_t, more general expressions for them will serve equally well. (We recall that a frieze pattern is not necessarily Ptolemaic.) Accordingly, if we can find a frieze pattern of order 6 such that

(5·44)
$$\frac{(-1,0)(1,2)}{(-1,1)(0,2)} = \cos^2\frac{\pi}{p},\quad \frac{(0,1)(2,3)}{(0,2)(1,3)} = \cos^2\frac{\pi}{q},$$
$$\frac{(1,2)(3,4)}{(1,3)(2,4)} = \cos^2\frac{\pi}{r},$$

then the angles ϕ, χ, ψ will be given by

$$\sin^2\psi = \frac{(2,3)(4,5)}{(2,4)(3,5)} = \frac{(-1,4)(2,3)}{(-1,3)(2,4)},$$
$$\cos^2\chi = \frac{(3,4)(5,6)}{(3,5)(4,6)} = \frac{(-1,0)(3,4)}{(-1,3)(0,4)},$$
$$\sin^2\phi = \frac{(4,5)(6,7)}{(4,6)(5,7)} = \frac{(-1,4)(0,1)}{(-1,1)(0,4)}.$$

In terms of the half-edge $l = O_0O_1$, we have the *circumradius*

(5·45) $$O_0O_4 = l\operatorname{cosec}\phi = l\sqrt{\frac{(-1,1)(0,4)}{(-1,4)(0,1)}}$$

and the *inradius*

(5·46) $$O_3O_4 = O_0O_4\cos\chi = l\sqrt{\frac{(-1,0)(-1,1)(3,4)}{(-1,3)(-1,4)(0,1)}}.$$

Here we are using the 'essential fragment' or 'fundamental region'

$$\begin{array}{ccccccccc}
(-1,0) & & (0,1) & & (1,2) & & (2,3) & & (3,4)\\
& (-1,1) & & (0,2) & & (1,3) & & (2,4) & \\
& & (-1,2) & & (0,3) & & (1,4) & & \\
& & & (-1,3) & & (0,4) & & & \\
& & & & (-1,4) & & & &
\end{array}$$

of the frieze pattern. For the polytopes $\{4, 3, 3\}$, $\{3, 3, 4\}$, $\{3, 4, 3\}$, the actual values may be taken to be

1 1 1 1   1 1 1 1   1 1 1 1 1 1 1
1 2 2 2   2 2 2 1   2 2 1 4
1 3 3   3 3 1   3 1 3
1 4   4 1   1 2
1   1   1

all extracted from (5·11); for the regular simplex $\{3, 3, 3\}$,

1 1 1 1 1
2 2 2 2
3 3 3
4 4
5

from (5·12); for $\{5, 3, 3\}$ and $\{3, 3, \frac{5}{2}\}$,

τ^3 1 1 1 1   1 1 1 1 τ^{-3}
2τ 2 2 2   2 2 2 $2\tau^{-1}$
$\sqrt{5}$ 3 3   and   3 3 $\sqrt{5}$
$2\tau^{-1}$ 4   4 2τ
τ^{-3} τ^3

from (5·13); for their reciprocals, $\{3, 3, 5\}$ and $\{\frac{5}{2}, 3, 3\}$, the same reversed (right to left), from the reversal of (5·13) (derived by reflecting in a vertical line); for $\{5, \frac{5}{2}, 5\}$ and $\{\frac{5}{2}, 5, \frac{5}{2}\}$,

τ τ^{-1} τ τ^{-1} τ τ^{-1} τ τ^{-1} τ τ^{-1}
2 2 2 2   2 2 2 2
$\sqrt{5}$ $\sqrt{5}$ $\sqrt{5}$ and $\sqrt{5}$ $\sqrt{5}$ $\sqrt{5}$
2 2   2 2
τ^{-1} τ

from (5·14); for $\{5, 3, \frac{5}{2}\}$, $\{3, \frac{5}{2}, 5\}$, $\{\frac{5}{2}, 5, 3\}$, fragments analogously extracted from (5·15); and for their reciprocals (which complete the list of sixteen polytopes $\{p, q, r\}$), the same reversed. (Compare Schläfli 1953, pp. 181, 260.)

EXERCISES

1. Obtain expressions for the 'intermediate' radii O_1O_4 and O_2O_4.
2. In terms of g, the order of the symmetry group $[p, q, r]$, obtain the 4-dimensional content ('hyper-volume') of $\{p, q, r\}$.

5·5 DIFFERENT PATTERNS FOR THE SAME POLYTOPE

When p, q, r are given, the equations (5·44) allow considerable freedom of choice in the formation of the corresponding frieze pattern. For instance (Coxeter 1963, pp. 160–2), we might choose $(t-1, t) = 1$ for $0 \leqslant t \leqslant 4$ and $(0, 2) = 2$, so that

$$(-1, 1) = \tfrac{1}{2}\sec^2\frac{\pi}{p}, \quad (1, 3) = \tfrac{1}{2}\sec^2\frac{\pi}{q}, \quad (2, 4) = 2\cos^2\frac{\pi}{q}\sec^2\frac{\pi}{r},$$

and so on.

In fact, all the cross-ratios will be unchanged if we replace each (s, t) by $a_s a_t (s, t)$ where a_s is *any function of s of period* 6 (that is, such that $a_{s+6} = a_s$).

The pattern chosen above for $\{3, 4, 3\}$ is aesthetically pleasing because it consists entirely of integers. However, this choice tends to obscure the fact that the 24-cell is self-reciprocal. A different choice, exhibiting bilateral symmetry like the pattern for $\{3, 3, 3\}$, is

1 1 1 1 1
$2^{\frac{3}{2}}$ $2^{\frac{1}{2}}$ $2^{\frac{1}{2}}$ $2^{\frac{3}{2}}$
3 1 3
$2^{\frac{1}{2}}$ $2^{\frac{1}{2}}$
1

given by $\quad a_{-1} = a_1 = a_3 = 2^{\frac{1}{4}}, \quad a_0 = a_2 = a_4 = 2^{-\frac{1}{4}}.$

EXERCISES

1. Does every $\{p, q, r\}$ admit a frieze pattern with $(t-1, t) = 1$ for every t, including $t = 5$?
2. Does any $\{p, q, r\}$ admit a Ptolemaic frieze pattern with $(s, t) = (O_s O_t)^2$ for $0 \leqslant s < t < 4$?

5·6 PATTERNS OF ORDER 6 AND PERIOD 3

If a Ptolemaic frieze pattern with $n = 6$ happens to have period 3 instead of 6, the hexagon $X_0X_1X_2X_3X_4X_5$ (Figure 5·6A) is centrally symmetric. We may conveniently take its circumradius to be $\frac{1}{2}$, so that

$$(0, 3) = (1, 4) = (2, 5) = 1.$$

Letting A, B, C denote the supplements of its angles, so that

$$A + B + C = \pi,$$

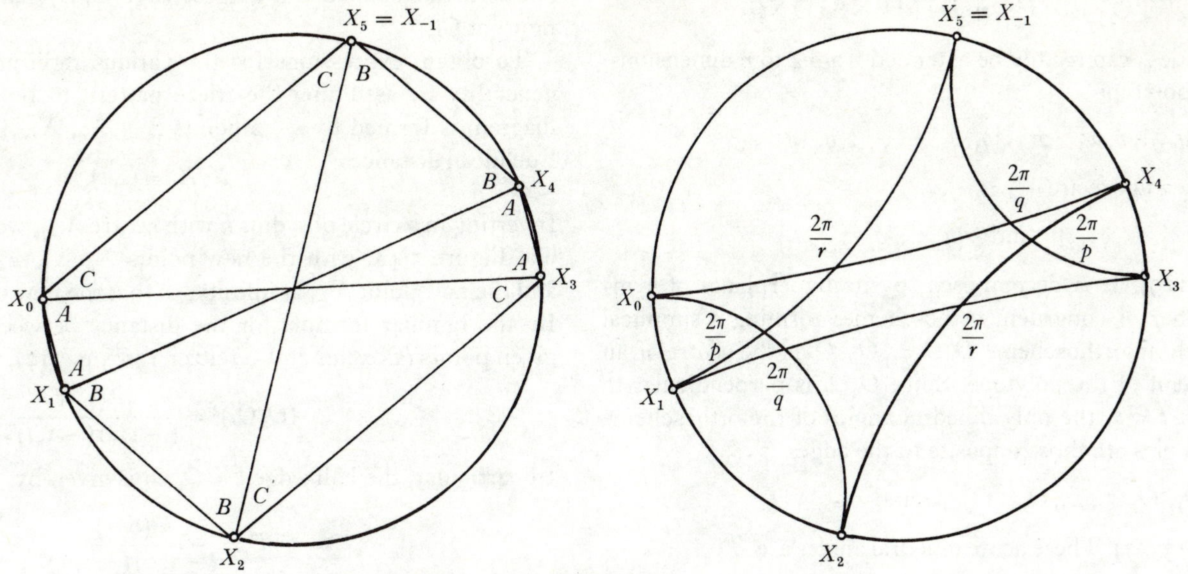

Figure 5·6A: A diagram for $\{5, 3, \frac{5}{2}\}$

we see that its sides are the cosines of A, B, C while its first diagonals are their sines. Thus the five rows of the pattern are

$$\begin{array}{ccccccccccccccc} & \cos C & & \cos A & & \cos B & & \cos C & & \cos A & & \cos B & & \cos C \ldots & \\ \sin A & & \sin B & & \sin C & & \sin A & & \sin B & & \sin C & & \sin A & & \sin B \\ & 1 & & 1 & & 1 & & 1 & & 1 & & 1 & & 1 & \ldots \\ \sin A & & \sin B & & \sin C & & \sin A & & \sin B & & \sin C & & \sin A & & \sin B \\ & \cos C & & \cos A & & \cos B & & \cos C & & \cos A & & \cos B & & \cos C \ldots & \end{array}$$

If this pattern belongs to a polytope $\{p, q, r\}$, and therefore also to $\{q, r, p\}$ and $\{r, p, q\}$, we have

$$\cos^2\frac{\pi}{p} = \cot B \cot C, \quad \cos^2\frac{\pi}{q} = \cot C \cot A, \quad \cos^2\frac{\pi}{r} = \cot A \cot B;$$

thus
$$\cos^2\frac{\pi}{p} + \cos^2\frac{\pi}{q} + \cos^2\frac{\pi}{r} = 1$$

(Hobson 1925, p. 47; Schläfli 1953, pp. 175, 252).

Since the cycle $\pi/p,\ \pi/q,\ \pi/r,\ \frac{1}{2}\pi - \psi,\ \chi,\ \frac{1}{2}\pi - \phi$ is now simply

$$\pi/p, \quad \pi/q, \quad \pi/r, \quad \pi/p, \quad \pi/q, \quad \pi/r,$$

such a polytope $\{p, q, r\}$ has

$$\psi = \tfrac{1}{2}\pi - \pi/p, \quad \chi = \pi/q, \quad \phi = \tfrac{1}{2}\pi - \pi/r.$$

(The last of these relations is clearly in agreement with (2·22) and (4·34).)

Since, in Figure 5·6A, the diameters $X_{-1}X_2$ and X_0X_3 make an angle $2C$ (corresponding to the arc X_0X_2), we have $C = \chi_{p,q}$. Similarly

$$A = \chi_{q,r}, \quad B = \chi_{r,p}.$$

In terms of $\kappa = \frac{1}{2}\operatorname{arc\,sec} 3$, $\lambda = \frac{1}{2}\arctan 2$, $\mu = \frac{1}{2}\arcsin\frac{2}{3}$, the actual cases are the permutations of

p	q	r	A	B	C
3	4	3	$\frac{1}{2}\pi - \kappa$	2κ	$\frac{1}{2}\pi - \kappa$
5	3	$\frac{5}{2}$	$\frac{1}{2}\pi - \lambda + \mu$	2λ	$\frac{1}{2}\pi - \lambda - \mu$

(Coxeter 1963, pp. 108–9, 293). The frieze patterns so constructed have the same effect as (5·11) and (5·15), respectively, though they are less convenient for practical computation.

EXERCISE

How do we know that (5·15) is not Ptolemaic?

5·7 REAL POLYTOPES IN n DIMENSIONS

Most of the above ideas can readily be extended from 4 to n dimensions. We have a regular polytope

$$\{p, q, \ldots, v, w\} \quad \text{or} \quad \{q_1, q_2, \ldots, q_{n-2}, q_{n-1}\},$$

whose cell and vertex figure are

$$\{q_1, q_2, \ldots, q_{n-2}\} \quad \text{and} \quad \{q_2, \ldots, q_{n-2}, q_{n-1}\}$$

(Coxeter 1963, p. 129). It is decomposed, by its hyperplanes of symmetry, into a number of congruent orthoschemes forming a simplical subdivision. In such an orthoscheme $O_0 O_1 \ldots O_n$, O_s is the centre of an s-dimensional element of the polytope. Since $O_s O_t$ is perpendicular to $O_t O_u$ whenever $s < t < u$, the only dihedral angles of the orthoscheme that are not right angles are those opposite to the edges

$$O_0 O_1, O_1 O_2, \ldots, O_{n-2} O_{n-1} \quad \text{and} \quad O_{n-1} O_n$$

(Coxeter 1963, pp. 133–7). These acute dihedral angles are

$$\pi/q_1, \pi/q_2, \ldots, \pi/q_{n-1} \quad \text{and} \quad \tfrac{1}{2}\pi - \psi,$$

the last being half the dihedral angle between two adjacent cells

$$\{q_1, q_2, \ldots, q_{n-2}\}$$

of the polytope $\{q_1, q_2, \ldots, q_{n-2}, q_{n-1}\}$. Thus ψ is one of the three 'principal angles'

(5·71) $\quad \phi = \angle O_0 O_n O_1, \quad \chi = \angle O_0 O_n O_{n-1}, \quad \psi = \angle O_{n-2} O_n O_{n-1}.$

In terms of the relevant frieze patterns of order $n+2$, we have

(5·72)
$$\cos^2\frac{\pi}{q_1} = \frac{(-1,0)(1,2)}{(-1,1)(0,2)}, \quad \cos^2\frac{\pi}{q_2} = \frac{(0,1)(2,3)}{(0,2)(1,3)}, \ldots,$$
$$\cos^2\frac{\pi}{q_{n-1}} = \frac{(n-3,n-2)(n-1,n)}{(n-3,n-1)(n-2,n)}.$$

Notice that we are now using the letter n for the number of dimensions, whereas at the beginning of this chapter we used it for the order of the pattern. No confusion need be caused as long as we remember that we are now using a frieze pattern of order $n+2$, representing a cycle of $n+2$ hyperplanes such that consecutive pairs are inclined at angles

(5·73) $\quad \dfrac{\pi}{q_1}, \ \dfrac{\pi}{q_2}, \ \ldots, \ \dfrac{\pi}{q_{n-1}}, \ \tfrac{1}{2}\pi - \psi, \ \chi, \ \tfrac{1}{2}\pi - \phi,$

while all other pairs are orthogonal. The first $n+1$ of these $n+2$ hyperplanes bound the characteristic orthoscheme $O_0 O_1 \ldots O_n$; the remaining one is orthogonal to the longest edge $O_0 O_n$, and contains the vertex figure at O_0.

To obtain expressions for the various radii and angles, we lose no generality by assuming the frieze pattern to be Ptolemaic, so that its diagram is formed by $n+2$ points $X_{-1}, X_0, X_1, \ldots, X_n$ on a circle, with Euclidean distances

$$X_s X_t = (s, t).$$

Inverting in a circle of radius k with centre X_{-1}, we obtain a new diagram like Figure 5·4A, with the new points $X_0, X_1, \ldots, X_n$ on a straight line and the new point X_{-1} at infinity, so that the new distances satisfy (5·43). By the familiar formula for the distance between the inverses of two given points (Coxeter and Greitzer 1967, p. 112), we have

$$(O_s O_t)^2 = \frac{k^2(s,t)}{(-1,s)(-1,t)}.$$

In particular, the half-edge $l = O_0 O_1$ is given by

$$l^2 = \frac{k^2(0,1)}{(-1,0)(-1,1)}.$$

Eliminating k, we deduce

(5·74) $$(O_s O_t)^2 = \frac{(-1,0)(-1,1)(s,t)}{(-1,s)(-1,t)(0,1)} l^2.$$

In particular, the polytope $\{q_1, q_2, \ldots, q_{n-2}, q_{n-1}\}$ (of edge $2l$) has circumradius and inradius

$$O_0 O_n = l\sqrt{\frac{(-1,1)(0,n)}{(-1,n)(0,1)}}, \quad O_{n-1} O_n = l\sqrt{\frac{(-1,0)(-1,1)(n-1,n)}{(-1,n-1)(-1,n)(0,1)}},$$

and its principal angles (using (5·26) with $n+2$ for n) are given by

$$\sin^2\psi = \left(\frac{O_{n-2}O_{n-1}}{O_{n-2}O_n}\right)^2 = \frac{(n-2,n-1)}{(-1,n-1)}\Big/\frac{(n-2,n)}{(-1,n)} = \frac{(n-2,n-1)(n,n+1)}{(n-2,n)(n-1,n+1)},$$
$$\cos^2\chi = \left(\frac{O_{n-1}O_n}{O_0 O_n}\right)^2 = \frac{(n-1,n)}{(-1,n-1)}\Big/\frac{(0,n)}{(-1,0)} = \frac{(n-1,n)(n+1,n+2)}{(n-1,n+1)(n,n+2)},$$
$$\sin^2\phi = \left(\frac{O_0 O_1}{O_0 O_n}\right)^2 = \frac{(0,1)}{(-1,1)}\Big/\frac{(0,n)}{(-1,n)} = \frac{(n,n+1)(n+2,n+3)}{(n,n+2)(n+1,n+3)},$$

completing the cycle (5·73) in the manner of (5·72). (Compare Coxeter 1963, p. 161.) As the only real polytopes $\{q_1, q_2, \ldots, q_{n-1}\}$ with $n > 4$ are

$$\{3, 3, \ldots, 3, 3\}, \quad \{4, 3, \ldots, 3, 3\}, \quad \{3, 3, \ldots, 3, 4\}$$

(Coxeter 1963, p. 136), with frieze patterns analogous to (5·12) and (5·11), the applications of these general expressions seem sadly meagre. However, their scope will be somewhat enlarged when we apply them to complex polytopes in §13·3.

EXERCISE

Letting g denote the order of the symmetry group $[q_1, q_2, \ldots, q_{n-1}]$ (generated by reflections in the bounding hyperplanes of the orthoscheme), obtain an expression for the n-dimensional content of the polytope $\{q_1, q_2, \ldots, q_{n-1}\}$.

5·8 REMARKS

In §3·5 we traced the origin of the frieze pattern to Gauss's *pentagramma mirificum*, which consists of a cycle of five planes, and to Wythoff's analogous cycle of six hyperplanes. However, it is unlikely that I would ever have thought of inventing the 'two-digit symbol' (s, t) without the stimulus of G. T. Bennett's diagram of intersecting circles orthogonal to a line (Figure 5·3B) or to a circle (Figure 5·3A). This diagram led to a simplified version[†] of Schläfli's trigonometry of the orthoscheme (Schläfli 1950, pp. 243–61; see also Schoute 1902, pp. 267–86, where such generalized trigonometry is called *polygonometry*).

The two-digit symbols have been fruitfully employed by Johannes Böhm,[‡] who quoted a proof by D. S. Mitrinović that the general solution of the equation (5·24) is

$$(s,t) = f_s g_t - f_t g_s,$$

where f_s and g_s are arbitrary functions of s. This expression (essentially the same as (5·22)) may have been inspired by the idea of the Plücker coordinates

$$p_{st} = f_s g_t - f_t g_s$$

for the line joining points (f_0, f_1, f_2, f_3) and (g_0, g_1, g_2, g_3) in projective 3-space.

Strangely, the simple idea of obtaining (5·74) by inversion did not occur to me till one early morning in January 1971.

† Coxeter, On Schläfli's generalization of Napier's Pentagramma Mirificum, *Bulletin of the Calcutta Mathematical Society* **28** (1936), 123–44; see especially p. 134.

‡ Zu Coxeters Integrationsmethode in gekrümmten Räumen, *Mathematische Annalen* **27** (1964), 179–214; see especially p. 189. See also Stenko Bilinski, Über Ptolemäische Sätze, *Monatschefte für Mathematik* **77** (1973), 193–205.

CHAPTER 6

The geometry of quaternions

Herod...proceeded further to take Peter...and delivered him to four quaternions...

Acts 12: 1–4

Although complex numbers were freely used in the eighteenth century, the mystery that surrounded them is revealed in the prevalence of such words as 'imaginary' and 'real'. This mystery lingered on till Gauss (and Hamilton, more explicitly) proposed the modern interpretation of complex numbers as pairs of real numbers which are added or multiplied according to certain formal rules. Sir William Rowan Hamilton (1805–65) adopted the term *quaternion* for an analogous set of *four* real numbers. By analogy with the expression $x+iy$ or $x+yi$ for the complex number (x,y), he denoted the quaternion (w,x,y,z) by $w+xi+yj+zk$ and represented the 'pure' quaternion $xi+yj+zk$ by the point (x,y,z) in Euclidean 3-space. A. S. Hathaway† extended this idea to four dimensions, representing the general quaternion by the point (w,x,y,z) in Euclidean 4-space. For some purposes it is convenient to express the quaternion as

$$x = x_0+x_1 i+x_2 j+x_3 k$$

and to name the point (x_0, x_1, x_2, x_3). This notation, replacing the (x_1, x_2, x_3, x_4) of Chapter 4, need not cause serious confusion.

Quaternions will be used to specify reflections in §6·3, and rotations in §6·4, although historically the order was reversed. The next two sections deal with an application of geometry to algebra: the enumeration of finite multiplicative groups of quaternions, and a convenient way to summarize the results. In §6·7 and §6·8, four-dimensional orthogonal transformations are expressed in terms of quaternions along the lines proposed by Du Val (1964, pp. 42, 58).

6·1 PAIRS OF COMPLEX NUMBERS

Segre (1961, p. 3) and Du Val (1964, p. 33) have proposed the convenient term *corpus* for a division ring or 'skew field', which satisfies all the usual laws of elementary algebra with the possible exception of the commutative law of multiplication.

From the field of complex numbers

$$u = x+yi \quad (x \text{ and } y \text{ real})$$

† Quaternions as numbers of four-dimensional space, *Bulletin of the American Mathematical Society* **4** (1897), 54–7.

the corpus of *quaternions*

$$a = u+vj \quad (u \text{ and } v \text{ complex})$$

can be derived by introducing a new unit j which interacts with the familiar i according to the simple rules

$$(6·11) \qquad jij = i, \quad iji = j.$$

These rules immediately imply $j^2 = (ij)^2 = i^2 = -1$ and

$$(6·12) \qquad ji = -ij.$$

(Thus we see already that this corpus is not a field.)

The new unit j has the effect of transforming each complex number $u = x+yi$ into its complex conjugate

$$j^{-1}uj = -juj = -xj^2-yjij = x-yi = \bar{u}.$$

Notice that, although $x+yi$ (where x and y are real) can just as well be written as $x+iy$, we must not transpose the v and j in $u+vj$, because

$$(6·13) \qquad jv = \bar{v}j.$$

Since $jvj = -\bar{v}$, we could define quaternions to be ordered pairs of complex numbers (u,v) or $u+vj$ which are added and multiplied according to the rules

$$(6·14) \qquad \begin{aligned}(s,t)+(u,v) &= (s+u, t+v),\\ (s,t)(u,v) &= (su-t\bar{v}, sv+t\bar{u}).\end{aligned}$$

These closely resemble the familiar rules for adding and multiplying complex numbers, and allow us just as easily to verify the laws of addition, the distributive laws, and the associative law of multiplication. The roles of the complex conjugate $\bar{u}$ and the squared modulus $|u|^2$ are taken over by the *quaternion conjugate*

$$(6·15) \qquad \tilde{a} = \bar{u}-vj$$

and the *norm* (a real number)

$$(6·16) \qquad \mathrm{N}a = \tilde{a}a = a\tilde{a} = u\bar{u}+v\bar{v} \geqslant 0.$$

In terms of these, any non-zero quaternion a has a unique inverse

$$(6·17) \qquad a^{-1} = \tilde{a}/\mathrm{N}a.$$

Since the quaternion conjugate of

$$(s+tj)(u+vj) = (su-t\bar{v})+(sv+t\bar{u})j$$

is
$$(\bar{s}\bar{u}-\bar{t}v)-(sv+t\bar{u})j = (\bar{u}-vj)(\bar{s}-tj),$$

the quaternion conjugate of ab is $\tilde{b}\tilde{a}$; therefore

$$\mathsf{N}(ab) = \tilde{b}\tilde{a}ab = \tilde{b}\mathsf{N}a\cdot b$$

(6·18)
$$= \mathsf{N}a\mathsf{N}b.$$

If $\mathsf{N}b = 1$, we call b a *unit* quaternion. To every non-zero quaternion a there corresponds a unit quaternion

(6·19)
$$\mathsf{U}a = a/\surd(\mathsf{N}a).$$

EXERCISES

1. Use (6·14) to check the associative law for multiplication.
2. Regarding i and j as the generators of an abstract group defined by (6·11), compute the order of this group and give a list of all its elements.

6·2 QUATERNIONS OF REAL NUMBERS

According to the definition in §6·1 a quaternion is an ordered pair of complex numbers, each of which is an ordered pair of real numbers. Clearly, we can combine the two steps and regard the quaternion as an ordered set of four real numbers a_0, a_1, a_2, a_3:

$$a = (a_0+a_1 i)+(a_2+a_3 i)j = a_0+a_1 i+a_2 j+a_3 ij$$
$$= a_0+a_1 i+a_2 j+a_3 k,$$

where k is an abbreviation for ij. This aspect is, of course, the origin of the name 'quaternion'. When taken along with the definition $k = ij$, the relations (6·11) are easily seen to be equivalent to

(6·21)
$$i^2 = j^2 = k^2 = ijk.$$

Other useful identities are

$$jk = i = -kj, \quad ki = j = -ik, \quad ij = k = -ji,$$
$$jij = i = kik, \quad kjk = j = iji, \quad iki = k = jkj.$$

Hamilton expressed the quaternion a as the sum of its *scalar* and *vector* parts:

$$\mathsf{S}a = a_0, \quad \mathsf{V}a = a_1 i+a_2 j+a_3 k,$$

so that $a = \mathsf{S}a+\mathsf{V}a$, $\tilde{a} = \mathsf{S}a-\mathsf{V}a$, and consequently

(6·22)
$$\mathsf{S}a = \tfrac{1}{2}(a+\tilde{a}), \quad \mathsf{V}a = \tfrac{1}{2}(a-\tilde{a}).$$

Thus the condition for a to be real ('scalar') is $\mathsf{V}a = 0$, or $a = \tilde{a}$. A quaternion x is said to be *pure* if $\mathsf{S}x = 0$; then $\tilde{x} = -x$ and $x^2 = -\mathsf{N}x$. Thus every unit pure quaternion is a square root of -1.

By writing $\mathsf{N}a$ as $a_0{}^2+(a_1{}^2+a_2{}^2+a_3{}^2)$, we see that any *unit* quaternion (such that $\mathsf{N}a = 1$) may be expressed as

$$a = \cos\alpha+y\sin\alpha,$$

where $\cos\alpha = \mathsf{S}a$ and y is a unit *pure* quaternion, namely $y = \mathsf{UV}a$. Since y is a square root of -1, de Moivre's theorem shows that, for any integer n,

$$a^n = \cos n\alpha+y\sin n\alpha.$$

It is clearly appropriate to use the same equation to define a^n for other real values of n and, by analogy with Euler's

$$\exp(\alpha i) = \cos\alpha+i\sin\alpha,$$

to write
$$\exp(\alpha y) = \cos\alpha+y\sin\alpha,$$

so that $\{\exp(\alpha y)\}^n = \exp(n\alpha y)$. In other words, the unit quaternion a can be expressed as $\exp(\alpha y)$ or $\exp(y\alpha)$, where α and y are given by $\mathsf{S}a = \cos\alpha$, $\mathsf{V}a = y\sin\alpha$; and *any* quaternion a can be expressed in the form

(6·23)
$$a = \surd(\mathsf{N}a)\exp(\alpha y), \quad \alpha = \operatorname{arc}\cos\mathsf{SU}a, \quad y = \mathsf{UV}a.$$

For instance, $1-k = \surd 2\exp(-k\pi/4)$, and any unit pure quaternion y can be expressed as

$$\exp(\tfrac{1}{2}\pi y).$$

However, since multiplication is not commutative, we cannot write $\exp(\alpha y)\exp(\beta z) = \exp(\alpha y+\beta z)$ unless $y = z$.

EXERCISES

1. Work out the products

(i) $(1+i)^4$, (ii) $(1+i+j+k)^3$,
(iii) $(i+k)^2$, (iv) $(1+i)(1+i+j+k)(i+k)$.

2. Among quaternions, the *only* square roots of -1 are the unit pure quaternions.
3. For any quaternion x and any unit quaternion a, the scalar part of $\tilde{a}xa$ is equal to the scalar part of x.
4. Any unit quaternion can be expressed as a power of a unit pure quaternion.
5. Any unit quaternion can be expressed as a product $i^a j^b i^c$, where a, b, c are real.

6·3 REFLECTIONS

One of the earliest applications of quaternions was to the theory of isometries in Euclidean 3-space. Each pure quaternion

$$x = x_1 i + x_2 j + x_3 k$$

represents a point (x_1, x_2, x_3) which we shall simply call 'the point x'. It also represents the vector $\mathbf{x}$ which goes from the origin to x. If

$$y = y_1 i + y_2 j + y_3 k$$

is another such quaternion, the distance between x and y, being the length of the vector $\mathbf{y} - \mathbf{x}$, is the square root of $\mathsf{N}(y-x)$. Since

$$xy = -(x_1 y_1 + x_2 y_2 + x_3 y_3) + \begin{vmatrix} x_2 & x_3 \\ y_2 & y_3 \end{vmatrix} i + \begin{vmatrix} x_3 & x_1 \\ y_3 & y_1 \end{vmatrix} j + \begin{vmatrix} x_1 & x_2 \\ y_1 & y_2 \end{vmatrix} k,$$

the inner (or 'scalar') product of two vectors $\mathbf{x}$ and $\mathbf{y}$ is

(6·31) $$\mathbf{x}\cdot\mathbf{y} = -\mathsf{S}(xy),$$

while the outer (or 'vector') product $\mathbf{x}\times\mathbf{y}$ is the vector represented by the quaternion $\mathsf{V}(xy)$. Since the conjugate of xy is $\tilde{y}\tilde{x} = (-y)(-x) = yx$, we have

(6·32) $$\mathsf{S}(xy) = \tfrac{1}{2}(xy+yx).$$

Thus the condition for the points x and y (where $xy \neq 0$) to lie in perpendicular directions from the origin is

(6·33) $$xy + yx = 0.$$

Any plane through the origin may be expressed in the form

(6·34) $$y_1 x_1 + y_2 x_2 + y_3 x_3 = 0,$$

where $y_1^2 + y_2^2 + y_3^2 = 1$. It is determined by its unit normal vector $\mathbf{y}$, which is represented by the unit pure quaternion

$$y = y_1 i + y_2 j + y_3 k \quad (\mathsf{N}y = 1).$$

Since the length of the projection of any vector $\mathbf{x}$ on the unit vector $\mathbf{y}$ is the inner product $\mathbf{x}\cdot\mathbf{y}$, the *reflection* in the plane (6·34) (see Figure 6·3A) reverses this normal vector $\mathbf{y}$ and transforms $\mathbf{x}$ into

$$\mathbf{x}' = \mathbf{x} - 2\mathbf{x}\cdot\mathbf{y}\,\mathbf{y} = \mathbf{x} + 2\mathsf{S}(xy)\,\mathbf{y}.$$

Since $y^2 = -1$, the corresponding transformation of quaternions is, by (6·32),

$$\begin{aligned} x' &= x + (xy+yx)y \\ &= yxy \qquad (\mathsf{S}y = 0, \mathsf{N}y = 1). \end{aligned}$$

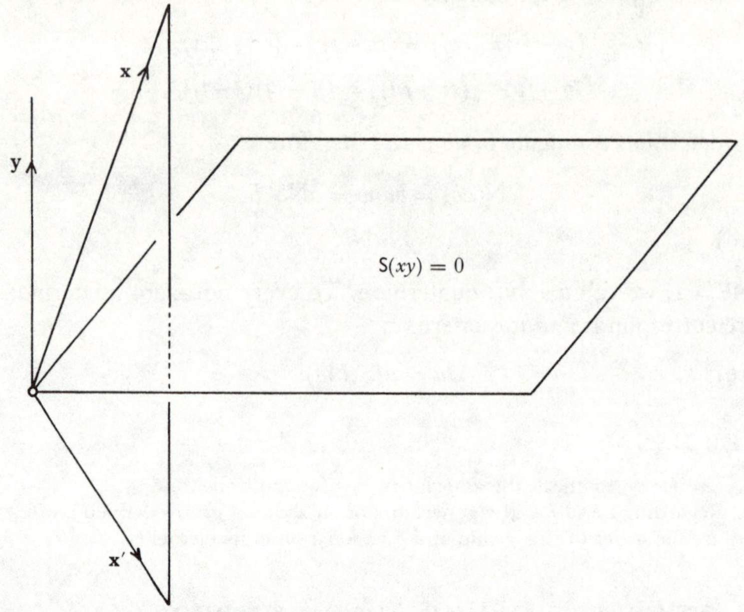

Figure 6·3A: Reflection

In such a case it is convenient to write, instead of 'the transformation $x' = yxy$' or 'the transformation $x \to yxy$', simply 'the transformation yxy'.

We have thus proved the following theorem:

(6·35) *The reflection in the plane*

$$y_1 x_1 + y_2 x_2 + y_3 x_3 = 0 \quad (\textit{where } y_1^2 + y_2^2 + y_3^2 = 1)$$

is represented by the quaternion transformation

$$yxy \quad (\mathsf{S}x = \mathsf{S}y = 0, \mathsf{N}y = 1).$$

EXERCISE

Given two quaternions, a and b, that have the same scalar part and the same norm, find quaternions y and z such that $a = y\tilde{z}$ and $b = \tilde{y}z$.

6·4 ROTATIONS

In three dimensions (see § 1·1) the central inversion in O can be expressed as the product of reflections in 3 mutually perpendicular planes through O, or as the product of one such reflection and the half-turn about the line through O perpendicular to the mirror. Since all these transformations are involutory, it follows that the half-turn can be expressed as the product of the reflection and the central inversion. Since the central inversion is the transformation $-x$, the following theorem has now been proved:

(6·41) *The half-turn about the line*

$$\frac{x_1}{y_1} = \frac{x_2}{y_2} = \frac{x_3}{y_3} \quad (y_1^2+y_2^2+y_3^2 = 1)$$

is the quaternion transformation

$$-yxy \quad (\mathsf{S}x = \mathsf{S}y = 0, \quad \mathsf{N}y = 1).$$

By regarding this as a transformation of the unit sphere $\Sigma x^2 = 1$, we may call it 'the half-turn about the point y'. Writing $-yxy$ in the form

$$e^{-\pi y/2}x\,e^{\pi y/2}$$

we exhibit it as a special case of $e^{-\alpha y}x\,e^{\alpha y}$ which, leaving every norm invariant, must be an isometry for any real α. Moreover, since this isometry leaves invariant both 0 and y, and is not a reflection, it must be a rotation about the line joining these two points or, let us say, a rotation about the point y. Since the nth power of $a^{-1}xa$ is $a^{-n}xa^n$, the angle of the rotation is proportional to α. But the angle is π when $\alpha = \frac{1}{2}\pi$ (by (6·41)). Hence the angle is always 2α, and we have proved:

(6·42) *The rotation through 2α about the point y (where $\mathsf{N}y = 1$, $\mathsf{S}y = 0$) is the quaternion transformation*

$$e^{-\alpha y}x\,e^{\alpha y},$$

that is, $(\cos\alpha - y\sin\alpha)\,x(\cos\alpha + y\sin\alpha)$.

It is important to notice that each unit quaternion $\exp(\alpha y)$ determines a unique rotation, namely the rotation through 2α about the point y; but each rotation is represented equally well by either of two opposite quaternions

$$\exp(\alpha y) \quad \text{and} \quad \exp\{(\alpha+\pi)y\} = -\exp(\alpha y).$$

For a given axis of rotation (or a given point y), a gradual increase of α from 0 to π (or of 2α from 0 to 2π) corresponds to a gradual change of the quaternion

$$a = \exp(\alpha y)$$

from 1 to -1. Since the product of transformations

$$\tilde{a}xa \quad \text{and} \quad \tilde{b}xb$$

is $\tilde{b}\tilde{a}xab = \tilde{c}xc$, where $c = ab$, it follows that

(6·43) *The multiplicative group of all unit quaternions is* 2:1 *homomorphic to the group of all rotations that leave the origin fixed. The kernel of the homomorphism is the group of order* 2 *generated by the quaternion* -1.

Considering finite subgroups of these continuous groups, we see that every finite group of rotations is 2:1 homomorphic to a finite group of unit quaternions including the 'central' quaternion -1.

EXERCISE

What restriction have we placed on our frame of coordinate axes by writing $e^{-\alpha y}x\,e^{\alpha y}$ (in Theorem (6·42)) rather than $e^{\alpha y}x\,e^{-\alpha y}$?

6·5 FINITE GROUPS OF QUATERNIONS

Since the norm of the nth power of a quaternion is equal to the nth power of the norm, any finite (multiplicative) group of quaternions must consist entirely of *unit* quaternions. If such a group contains the quaternion -1, Theorem (6·43) shows that it is 2:1 homomorphic to a finite group of rotations, each pair of 'opposite' quaternions $\pm\exp(\alpha y)$ being mapped on the rotation through 2α about a point y.

If a finite group of quaternions does *not* contain -1, each of its elements $\exp(\alpha y)$ represents uniquely the rotation through 2α about y, so there is an *isomorphic* group of rotations. In this, none of the rotations can be a half-turn, because then the corresponding quaternion would be pure, and its square would be -1. Looking through the list of finite rotation groups in §2·3 (or in Coxeter 1969, p. 275), we see that the only kind not containing a half-turn is the cyclic group $\mathfrak{C}_p$, where p is odd. The most obvious generator for this cyclic group is the rotation through $2\pi/p$ about some point, say $(1, 0, 0)$ or i. The corresponding quaternion is $\exp(i\pi/p)$, which generates $\mathfrak{C}_{2p}$ (since its pth power is $\exp(i\pi) = -1$). The same rotation group $\mathfrak{C}_p$ (with p odd) is equally well generated by the rotation through $4\pi/p$ about the same point, but now the corresponding quaternion $\exp(2i\pi/p)$ generates a cyclic group $\mathfrak{C}_p$ not containing -1. Hence

(6·51) *The only finite groups of quaternions not containing* -1 *are the cyclic groups of odd order.*

The remaining finite groups of quaternions, being 2:1 homomorphic to the finite groups of rotations

$$(p, p, 1), \quad (p, 2, 2), \quad (3, 3, 2), \quad (4, 3, 2), \quad (5, 3, 2),$$

are conveniently denoted by

$$\langle p,p,1\rangle,\quad \langle p,2,2\rangle,\quad \langle 3,3,2\rangle,\quad \langle 4,3,2\rangle,\quad \langle 5,3,2\rangle.$$

(Coxeter and Moser 1972, p. 68.)

We have now completed an *abstract* enumeration of the finite groups of quaternions. In the continuous group of *all* unit quaternions, each finite subgroup is one of an infinite class of conjugate subgroups derived from one another by inner automorphisms, that is, by multiplying on the right by any fixed unit quaternion and on the left by the inverse (or conjugate) quaternion. (The reader must forgive this use of 'conjugate' in two different senses!) For instance, instead of taking the generator $\exp(2i\pi/p)$ for $\mathfrak{C}_p$, we could equally well have taken $\exp(2\mathrm{P}\pi/p)$ where P is j or k or any other unit pure quaternion. We naturally use one convenient specimen from each class.

Each group $\langle p,q,r\rangle$ is generated by three quaternions

$$(6{\cdot}52)\qquad \mathrm{A}=\exp(\mathrm{P}\pi/p),\quad \mathrm{B}=\exp(\mathrm{Q}\pi/q),\quad \mathrm{C}=\exp(\mathrm{R}\pi/r),$$

which represent rotations through angles $2\pi/p$, $2\pi/q$, $2\pi/r$ about the vertices P, Q, R of a 'clockwise' spherical triangle $(p\,q\,r)$ having angles π/p at P, π/q at Q, π/r at R. Since the rotation group (p,q,r), of order $2s$ (see (2·34)), has the presentation (3·32), it follows that $\langle p,q,r\rangle$, of order $4s$, has the presentation

$$(6{\cdot}53)\qquad \mathrm{A}^p=\mathrm{B}^q=\mathrm{C}^r=\mathrm{ABC}=\mathrm{Z},\quad \mathrm{Z}^2=1,$$

where Z is the 'central' quaternion -1, which commutes with every quaternion and generates a normal subgroup of order 2 whose quotient group is (p,q,r). We shall see later that the relation $\mathrm{Z}^2=1$ is actually superfluous, being (in each case) a consequence of the preceding relations. For instance, the *cyclic group*

$$(6{\cdot}54)\qquad \langle p,p,1\rangle\cong\mathfrak{C}_{2p}$$

has the presentation $\quad \mathrm{A}^p=\mathrm{B}^p=\mathrm{C}=\mathrm{ABC},$

which implies $\mathrm{A}^{2p}=1$.

To sum up, the finite groups of quaternions are as follows:

the *cyclic* group	$\mathfrak{C}_n$,	of order n,
the *dicyclic* group	$\langle p,2,2\rangle$,	of order $4p$,
the *binary tetrahedral* group	$\langle 3,3,2\rangle$,	of order 24,
the *binary octahedral* group	$\langle 4,3,2\rangle$,	of order 48,
the *binary icosahedral* group	$\langle 5,3,2\rangle$,	of order 120.

EXERCISES

1. Express (6·53) in terms of the three generators A, B, Z.
2. Do the two relations $\mathrm{A}^p=\mathrm{B}^2=(\mathrm{AB})^2$ suffice for the dicyclic group?

6·6 GENERATORS FOR $\langle p,q,2\rangle$

Let us consider once more the characteristic triangle PQR formed by the axes of the three generating rotations for $(p,q,2)$. We recall, from pages 15 and 16, that the great circles QR, RP, PQ, or the planes that contain them, form a trihedral kaleidoscope in which the images of

$$P,\quad Q,\quad R$$

are the vertices of the regular or quasi-regular polyhedra

$$\{q,p\},\quad \{p,q\},\quad \begin{Bmatrix}p\\q\end{Bmatrix}.$$

In the notation of Figure 3·1B, the triangle PQR, being the fundamental region for $[p,q]$, is a spherical right-angled triangle whose catheti $QR=\phi$ and $RP=\psi$ are given by

$$\cos\phi=\cos\frac{\pi}{p}\Big/\sin\frac{\pi}{q},\qquad \sin\phi=\sin\frac{\pi}{h}\Big/\sin\frac{\pi}{q},$$

$$\cos\psi=\cos\frac{\pi}{q}\Big/\sin\frac{\pi}{p},\qquad \sin\psi=\sin\frac{\pi}{h}\Big/\sin\frac{\pi}{p},$$

where h, being the number of vertices of the equatorial polygon for $\begin{Bmatrix}p\\q\end{Bmatrix}$, or of the Petrie polygon for $\{p,q\}$, is given by (2·22).

Three quaternions A, B, C which generate $\langle p,q,2\rangle$ (as one explicit specimen of the infinite class of conjugate groups) may be obtained by locating the right-angled triangle $(p\,q\,2)$ at one corner of the trirectangular triangle (or octant) formed by the points that represent the units i,j,k (in clockwise order, because we are using a left-handed coordinate frame), as in Figure 6·6A. Accordingly, $\mathrm{R}=k$,

$$\mathrm{P}=k\cos\psi+i\sin\psi=\left(k\cos\frac{\pi}{q}+i\sin\frac{\pi}{h}\right)\Big/\sin\frac{\pi}{p},$$

$$\mathrm{Q}=k\cos\phi+j\sin\phi=\left(k\cos\frac{\pi}{p}+j\sin\frac{\pi}{h}\right)\Big/\sin\frac{\pi}{q},$$

and the three generators of $\langle p,q,2\rangle$ are, by (6·52),

$$\mathrm{A}=\cos\frac{\pi}{p}+\mathrm{P}\sin\frac{\pi}{p}=\cos\frac{\pi}{p}+k\cos\frac{\pi}{q}+i\sin\frac{\pi}{h},$$

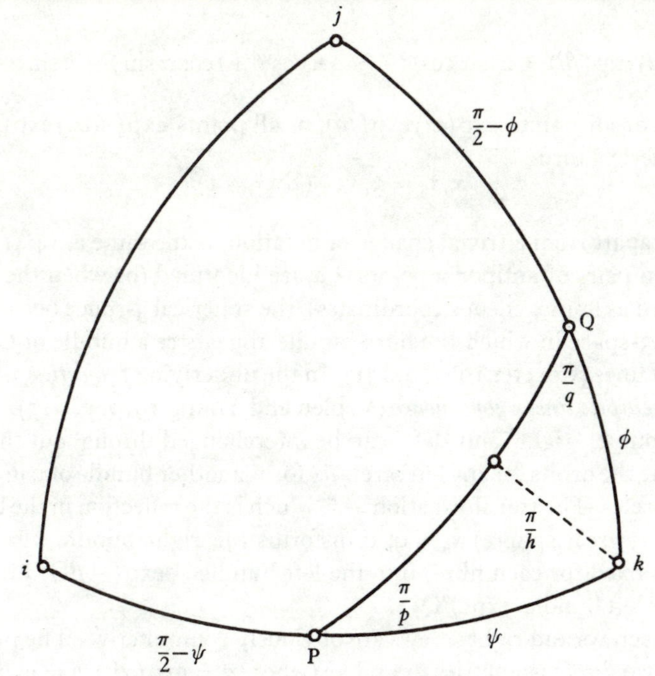

Figure 6·6A: The Möbius triangle

$$B = \cos\frac{\pi}{q}+Q\sin\frac{\pi}{q} = \cos\frac{\pi}{q}+k\cos\frac{\pi}{p}+j\sin\frac{\pi}{h},$$

$$C = \cos\frac{\pi}{2}+R\sin\frac{\pi}{2} = R = k.$$

Since
$$BC = \left(\cos\frac{\pi}{q}+k\cos\frac{\pi}{p}+j\sin\frac{\pi}{h}\right)k$$
$$= -\cos\frac{\pi}{p}+k\cos\frac{\pi}{q}+i\sin\frac{\pi}{h} = -A^{-1},$$

we can immediately verify the relations

(6·61) $$A^p = B^q = C^2 = ABC = -1$$

(which become Hamilton's $i^2 = j^2 = k^2 = ijk = -1$ in the case when $p = q = 2$). Since $C=AB$, we do not really need all three generators: $\langle p,q,2\rangle$ is equally well generated by any *two* of A, B, C. For instance, in terms of A and B the presentation (6·53) becomes

(6·62) $$A^p = B^q = (AB)^2 = Z, \quad Z^2 = 1.$$

We conclude that every finite group of quaternions is either cyclic or binary polyhedral, and in the latter case one possible pair of generators is

(6·63) $$A = \cos\frac{\pi}{p}+k\cos\frac{\pi}{q}+i\sin\frac{\pi}{h}, \qquad B = \cos\frac{\pi}{q}+k\cos\frac{\pi}{p}+j\sin\frac{\pi}{h}.$$

For instance, $\langle p, 2, 2\rangle$ (with $h = p$) is generated by

(6·64) $$A = \cos\frac{\pi}{p}+i\sin\frac{\pi}{p} = \exp(\pi i/p), \qquad B = k\cos\frac{\pi}{p}+j\sin\frac{\pi}{p} = k\exp(\pi i/p) = \exp(-\pi i/p)k$$

and, if we wish to include it, $C = AB = k$. Again, $\langle 5, 3, 2\rangle$ (with $h = 10$, so that $\sin\frac{\pi}{h} = \cos\frac{2\pi}{5} = \frac{1}{2}\tau^{-1}$) is generated by

(6·65) $$A = \cos\frac{\pi}{5}+k\cos\frac{\pi}{3}+i\cos\frac{2\pi}{5} = \tfrac{1}{2}(\tau+\tau^{-1}i+k), \qquad B = \cos\frac{\pi}{3}+k\cos\frac{\pi}{5}+j\cos\frac{2\pi}{5} = \tfrac{1}{2}(1+\tau^{-1}j+\tau k)$$

and $C = k$. The 120 elements $x_0+x_1i+x_2j+x_3k$ of this group are represented, in Euclidean 4-space, by the 120 vertices (x_0, x_1, x_2, x_3) of $\{3, 3, 5\}$. (See Coxeter 1963, pp. 156–7 (8·74).)

EXERCISE

Referring to Figure 6·6A, verify that the half-turn about R has the same effect on P as the reflection in the great circle QR, and has the same effect on Q as the reflection in PR.

6·7 SCREWS IN EUCLIDEAN 4-SPACE

From the general discussion in §1·1, we see that a reflection has, for its mirror, a hyperplane. In the four-dimensional case, then, the mirror is a 3-flat. A *rotation*, being the product of reflections in two intersecting hyperplanes, leaves invariant every point in its *axial plane*: the common plane of the two hyperplanes. Every direct orthogonal transformation, being the product of 2 or 4 reflections, is either a rotation or the product of two rotations.

The transformation $-x$, which is the *central inversion*, may be described as the product of reflections in any four mutually orthogonal hyperplanes through the origin. Combining these reflections in pairs, we see that $-x$ may just as well be described as the product of half-turns about two completely orthogonal planes (such as the planes $x_2 = x_3 = 0$ and $x_0 = x_1 = 0$).

When the quaternion $x = x_0+x_1 i+x_2 j+x_3 k$ is regarded as representing a point (x_0, x_1, x_2, x_3), the distance between two such points x and y is the square root of the norm

$$\mathsf{N}(y-x) = \Sigma(y_\nu - x_\nu)^2.$$

We see from (6·18) that the transformation xb, multiplying every quaternion x on the right by a fixed *unit* quaternion b, leaves this norm unchanged and is thus an isometry. By (6·23), we can write $b = \exp(\beta \mathrm{Q})$, where β is an angle and Q is a unit *pure* quaternion, namely $\mathrm{Q} = \mathsf{UV}b$. Continuous variation of β, beginning with $\beta = 0$, shows that the isometry xb is direct. Let us call it a *right screw*. Clearly, the set of all right screws is a continuous group, isomorphic to the group of all unit quaternions b.

Similarly, the same group of all unit quaternions is isomorphic also to the group of *left screws* $\tilde{a}x$, where $a = \exp(\alpha \mathrm{P})$ so that $\tilde{a} = \exp(-\alpha \mathrm{P})$. (The minus sign is desirable because the product of two such transformations $\tilde{a}x$ and $\tilde{a}'x$ is $\tilde{a}'\tilde{a}x$, and $\tilde{a}'\tilde{a}$ is the conjugate of aa'.)

Since the continuous group of unit quaternions is mapped continuously on the unit sphere $\mathsf{N}x = 1$, we have here a perfect instance of a 'topological group'. By keeping Q fixed while allowing β to increase continuously from 0 to 2π, we see that the orbit of the point 1 for the right screw xb (where $b = \exp(\beta\mathrm{Q})$ and $\mathrm{Q}^2 = -1$) is a great circle; for instance, if $\mathrm{Q} = j$, this circle (the locus of $\exp(\beta j)$ as β varies) is the section of the 3-sphere

$$x_0^2+x_1^2+x_2^2+x_3^2 = 1$$

by the plane $x_1 = x_3 = 0$. By applying a left screw to the circle, we see that the orbits of other points (for the same right screw) are congruent great circles; the orbits of two points are either coincident or entirely disjoint (Du Val 1964, p. 49). In other words, all the orbits form a *fibre bundle* filling the three-dimensional spherical space in much the same way that three-dimensional Euclidean space is filled with a bundle of parallel lines (e.g. vertical lines), the orbits for a translation. It is remarkable that any two circles in the bundle are interlocked, in the sense that any sphere through one has two points of intersection with the other; for instance, the sphere $x_3 = 0$ through the circle $\exp(\beta j)$ $(0 \leqslant \beta < 2\pi)$ intersects the circle $i\exp(\beta j)$ in the two points $\pm i$.

Since

$$\exp(\alpha i)\exp(\beta j) = \cos\alpha\cos\beta + i\sin\alpha\cos\beta + j\cos\alpha\sin\beta + k\sin\alpha\sin\beta,$$

the set of all points $\exp(\alpha i)\exp(\beta j)$, or all points $\exp(-\alpha i)\exp(\beta j)$, is the spherical torus

$$x_0 x_3 = x_1 x_2 \quad (\Sigma x_\nu^2 = 1)$$

which, apart from a trivial change of notation, is the same as (4·41).

When pairs of antipodal points $\pm x$ are identified (or when the x_ν are regarded as homogeneous coordinates), the spherical 3-space becomes an *elliptic* 3-space in which the fibre bundle appears as a bundle of *Clifford parallel* lines (Coxeter 1965, p. 151). In the underlying *projective* 3-space, it is an *elliptic linear congruence* (Veblen and Young 1910, p. 315).

Of course, 'right' and 'left' can be interchanged throughout this discussion: the orbits for any left screw $\tilde{a}x$ form another bundle of congruent great circles. The transformation $-\tilde{x}$, which is the reflection in the hyperplane (or great sphere) $x_0 = 0$, transforms the right bundle $x\exp(\beta\mathrm{Q})$ (with x fixed for each fibre) into the left bundle $-\exp(-\beta\mathrm{Q})\tilde{x}$, that is, into the left bundle $\exp(\beta\mathrm{Q})x$.

Left screws and right screws are obviously commutative. The product of the two direct isometries $\tilde{a}x$ and xb (where $a = \exp(\alpha\mathrm{P})$, $b = \exp(\beta\mathrm{Q})$, and $\mathrm{P}^2 = \mathrm{Q}^2 = -1$) is a direct isometry $\tilde{a}xb$. In the special case when $a = b$, Theorem (6·42) shows that this isometry $\tilde{a}xa$ is simply a rotation through 2α about the plane determined by the three points 0, 1, P (where $\mathrm{P} = \mathsf{UV}a$). In particular, Theorem (6·41) shows that the *half-turn* about the same plane is $-\mathrm{P}x\mathrm{P}$. This is also obvious from first principles, as the transformation $-\mathrm{P}x\mathrm{P}$ is of period 2 and leaves 0, 1, P all invariant.

Combining $-\mathrm{P}x\mathrm{P}$ with the central inversion $-x$, we find that the transformation $\mathrm{P}x\mathrm{P}$ (without the minus sign) is the half-turn about the plane through 0 completely orthogonal to the one through 1 and P. We easily verify that $\mathrm{P}x\mathrm{P}$ transforms 1 into -1, and P into $-\mathrm{P}$.

Let $\mathbf{x}$ denote the vector from $0 = (0, 0, 0, 0)$ to $x = (x_0, x_1, x_2, x_3)$. Since $\mathsf{S}a = \frac{1}{2}(a+\tilde{a})$ and the conjugate of ab is $\tilde{b}\tilde{a}$, the inner product of two such vectors is

(6·71) $$\mathbf{x}\cdot\mathbf{y} = x_0 y_0 + x_1 y_1 + x_2 y_2 + x_3 y_3 = \mathsf{S}(\tilde{x}y) = \mathsf{S}(x\tilde{y}) = \tfrac{1}{2}(x\tilde{y}+y\tilde{x})$$

and the condition for two non-zero vectors $\mathbf{x}$ and $\mathbf{y}$ to be perpendicular is

(6·72) $$x\tilde{y} + y\tilde{x} = 0.$$

This implies that the two conjugate quaternions $x\tilde{y}$ and $y\tilde{x}$ are pure.

We are now ready to prove the following theorem:

(6·73) *Every rotation through 2α can be expressed as* $\tilde{a}xb$, *where*

$$\mathrm{N}a = \mathrm{N}b = 1 \quad \textit{and} \quad \mathrm{S}a = \mathrm{S}b = \cos\alpha.$$

Every such quaternion transformation represents a rotation.

Let the axial plane for a given rotation be the plane through 0 spanned by two perpendicular unit vectors **y** and **z**, so that

$$y\tilde{y} = z\tilde{z} = 1, \quad y\tilde{z} + z\tilde{y} = 0.$$

Since $y\tilde{z}y\tilde{y}z = y$ and $y\tilde{z}z\tilde{y}z = z$ and $(y\tilde{z})^2 = (\tilde{y}z)^2 = -1$, the transformation

$$(6{\cdot}74) \qquad y\tilde{z}x\tilde{y}z$$

leaves invariant both y and z (as well as 0) and has period 2. These properties suffice to identify it with the half-turn about the given plane. More generally, the transformation

$$(6{\cdot}75) \qquad \begin{aligned} e^{\alpha y\tilde{z}}\,x\,e^{\alpha\tilde{y}z} &= (\cos\alpha + y\tilde{z}\sin\alpha)\,x(\cos\alpha + \tilde{y}z\sin\alpha) \\ &= x\cos^2\alpha + (x\tilde{y}z + y\tilde{z}x)\cos\alpha\sin\alpha + y\tilde{z}x\tilde{y}z\sin^2\alpha, \end{aligned}$$

being the product of a left screw and a right screw, is a direct isometry. Since it leaves y and z invariant, it is a rotation. Expressing it in the form $\tilde{a}xb$, we see that

$$\mathrm{N}a = \mathrm{N}b = 1 \quad \text{and} \quad \mathrm{S}a = \mathrm{S}b = \cos\alpha.$$

Since the nth power of $\tilde{a}xb$ is $\tilde{a}^n x b^n$, the angle of the rotation is proportional to α. But (6·74) shows that the angle is π when $\alpha = \frac{1}{2}\pi$. Hence it is always 2α.

In particular, the rotation through 2α about the plane spanned by 1 and P (where $\mathrm{P}^2 = -1$) is $e^{-\alpha\mathrm{P}}xe^{\alpha\mathrm{P}}$, in agreement with (6·42). Analogously, since the half-turn about the completely orthogonal plane is $\mathrm{P}x\mathrm{P}$, the rotation through 2α about that plane is

$$e^{\alpha\mathrm{P}}\,x\,e^{\alpha\mathrm{P}}.$$

Conversely, every transformation $\tilde{a}xb$, where $\mathrm{N}a = \mathrm{N}b = 1$ and $\mathrm{S}a = \mathrm{S}b = \cos\alpha$, is a rotation through 2α. In the special case when $\tilde{a} = b$, this follows from the above remark about $e^{\alpha\mathrm{P}}xe^{\alpha\mathrm{P}}$. In every other case we can identify $\tilde{a}xb$ with $e^{\alpha y\tilde{z}}xe^{\alpha\tilde{y}z}$ by writing $a = e^{\alpha\mathrm{P}}$, $b = e^{\alpha\mathrm{Q}}$,

$$(6{\cdot}76) \qquad y = \mathrm{U}(1 - \mathrm{PQ}), \quad z = \mathrm{U}(\mathrm{P} + \mathrm{Q}).$$

In fact, these expressions for y and z imply

$$\mathrm{P}z = \mathrm{U}(-1 + \mathrm{PQ}) = -y, \quad y\mathrm{Q} = \mathrm{U}(\mathrm{Q} + \mathrm{P}) = z,$$

whence

$$y\tilde{z} = -\mathrm{P}, \quad \tilde{y}z = \mathrm{Q}.$$

Thus, if $\mathrm{P} + \mathrm{Q} \neq 0$, $e^{-\alpha\mathrm{P}}xe^{\alpha\mathrm{Q}}$ is a rotation through 2α about the plane determined by the three points 0, $1 - \mathrm{PQ}$, $\mathrm{P} + \mathrm{Q}$. In particular, $-\mathrm{P}x\mathrm{Q}$ is the half-turn about this plane. Combining it with the central inversion $-x$, we deduce that $\mathrm{P}x\mathrm{Q}$ is the half-turn about the completely orthogonal plane, which is therefore, if $\mathrm{P} \neq \mathrm{Q}$, the plane through 0, $1 + \mathrm{PQ}$, $-\mathrm{P} + \mathrm{Q}$ (or through 0, $1 + \mathrm{PQ}$, $\mathrm{P} - \mathrm{Q}$). Hence, if $\mathrm{P} \neq \mathrm{Q}$, $e^{\alpha\mathrm{P}}xe^{\alpha\mathrm{Q}}$ is a rotation through 2α about the plane through 0, $1 + \mathrm{PQ}$, $\mathrm{P} - \mathrm{Q}$.

The most general direct orthogonal transformation, being the product of two rotations, is of the form $\tilde{a}xb$ where still $\mathrm{N}a = \mathrm{N}b = 1$ but no longer (in general) $\mathrm{S}a = \mathrm{S}b$. In fact, $e^{-\alpha\mathrm{P}}\,x\,e^{\beta\mathrm{Q}}$ is the (commutative) product of rotations

$$(6{\cdot}77) \qquad e^{-\frac{1}{2}(\alpha+\beta)\mathrm{P}}\,x\,e^{\frac{1}{2}(\alpha+\beta)\mathrm{Q}} \quad \text{and} \quad e^{\frac{1}{2}(-\alpha+\beta)\mathrm{P}}\,x\,e^{\frac{1}{2}(-\alpha+\beta)\mathrm{Q}}$$

through angles $\pm\alpha + \beta$ about two completely orthogonal planes. If $\alpha\beta \neq 0$, the two axial planes are uniquely determined. In general, they are the planes through 0, $1 \mp \mathrm{PQ}$, $\mathrm{P} \pm \mathrm{Q}$. If $\mathrm{P} = \pm\mathrm{Q}$, this description for one of the two is inadequate; but then this one is determined as being completely orthogonal to the other one. Thus the only direct orthogonal transformations having more than two invariant planes are the screws $\exp(-\alpha\mathrm{P})x$ and $x\exp(\beta\mathrm{Q})$.

Since $-\mathrm{P}(\mathrm{P} \pm \mathrm{Q}) = 1 \mp \mathrm{PQ}$ and $(\mathrm{P} \pm \mathrm{Q})\mathrm{Q} = \mp(1 \mp \mathrm{PQ})$, the plane through 0, $1 \mp \mathrm{PQ}$, $\mathrm{P} \pm \mathrm{Q}$ can be described either as the plane of the circle $\exp(\theta\mathrm{P})\,\mathrm{U}(\mathrm{P} \pm \mathrm{Q})$ or as the plane of the circle $\mathrm{U}(\mathrm{P} \pm \mathrm{Q})\exp(\theta\mathrm{Q})$. Setting $\beta = 0$ in (6·77), we see that the left screw $\exp(-\alpha\mathrm{P})x$, for which P is determined while Q is arbitrary, is the product of rotations through angles $\pm\alpha$ about two completely orthogonal planes containing circles

$$\exp(\theta\mathrm{P})\,\mathrm{U}(\mathrm{P} \pm \mathrm{Q}),$$

one such pair of planes for each choice of Q. Similarly, the right screw $x\exp(\beta\mathrm{Q})$ is the product of rotations through equal angles β about two completely orthogonal planes containing circles $\mathrm{U}(\mathrm{P} \pm \mathrm{Q})\exp(\theta\mathrm{Q})$, one such pair of planes for each choice of P. To sum up,

(6·78) *Every direct orthogonal transformation can be expressed as the product of a left screw and a right screw, that is,* $\tilde{a}xb$ *where* $\mathrm{N}a = \mathrm{N}b = 1$.

EXERCISES

1. If $\mathsf{N}a = \mathsf{N}b$, the two quaternions $a \pm b$ represent perpendicular vectors.
2. If $P^2 = Q^2 = -1$ and $P \neq Q \neq -P$, the four vectors represented by $1 \pm PQ$ and $P \pm Q$ are mutually perpendicular.
3. If $\mathsf{N}a = \mathsf{N}b = 1$ and $\mathsf{S}\tilde{a} = \mathsf{S}b = \cos\alpha$, $\tilde{a}xb$ is the rotation through 2α about the plane of 0, $1-ab$, $\tilde{a}-b$.

6·8 ROTATORY-REFLECTIONS

Having studied the direct orthogonal transformations in Euclidean 4-space, let us consider more briefly the opposite (or 'indirect') orthogonal transformations, for which the associated matrix has determinant -1. Being the product of m reflections where $m \leqslant 4$ and m is odd, such a transformation is either a reflection or the product of three reflections. In the latter case the three mirrors intersect in a line whose points are all invariant; thus their product leaves invariant the hyperplane perpendicular to this line, and in this hyperplane it appears as a *rotatory-reflection* (Coxeter 1963, p. 37) or 'compound reflection' (Du Val 1964, p.31), which is a rotation combined with the reflection in a plane perpendicular to the axis.

Let us consider first the reflection in the hyperplane

$$y_0x_0+y_1x_1+y_2x_2+y_3x_3 = 0 \quad (y_0{}^2+y_1{}^2+y_2{}^2+y_3{}^2 = 1),$$

whose unit normal vector $\mathbf{y}$ is represented by a unit quaternion y. Since the length of the projection of any vector $\mathbf{x}$ on the unit vector $\mathbf{y}$ is the inner product

$$\mathbf{x}\cdot\mathbf{y} = \mathsf{S}(x\tilde{y}),$$

this reflection transforms each vector $\mathbf{x}$ into

$$\mathbf{x}-2\mathbf{x}\cdot\mathbf{y}\mathbf{y} = \mathbf{x}-2\mathsf{S}(x\tilde{y})\mathbf{y}$$

and hence transforms each quaternion x into

$$x-(x\tilde{y}+y\tilde{x})y = -y\tilde{x}y.$$

We have thus proved

(6·81) *The reflection in the hyperplane* $\Sigma y_\nu x_\nu = 0$ (*where* $\Sigma y_\nu{}^2 = 1$) *is the transformation*

$$-y\tilde{x}y \quad (\mathsf{N}y = 1).$$

The product of any given opposite orthogonal transformation with the reflection $-\tilde{x}$ (in the hyperplane $x_0 = 0$) is a direct orthogonal transformation, say axb. Hence the opposite transformation itself is expressible as

$$-a\tilde{x}b \quad (\mathsf{N}a = \mathsf{N}b = 1).$$

When $a = b$, this is simply a reflection, and when $-a = b$ it is the product of a reflection with the central inversion, which we might reasonably call 'reflection in a line'. (The simplest instance is the transformation $\tilde{x}$, which is the reflection in the line $x_1 = x_2 = x_3 = 0$.) Setting these two cases aside, let us suppose $b \neq \pm a$.

Applying the transformation $-a\tilde{x}b$ to $a \pm b$, we obtain

$$-a(\tilde{a} \pm \tilde{b})b = -b \mp a.$$

Thus $a+b$ is reversed, while $a-b$ remains unchanged. This proves that $-a\tilde{x}b$ is a rotatory-reflection with its axis along the line of the vector $\mathbf{a}+\mathbf{b}$, acting in the hyperplane perpendicular to the vector $\mathbf{a}-\mathbf{b}$. If the angle of the rotation is α, the 'square' of the rotatory-reflection (i.e. its product with itself) must be a rotation through 2α. But the square of $-a\tilde{x}b$ is

$$a\tilde{b}x\tilde{a}b.$$

Hence, by Theorem (6·73), $\mathsf{S}(a\tilde{b}) = \cos\alpha$, which means that α is the angle between $\mathbf{a}$ and $\mathbf{b}$ (Du Val 1964, p. 58, with α for $\theta-\pi$). We have thus proved

(6·82) *If* $\mathsf{N}a = \mathsf{N}b = 1$ *and* $\pm a \neq b$, *the transformation* $-a\tilde{x}b$ *operates on the hyperplane perpendicular to* $\mathbf{a}-\mathbf{b}$ *as a rotatory-reflection whose axis and angle are the line of* $\mathbf{a}+\mathbf{b}$ *and the angle between* $\mathbf{a}$ *and* $\mathbf{b}$.

Letting $\mathbf{y}$ and $\mathbf{z}$ denote unit vectors along the diagonals of a rhombus whose pairs of opposite sides represent the vectors $\mathbf{a}$ and $\mathbf{b}$, we can write

$$a = -y\sin\tfrac{1}{2}\alpha+z\cos\tfrac{1}{2}\alpha, \quad b = y\sin\tfrac{1}{2}\alpha+z\cos\tfrac{1}{2}\alpha,$$

and conclude that, in the hyperplane perpendicular to a given unit vector $\mathbf{y}$ (that is, in the hyperplane $\Sigma y_\nu x_\nu = 0$), the rotatory-reflection involving rotation through α about the line of a given unit vector $\mathbf{z}$ is

$$(y\sin\tfrac{1}{2}\alpha-z\cos\tfrac{1}{2}\alpha)\,\tilde{x}\,(y\sin\tfrac{1}{2}\alpha+z\cos\tfrac{1}{2}\alpha).$$

EXERCISE

Consider the product of reflections in two hyperplanes $\Sigma y_\nu x_\nu = 0$ and $\Sigma z_\nu x_\nu = 0$. Express this product in the form $\tilde{a}xb$, where $a = y\tilde{z}$ and $b = \tilde{y}z$. Reconcile the result with (6·74).

6·9 REMARKS

Having dispelled the sense of mystery that formerly obscured the notion of a complex number $x+yi$, Hamilton tried for a long time to invent some kind of hypercomplex number $x+yi+zj$, where i and j would be essentially different square roots of -1. To add two such 'numbers' would be easy, but how should they be multiplied? (What is the product ij? Is it the same as ji?) The idea of writing $ij = k$, and treating the three symbols i, j, k all alike, occurred to him while walking beside a canal near Dublin. It pleased him so much that he scratched his famous equations on the supporting structure of a bridge, where a plaque commemorating the event still bears this inscription:

Here as he walked by
on the 16th of October 1843
Sir William Rowan Hamilton
in a flash of genius discovered
the fundamental formula for
quaternion multiplication

$$i^2 = j^2 = k^2 = ijk = -1$$

& cut it on a stone of this bridge

According to Blaschke (1954, p. 145), Hamilton's quaternions were anticipated by Euler in his letter of 4 May 1748 to C. Goldbach, and by Gauss,[†] who considered (about 1819) an algebra of symbols (a, b, c, d), adding them like vectors and multiplying them according to the rule

$$(a, b, c, d)\,(\alpha, \beta, \gamma, \delta) = (A, B, C, D),$$

where A, B, C, D are certain bilinear expressions in the other letters. All that is lacking for perfect agreement is a trivial matter of sign: Gauss's (a, b, c, d) is not $a+bi+cj+dk$ but $a-bi-cj-dk$. Another of Gauss's fragments of manuscript (quoted by van der Blij[‡]) came close to regarding a quaternion as a pair of complex numbers; in fact, Gauss's formula

$$(s\bar{s}+t\bar{t})\,(u\bar{u}+v\bar{v}) = \xi\bar{\xi}+\eta\bar{\eta},$$

where $\xi = su-tv$ and $\eta = sv+t\bar{u}$, is the result of equating norms of the two sides of our equation (6·14).

Theorems (6·42) and (6·78) were discovered by Cayley[§] in 1845 and 1855. In 1866 he gave the complete lists of quaternions forming the groups $\langle p, 2, 2\rangle$, $\langle 3, 3, 2\rangle$, $\langle 4, 3, 2\rangle$, $\langle 5, 3, 2\rangle$. Stringham[‖] proved that these (along with the cyclic groups) are the only finite groups of quaternions.

During the second half of the nineteenth century, quaternions became immensely popular. Every student of mathematics was expected to be familiar with them; they were 'fashionable' (like sets, vector spaces and functional analysis in more recent times). After 1900, interest waned for nearly forty years, but then a partial revival occurred. For instance, Theorem (6·81) was discovered by Witt[¶].

† C. F. Gauss, *Werke*, vol. 8 (Leipzig, 1900), p. 357.

‡ F. van der Blij, History of the octaves, *Simon Stevin* **34** (1960), 106–25; see especially p. 108.

§ Arthur Cayley, On certain results relating to quaternions, *Collected Mathematical Papers, I* (1889), pp. 123–6; Recherches ulterieures sur les déterminants gauches, *ibid. II*, pp. 202–15; Notes on polyhedra, *ibid. V* (1892), pp. 529–41.

‖ W. I. Stringham, Determination of the finite quaternion groups, *American Journal of Mathematics* **4** (1881), 345–57.

¶ Ernst Witt, Spiegelungsgruppen und Aufzählung halbeinfacher Liescher Ringe, *Abhandlungen aus dem Mathematischen Seminar der Hansischen Universität* **14** (1941), 289–322 (308). See also Coxeter, Quaternions and reflections, *American Mathematical Monthly* **53**, (1946), 136–46.

CHAPTER 7

The binary polyhedral groups

Tait once urged the advantage of Quaternions on Cayley (who never used them), saying: 'You know Quaternions are just like a pocket-map.' 'That may be,' replied Cayley, 'but you've got to take it out of your pocket, and unfold it, before it's of any use.' And he dismissed the subject with a smile.

Silvanus P. Thompson (1910, p. 1137)

We saw, in §6·6, that every finite non-cyclic group of quaternions is generated by two unit quaternions A and B which satisfy the relations (6·62) or

$$A^p = B^q = (AB)^2 = -1, \quad (p^{-1}+q^{-1} > \tfrac{1}{2}).$$

In the first four sections of the present chapter, we shall prove that a *presentation* of the group can be derived by the simple trick of deleting '$= -1$'. This trick was discovered by William Threlfall†. We shall justify it in most cases by observing that the quaternion group $\langle 2, 2, 2\rangle$ is a normal subgroup of index 3 in the binary tetrahedral group $\langle 3, 3, 2\rangle$, which is a subgroup of index 2 in the binary octahedral group $\langle 4, 3, 2\rangle$. Such information will be found useful later on, in connection with unitary groups. Since the icosahedral group $(5, 3, 2)$ is *simple*, the case of $\langle 5, 3, 2\rangle$ requires an entirely different procedure (§7·4; see also §11·7). In §7·5 we shall see that, except when $p = q = 3$, each group $\langle p, q, 2\rangle$ can be generated by two or three *pure* quaternions.

The remaining sections of this chapter deal with matrix representations of the same abstract groups and of the analogous groups with $p > 5$, $q = 3$, and one extra relation (7·67).

7·1 THE CYCLIC AND DICYCLIC GROUPS

In preparation for a closer look at the dicyclic group $\langle p, 2, 2\rangle$ (see (6·64)), let us first consider the cyclic group $\mathfrak{C}_n$. Any quaternion of period n will serve as a generator. Clearly, the most natural one to choose is the complex number

$$A = \exp(2\pi i/n) = \cos\frac{2\pi}{n} + i\sin\frac{2\pi}{n}.$$

† *Jahresbericht der deutschen Mathematiker–Vereinigung* **41** (1932), 6–8, and **46** (1936), 80.

The elements A^ν of this group $\mathfrak{C}_n$ are represented, in Euclidean 4-space, by the vertices

$$\left(\cos\frac{2\nu\pi}{n},\quad \sin\frac{2\nu\pi}{n},\ 0,\ 0\right) \quad (\nu = 0, \ldots, n-1)$$

of a regular n-gon $\{n\}$ in the plane $x_2 = x_3 = 0$.

For later application, we record the obvious fact that every cyclic group $\mathfrak{C}_{pq}$ has a normal subgroup $\mathfrak{C}_q$ whose quotient group is

(7·11) $$\mathfrak{C}_{pq}/\mathfrak{C}_q \simeq \mathfrak{C}_p.$$

In fact, the element $B = A^p$ of the group $A^{pq} = 1$ satisfies $B^q = 1$, and the quotient group is obtained by setting $B = 1$.

Let us consider the effect of extending the 'even' cyclic group

$$\mathfrak{C}_{2p} \simeq \langle p, p, 1\rangle$$

or

(7·12) $$A^{2p} = 1$$

by adjoining a second generator B, whose square is A^p and which transforms $\mathfrak{C}_{2p}$ according to the automorphism $A \to A^{-1}$ (Coxeter and Moser 1972, p. 7). Since the quaternion j, whose inverse is $-j$, transforms the quaternion

$$A = \exp(\pi i/p) = \cos\frac{\pi}{p} + i\sin\frac{\pi}{p}$$

into

$$\cos\frac{\pi}{p} - jij\sin\frac{\pi}{p} = \exp(-\pi i/p) = A^{-1},$$

we can take $B = j$ and exhibit the extended group, of order $4p$, as a group of quaternions.

The relations $B^2 = A^p$, $B^{-1}AB = A^{-1}$, which are satisfied by the quaternions

(7·121) $$A = \exp(\pi i/p), \quad B = j,$$

are equivalent to

(7·13) $$A^p = B^2 = (AB)^2$$

and imply $B^{-1}A^pB = A^{-p}$, making it unnecessary to mention the relation (7·12). In other words, the relations

$$A^p = B^2 = (AB)^2 = Z,$$

which obviously make Z commute with A and B, actually imply

$$Z^2 = 1.$$

Thus (7·13) is a complete presentation for the extended group of order $4p$. Introducing a redundant third generator

$$C = AB = \exp(\pi i/p)j = j\cos\frac{\pi}{p} + k\sin\frac{\pi}{p},$$

we can express (7·13) in the equivalent form

(7·131) $$A^p = B^2 = C^2 = ABC,$$

which serves to identify this group with the *dicyclic group* $\langle p, 2, 2\rangle$. In this representation, the $4p$ quaternions are

(7·14) $$A^\nu = \exp(\nu\pi i/p) = \cos\frac{\nu\pi}{p} + i\sin\frac{\nu\pi}{p},$$
$$A^\nu B = \exp(\nu\pi i/p)j = j\cos\frac{\nu\pi}{p} + k\sin\frac{\nu\pi}{p} \quad (\nu = 0, \ldots, 2p-1).$$

The $4p$ corresponding points are the vertices

(7·15) $$\left(\cos\frac{\nu\pi}{p}, \sin\frac{\nu\pi}{p}, 0, 0\right), \quad \left(0, 0, \cos\frac{\nu\pi}{p}, \sin\frac{\nu\pi}{p}\right)$$

of two $\{2p\}$'s in completely orthogonal planes, justifying the term 'dicyclic'. In the Euclidean 4-space, the convex hull of these $4p$ points is the generalized dipyramid whose cells (or facets) are $4p^2$ tetragonal disphenoids (see § 1·7) joining the $2p$ edges of one $2p$-gon to the $2p$ edges of the other. In the special case when $p = 2$, the two polygons are squares, the generalized dipyramid is the regular 16-cell β_4 or $\{3, 3, 4\}$, and $\langle 2, 2, 2\rangle$ is the quaternion group (6·21).

Since the relations (7·13) imply $B^{-1}AB = A^{-1}$, they imply $B^{-1}A^qB = A^{-q}$ for all q, whence

$$A^p = B^2 = (A^qB)^2.$$

Therefore $\langle p, 2, 2\rangle$ has a subgroup $\langle p/q, 2, 2\rangle$ (generated by A^q and B) for any divisor q of p. In particular, since every subgroup of index 2 is normal, $\langle 2p, 2, 2\rangle$ has a normal subgroup $\langle p, 2, 2\rangle$ and, of course,

(7·16) $$\langle 2p, 2, 2\rangle/\langle p, 2, 2\rangle \cong \mathfrak{C}_2.$$

Another consequence of $B^{-1}A^qB = A^{-q}$ is that, if q divides p, the element A^q of $\langle p, 2, 2\rangle$ generates a normal subgroup $\mathfrak{C}_{2p/q}$. The quotient group, obtained by combining (7·13) with $A^q = 1$, is the dihedral group

$$A^q = B^2 = (AB)^2 = 1$$

of order $2q$. Writing pq for p, we have

(7·17) $$\langle pq, 2, 2\rangle/\mathfrak{C}_{2p} \cong (q, 2, 2).$$

In particular,

(7·171) $$\langle p, 2, 2\rangle/\mathfrak{C}_{2p} \cong \mathfrak{C}_2$$

and, of course,

(7·18) $$\langle q, 2, 2\rangle/\mathfrak{C}_2 \cong (q, 2, 2).$$

Similarly, if p/q is odd, the element A^{2q} of $\langle p, 2, 2\rangle$ generates a normal subgroup $\mathfrak{C}_{p/q}$ whose quotient group is the dicyclic group

$$A^q = B^2 = (AB)^2$$

of order $4q$ (since $A^{2q} = 1$ implies $A^q = \{A^q\}^{p/q} = A^p$). Thus

(7·19) $$\langle pq, 2, 2\rangle/\mathfrak{C}_p \cong \langle q, 2, 2\rangle \quad (p \text{ odd}).$$

In particular, since $\langle 1, r, r\rangle \cong \mathfrak{C}_{2r}$,

(7·191) $$\langle p, 2, 2\rangle/\mathfrak{C}_p \cong \mathfrak{C}_4 \quad (p \text{ odd}).$$

EXERCISES

1. The quaternion group $\langle 2, 2, 2\rangle$ has the alternative presentations

(i) $jij = i$, $iji = j$; (ii) $jk = i$, $ki = j$, $ij = k$.

2. Is there a square root of -1 that will transform the quaternion group $\langle 2, 2, 2\rangle$ (defined by $jij = i$, $iji = j$) according to the automorphism which interchanges i and j? If so, what group of order 16 is obtained by adjoining this new element to $\langle 2, 2, 2\rangle$?

3. The dicyclic group $\langle pq, 2, 2\rangle$ has a dicyclic subgroup $\langle p, 2, 2\rangle$. When is this a normal subgroup?

4. For $\langle p, 2, 2\rangle$, what quaternion will transform the representation (6·64) into (7·121)?

7·2 THE BINARY TETRAHEDRAL GROUP

Since the relations (6·21) involve i, j, k symmetrically, it is natural to seek a quaternion l that might transform these generators of $\langle 2, 2, 2\rangle$ according to a cyclic permutation (Burnside 1911, p. 132). The corresponding points are permuted by a rotation through $-2\pi/3$ about the centre $3^{-\frac{1}{2}}(i+j+k)$ of the spherical octant ijk (marked a in Figure 7·2A). The desired quaternion is thus

$$l = \cos\frac{\pi}{3} - \frac{i+j+k}{\sqrt{3}}\sin\frac{\pi}{3} = \frac{1-i-j-k}{2}.$$

In fact,

$$l^{-1}il = \tfrac{1}{4}(1+i+j+k)i(1-i-j-k) = \tfrac{1}{4}(1+i+j+k)(1+i+j-k)$$
$$= j;$$

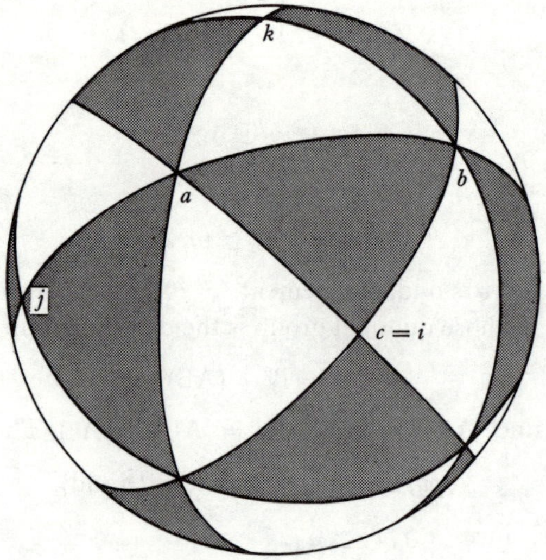

Figure 7·2A: (3, 3, 2) and ⟨3, 3, 2⟩

similarly, j is transformed into k, and k into i. Since $l^3 = -1$, we conclude that a group of quaternions, of order $3 \times 8 = 24$, can be derived from $\langle 2,2,2\rangle$ by adjoining l, whose cube is the central element Z and which cyclically permutes i, j, k. The relations

$$i^2 = j^2 = k^2 = ijk = Z = l^3, \quad l^{-1}il = j, \quad l^{-2}il^2 = k$$

imply (as we have seen) $Z^2 = 1$, and also

$$Z = ijk = i(l^{-1}il)(l^{-2}il^2) = (il^{-1})^3 l^3 = (Zi^{-1}l^{-1})^3 Z = (li)^{-3} Z^4 = (li)^{-3},$$

whence $(li)^3 = Z^{-1} = Z$. Conversely, the relations

$$(7\cdot21) \qquad l^{-3} = i^2 = (li)^3$$

(with l^{-3}, not l^3!), along with the definitions

$$Z = i^2, \quad j = l^{-1}il, \quad k = l^{-2}il^2,$$

imply

$$i^2 = j^2 = k^2 = Z = (li)^{-3} Z^2 = (Z^{-1}il^{-1})^3 Z^3 l^3 = i(l^{-1}il)(l^{-2}il^2) = ijk$$

and $Z^2 = 1$. Thus the two relations (7·21) suffice to define the group generated by the quaternions l and i. The substitution

$$(7\cdot22) \qquad a = l^{-1} = \tfrac{1}{2}(1+i+j+k), \quad b = li = \tfrac{1}{2}(1+i-j+k), \quad c = i$$

yields

$$(7\cdot23) \qquad a^3 = b^3 = c^2 = abc,$$

which serves to identify this group of order 24 with the *binary tetrahedral group* $\langle 3,2,2\rangle$. Since $c = ab$, it has the alternative presentation

$$(7\cdot24) \qquad a^3 = b^3 = (ab)^2,$$

involving a and b symmetrically. It follows, as before, that the relations $a^3 = b^3 = c^2 = abc = Z$ or

$$(7\cdot25) \qquad a^3 = b^3 = (ab)^2 = Z$$

imply $Z^2 = 1$.

Since the subgroup $\langle 2,2,2\rangle$, of index 3, has cosets $\langle 2,2,2\rangle a^{\pm 1}$, where a is given by (7·22), the 24 elements of $\langle 3,3,2\rangle$ can be recognized as the 24 unities

$$(7\cdot26) \qquad \pm 1,\ \pm i,\ \pm j,\ \pm k,\quad \frac{\pm 1 \pm i \pm j \pm k}{2}$$

in Adolf Hurwitz's system of *integral* quaternions (Hardy and Wright 1960, p. 303). The corresponding points are the vertices

$$(7\cdot27) \qquad (\pm 1, 0, 0, 0) \ \text{permuted, and} \ (\pm\tfrac{1}{2}, \pm\tfrac{1}{2}, \pm\tfrac{1}{2}, \pm\tfrac{1}{2})$$

of the regular 24-cell $\{3,4,3\}$ (Coxeter 1963, pp. 149, 156; 1968, p. 241) whose three inscribed β_4's represent the cosets of the subgroup $\langle 2,2,2\rangle$. This subgroup, and the $\mathfrak{C}_2$ generated by Z (or -1), are the only normal subgroups. Clearly,

$$(7\cdot28) \qquad \langle 3,3,2\rangle/\langle 2,2,2\rangle \cong \mathfrak{C}_3 \quad \text{and} \quad \langle 3,3,2\rangle/\mathfrak{C}_2 \cong (3,3,2).$$

EXERCISES

1. The binary tetrahedral group ⟨3, 3, 2⟩ has the representation (by permutations of degree 8):

$$a = (1\ 2\ 3\ 1'\ 2'\ 3')\ (4\ 4'),$$
$$b = (3\ 4\ 2\ 3'\ 4'\ 2')\ (1\ 1'),$$
$$c = (3\ 1\ 3'\ 1')\ (2\ 4\ 2'\ 4'),$$

and the presentations

(i) $bab = a^2, \quad aba = b^2$;
(ii) $ml^2m = l, \quad lm^2l = m$;
(iii) $nm = ln = ml, \quad lmn = 1$;
(iv) $lmn = mkn = nkl = lkm = 1$;
(v) $NL = K, \quad KM = L, \quad LN = M, \quad MK = N$.

2. What group is defined by

$$mn = k, \quad nk = l, \quad kl = m, \quad lm = n?$$

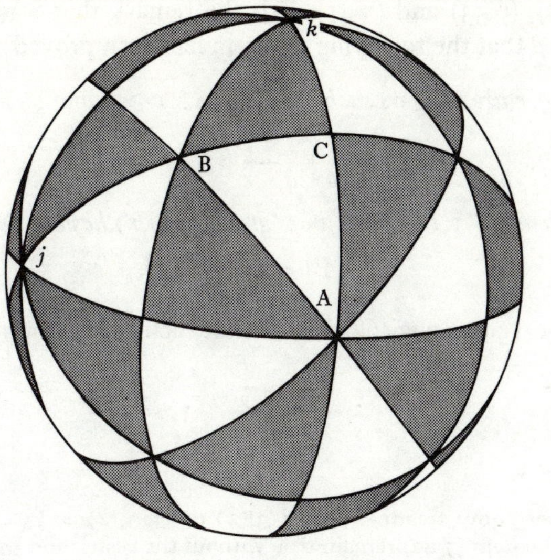

Figure 7·3A: (4, 3, 2) and ⟨4, 3, 2⟩

7·3 THE BINARY OCTAHEDRAL GROUP

Since the relations (7·24) involve a and b symmetrically, it is natural to seek a quaternion C that might interchange these generators of ⟨3, 3, 2⟩, transforming each into the other. The corresponding points are interchanged by a half-turn about the midpoint of the arc ab in Figure 7·2A, that is, about the point marked C in Figure 7·3A. Thus the desired quaternion is a numerical multiple of

$$(i+j+k)+(i-j+k) = 2(i+k),$$

namely, the unit quaternion

(7·31) $$\mathrm{C} = 2^{-\frac{1}{2}}(i+k) = i\exp(j\pi/4).$$

In fact, $\mathrm{C}^{-1}a\mathrm{C} = \frac{1}{2}(1+i-j+k) = b$. Since $\mathrm{C}^2 = -1$, we conclude that a group of quaternions, of order $2\times 24 = 48$, can be derived from ⟨3, 3, 2⟩ by adjoining C, whose square is Z and which transforms a into b. The relations

$$a^3 = b^3 = (ab)^2 = \mathrm{Z} = \mathrm{C}^2, \quad \mathrm{C}^{-1}a\mathrm{C} = b$$

imply (as we have seen) $\mathrm{Z}^2 = 1$, and also

$$(\mathrm{C}a^{-1})^4 = (a\mathrm{C}^{-1})^{-4} = (a\mathrm{C}^{-1}a\mathrm{C}\mathrm{Z}^{-1})^{-2} = (ab)^{-2}\mathrm{Z}^2 = \mathrm{Z}.$$

Conversely, the relations

(7·311) $$\mathrm{C}^2 = a^3 = (\mathrm{C}a^{-1})^4,$$

along with the definitions $\mathrm{Z} = \mathrm{C}^2$ and $b = \mathrm{C}^{-1}a\mathrm{C}$, imply

$$a^3 = b^3 = \mathrm{Z} = \mathrm{Z}^{-1+2} = (a\mathrm{C}^{-1})^4\mathrm{Z}^2 = (a\mathrm{C}^{-1}a\mathrm{C}\mathrm{Z}^{-1})^2\mathrm{Z}^2 = (a\mathrm{C}^{-1}a\mathrm{C})^2 = (ab)^2$$

and $\mathrm{Z}^2 = 1$. Thus the two relations (7·311) suffice to define the group generated by the quaternions C and a. The substitution

(7·32) $$\mathrm{A} = \mathrm{C}a^{-1} = 2^{-\frac{1}{2}}(1+i) = \exp(i\pi/4), \quad \mathrm{B} = a = \tfrac{1}{2}(1+i+j+k)$$

yields the presentation

(7·33) $$\mathrm{A}^4 = \mathrm{B}^3 = \mathrm{C}^2 = \mathrm{ABC},$$

which serves to identify this group of order 48 with the *binary octahedral group* ⟨4, 3, 2⟩. In terms of A and B alone, we have simply

(7·34) $$\mathrm{A}^4 = \mathrm{B}^3 = (\mathrm{AB})^2.$$

It follows that the relations $\mathrm{A}^4 = \mathrm{B}^3 = (\mathrm{AB})^2 = \mathrm{Z}$ imply $\mathrm{Z}^2 = 1$.

Since the subgroup ⟨3, 3, 2⟩, of index 2, has a coset ⟨3, 3, 2⟩C, where C is given by (7·31), the 48 elements of ⟨4, 3, 2⟩ are simply those of ⟨3, 3, 2⟩ along with

(7·35) $$\begin{matrix} 2^{-\frac{1}{2}}(\pm 1 \pm i), & 2^{-\frac{1}{2}}(\pm 1 \pm j), & 2^{-\frac{1}{2}}(\pm 1 \pm k), \\ 2^{-\frac{1}{2}}(\pm j \pm k), & 2^{-\frac{1}{2}}(\pm k \pm i), & 2^{-\frac{1}{2}}(\pm i \pm j). \end{matrix}$$

The corresponding points are the vertices of two reciprocal {3, 4, 3}'s: (7·27) and

(7·36) $$(\pm 2^{-\frac{1}{2}}, \pm 2^{-\frac{1}{2}}, 0, 0) \quad \text{permuted.}$$

(Coxeter 1963, p. 156 (8·72).)

In addition to the binary tetrahedral subgroup ⟨3, 3, 2⟩, for which

(7·37) $$\langle 4,3,2\rangle/\langle 3,3,2\rangle \cong \mathfrak{C}_2,$$

and the centre $\mathfrak{C}_2$ generated by Z, for which

(7·38) $$\langle 4,3,2\rangle/\mathfrak{C}_2 \cong (4,3,2),$$

there is a subgroup ⟨2, 2, 2⟩ which is normal because it is generated by $\mathrm{A}^2 = i$ and its conjugate $\mathrm{B}i\mathrm{B}^{-1} = aia^{-1} = l^{-1}il = j$. Combining the relations (7·34) with $\mathrm{A}^2 = 1$, we obtain the dihedral quotient group

$$\mathrm{A}^2 = \mathrm{B}^3 = (\mathrm{AB})^2 = 1;$$

thus

(7·39) $$\langle 4,3,2\rangle/\langle 2,2,2\rangle \cong (3,2,2).$$

EXERCISES

1. The binary octahedral group $\langle 4, 3, 2\rangle$ has the representation (by permutations of degree 16)

$$\begin{aligned} A &= (1\,2\,3\,4\,1'\,2'\,3'\,4')\,(5\,6\,7\,8\,5'\,6'\,7'\,8'),\\ B &= (5\,4\,2\,5'\,4'\,2')\,(8\,6\,1\,8'\,6'\,1')\,(3\,3')\,(7\,7'),\\ C &= (5\,1\,5'\,1')\,(3\,2\,3'\,2')\,(4\,8\,4'\,8')\,(7\,6\,7'\,6'), \end{aligned}$$

and the presentations

(i) $AA'A = A'AA', \quad A'A^2A' = A^2;$

(ii) $A'AA'' = A, \quad A''A'A = A', \quad AA''A' = A''.$

2. Do the relations $A'A^2A' = A^2$, $AA'^2A = A'^2$ define a finite group?

7·4 THE BINARY ICOSAHEDRAL GROUP

This section supplies the one remaining step that is needed to prove the following theorem:

(7·41) *If p and q are integers such that $p \geqslant q \geqslant 2$ and*

$$p^{-1}+q^{-1} > \tfrac{1}{2},$$

then the relations $A^p = B^q = (AB)^2 = Z$ *imply* $Z^2 = 1$.

The case $q = 2$ has already been proved in §7·1, the case $p = q = 3$ in §7·2, and the case $p = 4$, $q = 3$ in §7·3. Thus the only case still to be considered is the *binary icosahedral group* $\langle 5, 3, 2\rangle$, of order 120. Since the ordinary icosahedral group $(5, 3, 2)$ is simple, the indirect procedure of §7·2 and §7·3 is no longer available. Instead, we shall use a direct approach, as in the derivation of (7·13). The reader may like to pause here, and test his own skill before reading on! The recommended procedure is to express Z in various ways, until one of the expressions is recognized as being equal to its own inverse.

One way to accomplish this is to observe that the given relations imply

$$A = B^{-1}A^{-1}B^2 \quad \text{and} \quad B = A^4B^{-1}A^{-1},$$

whence

$$\begin{aligned} Z = A^5 = BA^5B^{-1} &= (B^{-1}A^{-1}B^2)^5\,B^{-1} = (A^{-1}B)^5 = (A^3B^{-1}A^{-1})^5 = (A^2B^{-1})^5\\ &= \{(B^{-1}A^{-1}B^2)^2\,B^{-1}\}^5 = (B^{-1}A^{-1}BA^{-1}B)^5 = (A^{-1}BA^{-1})^5\\ &= (BA^{-2})^5 = (A^2B^{-1})^{-5}. \end{aligned}$$

Having seen (on the line two above the last one) that $Z = (A^2B^{-1})^5$, we conclude that $Z = Z^{-1}$, as desired.

Combining (6·53) and (7·41) with the remark that was made about (6·54), we see that the following theorem has been proved:

(7·42) *If p, q, r are integers such that $p \geqslant q \geqslant r \geqslant 1$ and*

$$\frac{1}{p}+\frac{1}{q}+\frac{1}{r} > 1$$

and $p = q$ when $r = 1$, there is a finite group $\langle p, q, r\rangle$ having the presentation

(7·421) $$A^p = B^q = C^r = ABC.$$

As we observed on page 68, the order of each finite group $\langle p, q, r\rangle$ is $4s$, where

$$\frac{1}{s} = \frac{1}{p}+\frac{1}{q}+\frac{1}{r}-1.$$

EXERCISES

1. Adapt the above treatment of $\langle 5, 3, 2\rangle$ to $\langle 3, 3, 2\rangle$ and $\langle 4, 3, 2\rangle$.
2. Does Theorem (7·42) remain true without the restriction that $p = q$ when $r = 1$?
3. Does Theorem (7·42) remain true when the sign $>$ in

$$\frac{1}{p}+\frac{1}{q}+\frac{1}{r} > 1$$

is replaced by $=$?

4. The binary icosahedral group $\langle 5, 3, 2\rangle$ has the presentation

$$cb^{-1}c^{-1}b = a, \quad ac^{-1}a^{-1}c = b, \quad ba^{-1}b^{-1}a = c$$

(R. G. de Buda).

7·5 FINITE GROUPS GENERATED BY PURE QUATERNIONS

In §3·4 we saw that, apart from the cyclic groups $\mathfrak{C}_p$ with $p \neq 2$ and the tetrahedral group $(3, 3, 2)$, every finite rotation group is generated by its half-turns. It follows that, apart from the cyclic groups $\mathfrak{C}_p$ with $p \neq 4$ and the binary tetrahedral group $\langle 3, 3, 2\rangle$, every finite group of quaternions is generated by its *pure* quaternions. The only pure quaternions in $\langle 3, 3, 2\rangle$ are $\pm i$, $\pm j$, $\pm k$, which generate the subgroup $\langle 2, 2, 2\rangle$. For the rest, the details are as follows.

The two half-turns that generate the dihedral group $(p, 2, 2)$ may be represented by unit pure quaternions

$$U_1 = j, \quad U_2 = j\cos\frac{\pi}{p}+k\sin\frac{\pi}{p}$$

($= -jA$ (in the notation of (7·121)), which generate the dicyclic group $\langle p, 2, 2\rangle$ and satisfy the relations

(7·51) $$U_1{}^2 = U_2{}^2 = (U_1U_2)^p = -1.$$

Similarly, in the notation of (6·63) and Figure 6·6A, the three great circles *RP*, *PQ*, *QR* that bound the triangle $(p\,q\,2)$ have poles represented by the three unit pure quaternions

$$U_1 = j, \quad U_2 = -jA = -Bi = -i\cos\frac{\pi}{q} - j\cos\frac{\pi}{p} + k\sin\frac{\pi}{h}, \quad U_3 = i,$$

which satisfy

$$(7\cdot52) \qquad U_1U_2 = A, \quad U_2U_3 = B, \quad U_3U_1 = k = C$$

and

$$U_1U_2U_3 = Ai = jB = -\sin\frac{\pi}{h} + i\cos\frac{\pi}{p} + j\cos\frac{\pi}{q}$$

$$= -\sin\frac{\pi}{h} + J\cos\frac{\pi}{h},$$

for some unit pure quaternion J.

When $[p, q]$ is $[4, 3]$ or $[5, 3]$, h is even but $\frac{1}{2}h$ is odd, so that

$$(U_1U_2U_3)^{\frac{1}{2}h} = \cos\frac{h}{2}\left(\frac{\pi}{2} + \frac{\pi}{h}\right) = -\sin\frac{h\pi}{4} = (-1)^p.$$

Replacing all the quaternions U_ν by $-U_\nu$ if $p = 4$ (but letting them stand if $p = 5$), we deduce the presentation

$$(7\cdot53) \qquad U_\nu^2 = (U_1U_2)^p = (U_2U_3)^3 = (U_3U_1)^2 = (U_1U_2U_3)^{\frac{1}{2}h}$$

for $\langle p, 3, 2\rangle$, where $p = 4$ or 5 (and $\frac{1}{2}h = 3$ or 5, respectively). Since the products (7·52) satisfy

$$A^p = B^3 = C^2 = ABC,$$

they generate the whole group (and the U_ν must be expressible in terms of them).

When $p = 5$, the above expressions for the U_ν as quaternions become

$$(7\cdot54) \qquad U_1 = j, \quad U_2 = -\tfrac{1}{2}(i + \tau j - \tau^{-1}k), \quad U_3 = i.$$

But the corresponding expressions when $p = 4$ are rather clumsy; it is preferable to use

$$(7\cdot55) \qquad U_1 = j, \quad U_2 = 2^{-\frac{1}{2}}(k-j), \quad U_3 = 2^{-\frac{1}{2}}(i-k),$$

which are poles of the great circles that bound the left half of the triangle shaded in Figure 3·4A or 7·5A.

When $[p, q]$ is $[p, 2]$, we have

$$(U_1U_2)^{\frac{1}{2}p} = A^{\frac{1}{2}p} = \left(\cos\frac{\pi}{p} + i\sin\frac{\pi}{p}\right)^{\frac{1}{2}p} = i,$$

so that the quaternions $U_1 = j$, $U_2 = -j\exp(\pi i/p)$, $U_3 = i$ generate $\langle p, 2, 2\rangle$ (p even) in the form

$$(7\cdot56) \qquad U_\nu^2 = (U_1U_2)^p = (U_2U_3)^2 = (U_3U_1)^2 = (U_1U_2)^{\frac{1}{2}p}U_3,$$

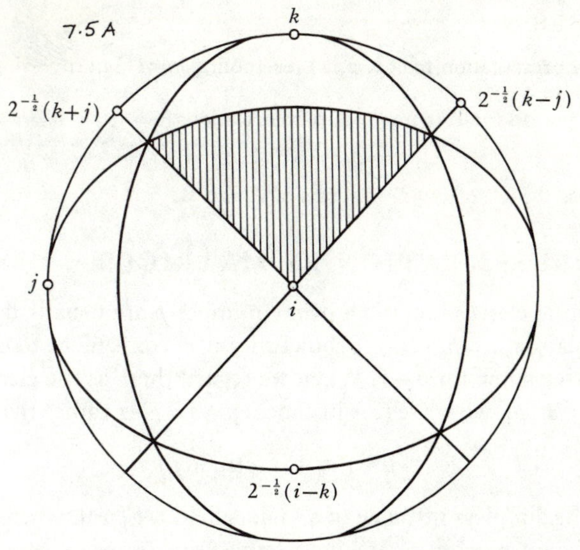

Figure 7·5A: Pure quaternions that generate $\langle 4, 3, 2\rangle$

analogous to (3·49). When p is odd, the same three quaternions generate $\langle 2p, 2, 2\rangle$ in the form

$$(7\cdot57) \qquad U_\nu^2 = (U_1U_2)^p = (U_2U_3)^2 = (U_3U_1)^2,$$

analogous to (3·42). Taking these last two results together, we notice that we have found two distinct sets of three pure quaternions which generate $\langle 2p, 2, 2\rangle$ (p odd):

$$j, \; -j\exp(\pi i/2p), \; i \quad \text{and} \quad j, \; -j\exp(\pi i/p), \; i.$$

It is only in the former set that the generator i is redundant.

Looking again at Figure 7·5A, we see that the three quaternions

$$(7\cdot58) \qquad U_1 = 2^{-\frac{1}{2}}(k-j), \quad U_2 = 2^{-\frac{1}{2}}(i-k), \quad U_3 = 2^{-\frac{1}{2}}(k+j)$$

generate $\langle 4, 3, 2\rangle$ and satisfy

$$(7\cdot59) \qquad U_\nu^2 = (U_1U_2)^3 = (U_2U_3)^3 = (U_3U_1)^2.$$

The products $a = U_1U_2$, $b = U_2U_3$, $c = U_3U_1$ satisfy (7·23) (see Figure 7·2A) and thus generate the subgroup $\langle 3, 3, 2\rangle$. It is interesting to compare our two presentations for $\langle 4, 3, 2\rangle$:

$$U_\nu^2 = (U_1U_2)^4 = (U_2U_3)^3 = (U_3U_1)^2 = (U_1U_2U_3)^3$$

(see (7·53)) with generators (7·55), and (7·59) with generators (7·58).

EXERCISES

1. Give a presentation for $\langle p, 2, 2\rangle$ resembling (7·51) but involving only two relations.
2. What period for Z is implied by the relations
$$U_1^2 = U_2^2 = (U_1U_2)^p = Z?$$
3. Express the U_ν of (7·53) in terms of A and B.

7·6 REPRESENTATION BY MATRICES

The p residue classes modulo a prime number p are usually denoted by the symbols 0, 1, ..., $p-1$, but sometimes it is convenient to use -1 as an alternative name for $p-1$. When we regard them as the elements of a finite *field* $GF[p]$, we write (for instance) $p-1 = -1$ rather than

$$p-1 \equiv -1 \pmod{p}.$$

$GF[p]$ is the simplest instance of a Galois field $GF[p^m]$, where m can be any positive integer but p remains a *prime* number.

If a, b, c, d are elements of $GF[p]$ satisfying $ad-bc \neq 0$, the matrices

$$\begin{bmatrix} a & c \\ b & d \end{bmatrix}$$

form the *general linear homogeneous group* $GL(2,p)$, of order

$$(p^2-1)(p^2-p) = p(p+1)(p-1)^2$$

(Coxeter and Moser 1972, p. 92). For any *odd* prime p, the condition $ad-bc \neq 0$ can be replaced by $ad-bc = \pm 1$ to yield a normal subgroup of index $\frac{1}{2}(p-1)$: the *unimodular group*, of order

$$2p(p^2-1) = 2(p-1)p(p+1).$$

(If $p = 3$, it coincides with $GL(2,p)$.) This, in turn, has a subgroup of index 2 consisting of the matrices for which $ad-bc = 1$: the *special linear homogeneous group* $SL(2,p)$, of order

$$p(p^2-1) = (p-1)p(p+1).$$

In particular, the orders of $SL(2,3)$ and $SL(2,5)$ are 24 and 120. The isomorphism

(7·61) $$SL(2,p) \cong \langle p, 3, 2\rangle \quad (p = 3 \text{ or } 5),$$

can be established by finding matrices A, B, C that satisfy the relations

(7·62) $$A^p = B^3 = C^2 = ABC \neq \begin{bmatrix} 1 & 0 \\ 0 & 1 \end{bmatrix}.$$

In fact, for *any* odd prime p, the matrices

(7·63) $$A = \begin{bmatrix} -1 & 0 \\ -1 & -1 \end{bmatrix}, \quad B = \begin{bmatrix} 0 & -1 \\ 1 & 1 \end{bmatrix}, \quad C = \begin{bmatrix} 0 & 1 \\ -1 & 0 \end{bmatrix} \pmod{p}$$

generate $SL(2,p)$ and satisfy

(7·64) $$A^p = B^3 = C^2 = ABC = \begin{bmatrix} -1 & 0 \\ 0 & -1 \end{bmatrix}.$$

Thus *$SL(2,p)$ is a factor group of $\langle p,3,2\rangle$*, and the agreement of order when $p = 3$ or 5 shows that, in these first two cases, the homomorphism is an isomorphism. However, this can never happen again, for, by Theorem (3·33), $\langle p,3,2\rangle$ has an infinite factor group whenever $p > 5$. In other words, if $p > 5$, a presentation for $SL(2,p)$ based on

$$A^p = B^3 = C^2 = ABC$$

or

(7·65) $$A^p = B^3 = (AB)^2$$

must involve at least one further relation.

J. G. Sunday[†] established the presentation

$$S^p = T^2 = (ST)^3 = (S^{(p+1)/2}TS^4T)^2$$

for $SL(2,p)$, where p is any odd prime. The same procedure serves to establish the closely related presentation

(7·66) $$A^p = B^3 = (AB)^2 = (A^{(p+1)/2}BA^{-3}B)^2.$$

The essential step, suggested by J. Schur,[‡] is the observation that these relations, along with $S \rightleftarrows T$ (or $A \rightleftarrows B$), imply $T = 1$ (or $AB = 1$), so that T (or AB) is a commutator.

Sunday observed, further, that the central quotient group $PSL(2,p)$ (the *linear fractional group* mod p, which is $(p,3,2)$ when $p = 3$ or 5) has likewise a three-generator presentation:

$$S^p = 1, \quad T^2 = (ST)^3, \quad (S^{(p+1)/2}TS^4T)^2 = 1.$$

Setting $S = A$, $T = (AB)^{-1}$, we may express this as

(7·67) $$A^p = 1, \quad B^3 = (AB)^2, \quad (A^{(p+1)/2}BA^{-3}B)^2 = 1.$$

EXERCISES

1. Referring to (7·66), deduce $(A^{(p+1)/2}BA^{-3}B)^2 = Z$ from
$$A^p = B^3 = (AB)^2 = Z$$
when $p = 3$ or 5.
2. What is the group $A^p = 1$, $B^3 = (AB)^2$ when $p = 3$ or 5? (Compare (7·67).)

[†] Presentations of the groups $SL(2, m)$ and $PSL(2, m)$, *Canadian Journal of Mathematics* **24** (1972), 1129–31.

[‡] Untersuchungen über die Darstellung der endlichen Gruppen durch gebrochene lineare Substitutionen, *Journal für die reine und angewandte Mathematik* **122** (1907), 85–137; see especially p. 96.

3. Use Sunday's trick to establish

$$A^5 = B^3 = (AB)^2$$

as a presentation for $SL(2, 5)$. (Compare §7·4.)

4. For any odd $m > 1$, the group of 2×2 matrices of determinant 1 over the ring of residues modulo m is generated by

$$S = \begin{bmatrix} 1 & 0 \\ 1 & 1 \end{bmatrix}, \quad T = \begin{bmatrix} 0 & 1 \\ -1 & 0 \end{bmatrix} \pmod{m}.$$

Find the order of this group $SL(2, m)$ when m has the complete factorization Πp^e. (Gunning 1962, p. 9.)

7·7 THE UNIMODULAR GROUP

From the presentation (7·66) for $SL(2,p)$, we can derive the unimodular group by adjoining the further generator

$$R = \begin{bmatrix} -1 & 0 \\ 1 & 1 \end{bmatrix},$$

of period 2 (and determinant -1), which transforms each of

$$A = \begin{bmatrix} -1 & 0 \\ -1 & -1 \end{bmatrix}, \quad B = \begin{bmatrix} 0 & -1 \\ 1 & 1 \end{bmatrix}$$

into its inverse. Thus the unimodular group, of order $2p(p^2-1)$, is given by (7·66) along with $R^2 = 1$, $RAR = A^{-1}$, $RBR = B^{-1}$.

Defining

$$R_1 = AR = \begin{bmatrix} 1 & 0 \\ 0 & -1 \end{bmatrix}, \quad R_2 = R = \begin{bmatrix} -1 & 0 \\ 1 & 1 \end{bmatrix}, \quad R_3 = RB = \begin{bmatrix} 0 & 1 \\ 1 & 0 \end{bmatrix},$$

so that $A = R_1R_2$ and $B = R_2R_3$, we deduce the presentation

(7·71) $$R_\nu^2 = 1, \quad (R_1R_2)^p = (R_2R_3)^3 = (R_1R_3)^2 = \{(R_1R_2)^{(p-1)/2}R_3(R_1R_2)^4R_3\}^2.$$

Of course, the last relation is superfluous when $p < 6$. Thus the unimodular group modulo 3, which coincides with $GL(2, 3)$, and the unimodular group modulo 5, of order 240, are given by

(7·72) $$R_\nu^2 = 1, \quad (R_1R_2)^p = (R_2R_3)^3 = (R_1R_3)^2 \quad (p = 3 \text{ or } 5).$$

This compares interestingly with (7·53) and (7·59).

In the case when $p = 3$, it is remarkable that the presentation of the unimodular group can be reduced to two generators and two relations:

(7·73) $$a^3 = c^4, \quad (a^{-1}c)^2 = 1.$$

This can be proved by observing that, when $p = 3$, the special linear group has the symmetrical presentation (7·24), making it natural to represent the generators as

$$a = \begin{bmatrix} -1 & 0 \\ -1 & -1 \end{bmatrix}, \quad b = \begin{bmatrix} -1 & -1 \\ 0 & -1 \end{bmatrix}.$$

Since these are transformed into each other by the matrix

$$R_3 = \begin{bmatrix} 0 & 1 \\ 1 & 0 \end{bmatrix}$$

of determinant -1, the unimodular group $GL(2,3)$ is generated by a (or b) and R_3. The relations

$$a^3 = b^3 = (ab)^2 \quad \text{and} \quad R_3^2 = 1, \quad R_3aR_3 = b$$

combine to yield $$a^3 = (aR_3)^4, \quad R_3^2 = 1,$$

and we obtain (7·73) by defining

$$c = aR_3 = \begin{bmatrix} 0 & -1 \\ -1 & -1 \end{bmatrix}.$$

This presentation shows that

$$GL(2,3) \cong \langle -3, 4 \mid 2 \rangle$$

in the notation of Coxeter and Moser (1972, pp. 75, 116, 140). In other words, $GL(2,3)$ is the largest finite group of automorphisms for a Riemann surface of genus 2, namely the Riemann surface for the curve

$$y^2 = x^5 - x.$$

EXERCISE

By (7·72), $GL(2, 3)$ has the presentation

$$R_\nu^2 = 1, \quad (R_1R_2)^3 = (R_2R_3)^3 = (R_1R_3)^2.$$

How can this be reconciled with (7·73)?

7·8 A REPRESENTATION USING RESIDUES MODULO $h+1$

The matrix representation (7·63) for $\langle p, 3, 2\rangle$ with $p = 3$ or 5 makes us wonder whether anything like this can be done for the binary *octahedral* group $\langle 4, 3, 2\rangle$. The simple group $PSL(2,7)$, of order 168, is known to have octahedral subgroups (Burnside 1911, p. 216). Accordingly, we may expect $SL(2,7)$, of order 336, to have *binary* octahedral subgroups. In fact, the relations

$$A^4 = B^3 = C^2 = ABC = \begin{bmatrix} -1 & 0 \\ 0 & -1 \end{bmatrix}$$

are satisfied by the following matrices over $GF[7]$:

(7·81) $$A = \begin{bmatrix} 1 & 1 \\ 2 & 3 \end{bmatrix}, \quad B = \begin{bmatrix} -1 & -3 \\ 1 & 2 \end{bmatrix}, \quad C = \begin{bmatrix} 0 & -1 \\ 1 & 0 \end{bmatrix}.$$

Thus $\langle 4, 3, 2\rangle$ *is a subgroup of index 7 in* $SL(2,7)$. It is interesting to observe that the equations

$$B^3 = C^2 = ABC = \begin{bmatrix} -1 & 0 \\ 0 & -1 \end{bmatrix}$$

remain valid when we take the elements of the matrices (7·81) to be ordinary integers instead of residues modulo 7. Accordingly, we may expect to find representations for other groups $\langle p, 3, 2\rangle$ by using different moduli. Since the powers of A include

$$A^2 = \begin{bmatrix} 3 & 4 \\ 8 & 11 \end{bmatrix}, \quad A^3 = \begin{bmatrix} 11 & 15 \\ 30 & 41 \end{bmatrix},$$

$$A^4 = \begin{bmatrix} 41 & 56 \\ 112 & 153 \end{bmatrix}, \quad A^5 = \begin{bmatrix} 153 & 209 \\ 418 & 571 \end{bmatrix},$$

the result of working modulo 3, 4, 7 or 11 is to obtain

$$\langle 3, 3, 2\rangle, \quad \langle 2, 3, 2\rangle, \quad \langle 4, 3, 2\rangle \quad \text{or} \quad \langle 5, 3, 2\rangle.$$

The second of these is the dicyclic group $\langle 3, 2, 2\rangle$ of order 12. (Of course, the residues modulo 4 do *not* form a *field*.)

Working modulo 5, we find $A^3 = 1$, making it less obvious that this is another representation for the binary tetrahedral group. However, the substitution

$$a = -A, \quad b = B, \quad c = -C = C^{-1}$$

yields $a^3 = b^3 = c^2 = abc$, allowing us to combine four of these five results into the following theorem:

(7·82) *The binary polyhedral group* $\langle p, 3, 2\rangle$ $(p = 2, 3, 4, 5;\ h = 3, 4, 6, 10)$ *is generated by*

$$\begin{bmatrix} -1 & -3 \\ 1 & 2 \end{bmatrix} \quad \textit{and} \quad \begin{bmatrix} 0 & -1 \\ 1 & 0 \end{bmatrix} \quad (\text{mod } h+1).$$

Since these are matrices of determinant 1, it follows that $\langle 5, 3, 2\rangle$, which is $SL(2, 5)$, is a subgroup of index 11 in $SL(2, 11)$.

EXERCISES

1. When $p = 3$, 4, or 5, the matrices B and C of Theorem (7·82) satisfy

$$(B^{-1}C^{-1}BC)^{h/2} = \begin{bmatrix} -1 & 0 \\ 0 & -1 \end{bmatrix}.$$

2. What groups are generated by the matrices (7·81)
   (i) mod 2, (ii) mod 13?

7·9 REMARKS

As an excuse for beginning this chapter with binary polyhedral groups and ending it with special linear groups, it is appropriate to remark that these two families of groups share an interesting property: each group contains exactly one subgroup of order 2. In fact, since -1 is the only quaternion of period 2, every finite group $\langle p, q, r\rangle$ has this property; and since

$$\begin{bmatrix} -1 & 0 \\ 0 & -1 \end{bmatrix}$$

is the only two-by-two matrix of determinant 1 and period 2, every group $SL(2, p^m)$ (p an odd prime) has this same property. So also does the direct product $\mathfrak{C}_2 \times \mathfrak{G}$ where $\mathfrak{G}$ is any group of odd order. Whether any other finite groups have it seems to be an open question. Burnside (1911, p. 132) proved that the only groups of order 2^m having exactly one subgroup of order 2 are the cyclic group $\langle 2^{m-1}, 2^{m-1}, 1\rangle$ and the dicyclic group $\langle 2^{m-2}, 2, 2\rangle$. He remarked that the latter with $m = 3$ is the quaternion group. Accordingly, many authors call $\langle 2^{m-2}, 2, 2\rangle$ the *generalized quaternion* group, although the name *dicyclic* group for $\langle p, 2, 2\rangle$ was coined by G. A. Miller (see Miller, Blichfeldt and Dickson 1916, p. 62).

The term *binary* groups was used as long ago as 1905 by L. E. Dickson.[†] He denoted the groups

$$\langle k, 2, 2\rangle, \quad \langle 3, 3, 2\rangle, \quad \langle 4, 3, 2\rangle, \quad \langle 5, 3, 2\rangle$$

by $\mathfrak{D}_{4k}$, $\mathfrak{T}_{24}$, $\mathfrak{O}_{48}$, $\mathfrak{J}_{120}$, gave complicated presentations for them, and represented the last three by matrices over $GF[5]$, $GF[7]$, $GF[11]$, respectively. Thus, he very nearly anticipated our Theorem (7·82).

In 1916 Steinitz[‡] noticed that the elements of $\langle 5, 3, 2\rangle$, $\langle 3, 3, 2\rangle$ and $\langle 4, 3, 2\rangle$, regarded as points in Euclidean 4-space, are the vertices of certain polytopes, namely $\{3, 3, 5\}$, $\{3, 4, 3\}$, and two reciprocal $\{3, 4, 3\}$'s.

The elegant presentation (7·42) was discovered by William Threlfall[§] in the course of his topological research. He encouraged me to look for a direct algebraic proof that these relations suffice.

At about the same time, Hans Zassenhaus[‖] investigated some algebras which he called *near-fields*. Three of them (his I, III, V) have multiplicative groups of orders 24, 48, 120. These groups are, in fact,

$$\langle 3, 3, 2\rangle, \quad \langle 4, 3, 2\rangle, \quad \langle 5, 3, 2\rangle.$$

The generators that he gives for them are easily seen to agree with Theorem (7·82), but his presentations are more complicated, and instead of $SL(2, p)$ he writes $\mathfrak{M}(2, p)$.

† On finite algebras, *Nachrichten der Königlichen Gesellschaft der Wissenschaften in Göttingen* (*Math.-Phys. Klasse*) 1905, 358–93; especially, p. 364.

‡ Ernst Steinitz, *Polyeder und Raumeinteilungen*, *Encyklopädie der Mathematischen Wissenschaften*, III 1,B (Leipzig, 1914–31), pp. 125–6.

§ See the reference cited on page 74. Miller, Blichfeldt and Dickson (1916, p. 188) presented the binary icosahedral group in the form

$$s_1^3 = s_2^5 = (s_1 s_2)^2, \quad s_1^6 = 1$$

without realizing that the last relation is superfluous.

‖ Über endliche Fastkörper, *Abhandlungen aus dem Mathematischen Seminar der Hansischen Universität* 11 (1936), 187–220; especially p. 217.

CHAPTER 8

Unitary space

> In this way we are able to create an infinite set of new universes, the laws of which are within our reach, though we can never set foot in them. László Fejes Tóth (1964, p. 125)

In §4·4 and §4·8 we saw how four real coordinates can be paired to form two complex coordinates, so that the real 4-space is reduced to the unitary plane. In the present chapter this idea is extended to unitary *n*-space and to 'oblique' coordinates, the coordinate axes being normal to the mirrors of a complex kaleidoscope (§8·8).

The words of Chesterton, at the beginning of chapter 3, suggest that a reflection should be of period 2. In complex geometry there is no need for such a restriction: for instance, instead of the real reflection $x' = -x$, of period 2, we can have a complex reflection $x' = \exp(2\pi i/p)\,x$, of period p.

8·1 AFFINE COORDINATES

In any affine *n*-space, as in the real affine plane (Coxeter 1969, p. 203), a projective collineation of the whole space onto itself is called an *affinity*. A *translation* may be defined as an affinity that transforms each line into a parallel line while leaving no point invariant (or else leaving every point invariant). Thus the group of all affinities has an Abelian subgroup consisting of all the translations. By regarding this subgroup as an additive group, we may identify the translations with vectors. For a fixed point O, the *origin*, and an arbitrary point P, the translation that takes O to P is called the *vector* $\overrightarrow{OP}$. Such vectors can be added and subtracted in the usual way, and can be multiplied by the elements of a given field. Two non-zero vectors $\mathbf{x}$ and $\mathbf{y}$ are said to be *parallel* (or 'dependent') if they act along parallel lines, that is, if $\mathbf{x} = c\mathbf{y}$ for some non-zero element c of the field.

Any n independent vectors $\mathbf{e}_1, \mathbf{e}_2, \ldots, \mathbf{e}_n$ can be used as a basis for *affine coordinates*, the point $(x_1, x_2, \ldots, x_n)$ being derived from the origin by applying the vector

(8·11) $$\mathbf{x} = x_1\mathbf{e}_1 + x_2\mathbf{e}_2 + \ldots + x_n\mathbf{e}_n.$$

For instance, the vector $\mathbf{e}_1$ takes (0, 0, ..., 0) to (1, 0, ..., 0). Any N points, given by vectors $\mathbf{x}, \mathbf{y}, \ldots$, have a *centroid* whose vector is

$$N^{-1}(\mathbf{x}+\mathbf{y}+\ldots).$$

Any *finite* group of affinities leaves invariant at least one point, namely the centroid of all the transforms of an arbitrary point. (Compare Coxeter 1969, pp. 214, 224). By choosing such an invariant point to be the origin for our system of affine coordinates, we may express the affinities as non-singular homogeneous linear transformations

(8·12) $$x_\nu' = \Sigma b_{\mu\nu} x_\mu \quad \text{(summed over } \mu)$$

or, in matrix notation,

(8·13) $$\mathbf{x}' = \mathbf{x}\mathrm{B},$$

where $\mathbf{x}$ stands for the 'one row' matrix $[x_1\ x_2\ \ldots\ x_n]$ and

$$\mathrm{B} = \begin{bmatrix} b_{11} & b_{12} & \cdots & b_{1n} \\ \cdots & \cdots & \cdots & \cdots \\ b_{n1} & b_{n2} & \cdots & b_{nn} \end{bmatrix}$$

(Birkhoff and MacLane 1965, p. 193). Since the affinity is a transformation of the whole space onto itself, the determinant of B is not zero.

EXERCISE

Write down coordinates for the centroid of three points

$$(x_1, \ldots, x_n), \quad (y_1, \ldots, y_n), \quad (z_1, \ldots, z_n).$$

8·2 HERMITIAN FORMS

Except in §4·4 and §4·8, our geometry has usually been real: the field of coordinates has consisted of the real numbers. To make things more exciting, let us now use the field of complex numbers and consider groups of complex affinities.

A bilinear form

(8·21) $$\Sigma\Sigma a_{\lambda\mu} x_\lambda \bar{x}_\mu,$$

in n complex variables $(x_1, \ldots, x_n)$ and their complex conjugates $(\bar{x}_1, \ldots, \bar{x}_n)$, is said to be *Hermitian* if

(8·22) $$a_{\mu\lambda} = \bar{a}_{\lambda\mu},$$

that is, if every 'diagonal' term $a_{\lambda\lambda} x_\lambda \bar{x}_\lambda$ is real while each 'mixed' term $a_{\lambda\mu} x_\lambda \bar{x}_\mu$ $(\lambda \neq \mu)$ is balanced by its complex conjugate

$$a_{\mu\lambda} x_\mu \bar{x}_\lambda = \bar{a}_{\lambda\mu} \bar{x}_\lambda x_\mu,$$

with the result that, whatever (complex) values are assigned for the x_λ, the value of whole expression is real (Burnside 1911, p. 253). The coefficients constitute a Hermitian matrix

$$\text{(8·23)} \qquad \mathrm{A} = \|a_{\lambda\mu}\|, \quad a_{\mu\lambda} = \bar{a}_{\lambda\mu},$$

in terms of which the form can be expressed as

$$\text{(8·24)} \qquad \mathbf{x}\mathrm{A}\bar{\mathbf{x}}^{\mathrm{T}},$$

where $\bar{\mathbf{x}}^{\mathrm{T}}$ is the 'one column' matrix with entries $\bar{x}_1, \ldots, \bar{x}_n$.

A Hermitian form is said to be *positive definite* if it is positive for all values of the variables except $(0, \ldots, 0)$, and to be *positive semidefinite* if it is never negative but is zero for some values not all zero. It is said to be *indefinite* if it is positive for some choices of the x_μ and negative for others. Thus, when $n = 2$, $x_1\bar{x}_1 + x_2\bar{x}_2$ is definite, $x_1\bar{x}_1$ and

$$(x_1 + x_2)(\bar{x}_1 + \bar{x}_2)$$

are semidefinite, and $x_1\bar{x}_1 - x_2\bar{x}_2$ is indefinite. Clearly, the sum of any finite number of Hermitian forms (in the same variables) is Hermitian, and if each is positive definite, the sum is positive definite.

For instance, any finite group of linear transformations leaves invariant at least one positive definite Hermitian form, namely the sum of all the transforms of the *primitive* form

$$\text{(8·25)} \qquad x_1\bar{x}_1 + x_2\bar{x}_2 + \ldots + x_n\bar{x}_n$$

(Burnside 1911, p. 256).

EXERCISES

Which of the following forms are definite, semidefinite or indefinite?

1. $x_1\bar{x}_1 - \sqrt{2}(x_1\bar{x}_2 + x_2\bar{x}_1) + x_2\bar{x}_2$.
2. $x_1\bar{x}_1 - \sqrt{\tfrac{1}{3}}(x_1\bar{x}_2 + x_2\bar{x}_1) + x_2\bar{x}_2 - \sqrt{\tfrac{1}{3}}(x_2\bar{x}_3 + x_3\bar{x}_2) + x_3\bar{x}_3$.
3. $x_1\bar{x}_1 - \sqrt{\tfrac{1}{3}}(x_1\bar{x}_2 + x_2\bar{x}_1) + x_2\bar{x}_2 - \sqrt{\tfrac{1}{3}}(x_2\bar{x}_3 + x_3\bar{x}_2) + x_3\bar{x}_3 - \sqrt{\tfrac{1}{3}}(x_3\bar{x}_4 + x_4\bar{x}_3) + x_4\bar{x}_4$.
4. $x_1\bar{x}_1 - \sqrt{\tfrac{1}{3}}(x_1\bar{x}_2 + x_2\bar{x}_1) + x_2\bar{x}_2 - \sqrt{\tfrac{1}{3}}(x_2\bar{x}_3 + x_3\bar{x}_2) + x_3\bar{x}_3 - \sqrt{\tfrac{1}{3}}(x_3\bar{x}_4 + x_4\bar{x}_3) + x_4\bar{x}_4 - \sqrt{\tfrac{1}{3}}(x_4\bar{x}_5 + x_5\bar{x}_4) + x_5\bar{x}_5$.

8·3 INNER PRODUCTS

Given a Hermitian matrix A, any two vectors $\mathbf{x}$ and $\mathbf{y}$ have an *inner product*

$$\text{(8·31)} \qquad \mathbf{x}\cdot\mathbf{y} = \mathbf{x}\mathrm{A}\bar{\mathbf{y}}^{\mathrm{T}} = \Sigma\Sigma a_{\lambda\mu} x_\lambda \bar{y}_\mu$$

such that

$$\text{(8·32)} \qquad (\mathbf{w}+\mathbf{x})\cdot\mathbf{y} = \mathbf{w}\cdot\mathbf{y} + \mathbf{x}\cdot\mathbf{y}, \quad (c\mathbf{x})\cdot\mathbf{y} = c(\mathbf{x}\cdot\mathbf{y}), \quad \mathbf{y}\cdot\mathbf{x} = \overline{\mathbf{x}\cdot\mathbf{y}}$$

and consequently $\mathbf{y}\cdot(\mathbf{w}+\mathbf{x}) = \mathbf{y}\cdot\mathbf{w} + \mathbf{y}\cdot\mathbf{x}$, $\mathbf{x}\cdot(c\mathbf{y}) = \bar{c}(\mathbf{x}\cdot\mathbf{y})$. Instead of Gibbs's $\mathbf{x}\cdot\mathbf{y}$, some authors prefer the notation (x, y) or $\langle x|y\rangle$.

Clearly, the Hermitian form (8·21) is a special inner product, namely $\mathbf{x}\cdot\mathbf{x}$, which we naturally abbreviate to $\mathbf{x}^2$. Conversely, the general inner product $\mathbf{x}\cdot\mathbf{y}$ is expressible in terms of the Hermitian form as follows:

$$\begin{aligned} 4\mathbf{x}\cdot\mathbf{y} &= (\mathbf{x}^2 + \mathbf{x}\cdot\mathbf{y} + \mathbf{y}\cdot\mathbf{x} + \mathbf{y}^2) + i(\mathbf{x}^2 - i\mathbf{x}\cdot\mathbf{y} + i\mathbf{y}\cdot\mathbf{x} + \mathbf{y}^2) \\ &\quad - (\mathbf{x}^2 - \mathbf{x}\cdot\mathbf{y} - \mathbf{y}\cdot\mathbf{x} + \mathbf{y}^2) - i(\mathbf{x}^2 + i\mathbf{x}\cdot\mathbf{y} - i\mathbf{y}\cdot\mathbf{x} + \mathbf{y}^2) \\ \text{(8·33)} \qquad &= (\mathbf{x}+\mathbf{y})^2 + i(\mathbf{x}+i\mathbf{y})^2 - (\mathbf{x}-\mathbf{y})^2 - i(\mathbf{x}-i\mathbf{y})^2 \end{aligned}$$

(Smirnoff 1964, p. 383).

For geometric reasons we shall be interested in those linear transformations (8·12) or (8·13) which preserve the Hermitian form (8·21) or (8·24), so that

$$\Sigma\Sigma a_{\lambda\mu} x_\lambda \bar{x}_\mu = \Sigma\Sigma a_{\kappa\nu} x_\kappa' \bar{x}_\nu'.$$

As we have just seen, the inner product (8·31) is a linear combination of Hermitian forms $(\mathbf{x} \pm \mathbf{y})^2$ and $(\mathbf{x} \pm i\mathbf{y})^2$, involving the same coefficients $a_{\lambda\mu}$. Thus the preservation of the Hermitian form is equivalent to the preservation of the inner product. We proceed to establish the necessary and sufficient condition

$$\text{(8·34)} \qquad \Sigma\Sigma a_{\kappa\nu} b_{\lambda\kappa} \bar{b}_{\mu\nu} = a_{\lambda\mu} \quad (\lambda, \mu = 1, \ldots, n)$$

or

$$\text{(8·35)} \qquad \mathrm{B}\mathrm{A}\bar{\mathrm{B}}^{\mathrm{T}} = \mathrm{A}.$$

(This is a condition on B; A is supposed to have been given in advance.) In fact, if the transformation $\mathbf{x}' = \mathbf{x}\mathrm{B}$ preserves the inner product, we have

$$\mathbf{x}\mathrm{A}\bar{\mathbf{y}}^{\mathrm{T}} = \mathbf{x}'\mathrm{A}\bar{\mathbf{y}}'^{\mathrm{T}} = \mathbf{x}\mathrm{B}\mathrm{A}\overline{\mathbf{y}\mathrm{B}}^{\mathrm{T}} = \mathbf{x}\mathrm{B}\mathrm{A}\bar{\mathrm{B}}^{\mathrm{T}}\bar{\mathbf{y}}^{\mathrm{T}}.$$

Since this equation must hold for all $\mathbf{x}$ and $\mathbf{y}$, it implies (8·35). Conversely (8·35) implies

$$\mathbf{x}\mathrm{A}\bar{\mathbf{y}}^{\mathrm{T}} = \mathbf{x}\mathrm{B}\mathrm{A}\bar{\mathrm{B}}^{\mathrm{T}}\bar{\mathbf{y}}^{\mathrm{T}} = \mathbf{x}\mathrm{B}\mathrm{A}\overline{\mathbf{y}\mathrm{B}}^{\mathrm{T}} = \mathbf{x}'\mathrm{A}\bar{\mathbf{y}}'^{\mathrm{T}}.$$

8·4 LENGTHS AND ANGLES

If the invariant Hermitian form

$$\text{(8·41)} \qquad \mathbf{x}^2 = \mathbf{x}\mathrm{A}\bar{\mathbf{x}}^{\mathrm{T}}$$

is positive definite, it enables us to introduce a *metric* into the affine space by calling the number

$$|\mathbf{x}| = (\mathbf{x}^2)^{\frac{1}{2}}$$

the *length* of any vector $\mathbf{x}$. This number is positive unless $\mathbf{x}$ is the zero vector.

In preparation for a useful definition of angle, let us prove, for non-zero vectors $\mathbf{x}$ and $\mathbf{y}$, Schwarz's inequality

$$\mathbf{x}\cdot\mathbf{y}\,\mathbf{y}\cdot\mathbf{x} \leqslant \mathbf{x}^2\mathbf{y}^2, \tag{8·42}$$

with equality only when $\mathbf{x}$ and $\mathbf{y}$ are parallel (Halmos 1958, p. 91). Since

$$(\mathbf{y}^2\mathbf{x} - \mathbf{x}\cdot\mathbf{y}\,\mathbf{y})^2 = \mathbf{y}^2(\mathbf{x}^2\mathbf{y}^2 - |\mathbf{x}\cdot\mathbf{y}|^2),$$

we have $\mathbf{x}^2\mathbf{y}^2 \geqslant |\mathbf{x}\cdot\mathbf{y}|^2$, as desired, with equality only when

$$\mathbf{y}^2\mathbf{x} = \mathbf{x}\cdot\mathbf{y}\,\mathbf{y}.$$

If $\mathbf{x}$ and $\mathbf{y}$ are parallel, we have $\mathbf{x} = c\mathbf{y}$ for some c; then $\mathbf{x}\cdot\mathbf{y} = c\mathbf{y}^2$ and $\mathbf{y}^2\mathbf{x} = c\mathbf{y}^2\mathbf{y} = \mathbf{x}\cdot\mathbf{y}\,\mathbf{y}$.

Taking (8·42) into consideration, we naturally define the *angle* θ between any two lines so that $0 \leqslant \theta \leqslant \frac{1}{2}\pi$ and

$$\cos\theta = \left(\frac{\mathbf{x}\cdot\mathbf{y}\,\mathbf{y}\cdot\mathbf{x}}{\mathbf{x}^2\mathbf{y}^2}\right)^{\frac{1}{2}} = \frac{|\mathbf{x}\cdot\mathbf{y}|}{|\mathbf{x}|\,|\mathbf{y}|}, \tag{8·43}$$

where $\mathbf{x}$ and $\mathbf{y}$ are non-zero vectors along the lines. The angle θ so defined is zero when the lines are parallel. If $\mathbf{x}\cdot\mathbf{y} = 0$, $\theta = \frac{1}{2}\pi$; then we naturally say that the lines (and the vectors) are *perpendicular*.

We can replace $\mathbf{x}$ by $c\mathbf{x}$, or $\mathbf{y}$ by $c\mathbf{y}$, without altering the angle between the lines of action of $\mathbf{x}$ and $\mathbf{y}$. In particular, for any two lines, we can choose vectors of length 1 along them so that $\mathbf{x}\cdot\mathbf{y}$ is real and therefore

$$\mathbf{x}\cdot\mathbf{y} = \pm\cos\theta. \tag{8·44}$$

(We could have insisted on the plus sign, but the minus sign will be found useful later.) For three lines, we cannot always choose vectors of length 1 along them so that all three products are real, but we can do so if two of the lines are perpendicular. In fact, if $\mathbf{x}\cdot\mathbf{z} = 0$, we can choose any vector $\mathbf{y}$ of length 1 along the remaining line and multiply $\mathbf{x}$ and $\mathbf{z}$ by suitable powers of $\exp i$ to make $\mathbf{x}\cdot\mathbf{y}$ and $\mathbf{y}\cdot\mathbf{z}$ real. Similarly,

(8·45) *If a sequence of n lines has the property that every two non-adjacent members are perpendicular, then vectors of length 1 can be chosen along them so that all the inner products are real.*

Expressing $\mathbf{x}$ and $\mathbf{y}$ as linear combinations of a basis $\mathbf{e}_1, \mathbf{e}_2, \ldots, \mathbf{e}_n$ in the manner of (8·11), we may compare the expressions $\mathbf{x}\cdot\mathbf{y} = \Sigma\Sigma a_{\lambda\mu}x_\lambda\bar{y}_\mu$ and

$$\mathbf{x}\cdot\mathbf{y} = \Sigma x_\lambda\mathbf{e}_\lambda\cdot\Sigma y_\mu\mathbf{e}_\mu = \Sigma\Sigma x_\lambda\bar{y}_\mu\mathbf{e}_\lambda\cdot\mathbf{e}_\mu,$$

and conclude that

$$a_{\lambda\mu} = \mathbf{e}_\lambda\cdot\mathbf{e}_\mu: \tag{8·46}$$

the coefficients $a_{\lambda\mu}$ are *inner products of the basic vectors*. Thus the equation (8·22) appears as a special case of $\mathbf{y}\cdot\mathbf{x} = \overline{\mathbf{x}\cdot\mathbf{y}}$. Each 'diagonal' coefficient $a_{\lambda\lambda}$, being the square of the length of the basic vector, is positive.

EXERCISE

Let n lines be represented by the nodes of a *graph* whose branches join nodes representing non-perpendicular lines. If the graph is a *tree*, vectors of length 1 can be chosen along the lines so that all the inner products are real.

8·5 UNITARY TRANSFORMATIONS

In order to relate a given positive definite Hermitian form (8·21) to a primitive form

$$\mathbf{u}\bar{\mathbf{u}}^{\mathrm{T}} = \Sigma u_\lambda\bar{u}_\lambda, \tag{8·51}$$

we choose an *orthonormal basis* $\mathbf{i}_1, \ldots, \mathbf{i}_n$, consisting of n mutually perpendicular vectors of length 1, so that

$$\mathbf{i}_\lambda\cdot\mathbf{i}_\mu = \delta_{\lambda\mu} = \begin{cases} 1 & \text{if } \lambda = \mu, \\ 0 & \text{if } \lambda \neq \mu. \end{cases} \tag{8·52}$$

For this purpose, we choose any unit vector $\mathbf{i}_1$. The equation $\mathbf{x}\cdot\mathbf{i}_1 = 0$ determines an $(n-1)$-flat not containing $\mathbf{i}_1$, in which we choose another unit vector $\mathbf{i}_2$. The equation $\mathbf{x}\cdot\mathbf{i}_2 = 0$, along with the previous one, determines an $(n-2)$-flat containing neither $\mathbf{i}_1$ nor $\mathbf{i}_2$, and so on (van der Waerden 1950, pp. 126–9).

When referred to this new basis, the general vector $\mathbf{x}$ is more naturally denoted by $\mathbf{u}$, so that

$$\mathbf{u} = \Sigma u_\lambda\mathbf{i}_\lambda. \tag{8·53}$$

Its inner product with itself is

$$\mathbf{u}^2 = \Sigma u_\lambda\mathbf{i}_\lambda\cdot\Sigma u_\mu\mathbf{i}_\mu = \Sigma\Sigma u_\lambda\bar{u}_\mu\delta_{\lambda\mu} = \Sigma u_\lambda\bar{u}_\lambda = \mathbf{u}\bar{\mathbf{u}}^{\mathrm{T}}.$$

For two vectors $\mathbf{u}$ and $\mathbf{v} = \Sigma v_\mu\mathbf{i}_\mu$, we have similarly

$$\mathbf{u}\cdot\mathbf{v} = \Sigma u_\lambda\bar{v}_\lambda = \mathbf{u}\bar{\mathbf{v}}^{\mathrm{T}}. \tag{8·54}$$

In particular,

$$\begin{aligned} \mathbf{u}\cdot\mathbf{i}_\mu &= \Sigma u_\lambda\mathbf{i}_\lambda\cdot\mathbf{i}_\mu = \Sigma u_\lambda\delta_{\lambda\mu} \\ &= u_\mu, \end{aligned} \tag{8·55}$$

but $\mathbf{i}_\mu\cdot\mathbf{u} = \bar{u}_\mu$.

A linear transformation $\mathbf{u}' = \mathbf{u}\mathrm{B}$ is said to be *unitary* if it preserves the primitive Hermitian form $\Sigma u_\lambda\bar{u}_\lambda$. Setting $a_{\lambda\mu} = \delta_{\lambda\mu}$ in (8·34), we see that a necessary and sufficient condition for this is

$$\Sigma b_{\lambda\nu}\bar{b}_{\mu\nu} = \delta_{\lambda\mu} \quad (\lambda, \mu = 1, \ldots, n). \tag{8·56}$$

In other words, B is a *unitary matrix* if and only if

$$\mathrm{B}\overline{\mathrm{B}}^{\mathrm{T}} = \mathrm{I}_n \tag{8·561}$$

(meaning the $n \times n$ identity matrix; see Miller, Blichfeldt and Dickson 1916, p. 210, or Birkhoff and MacLane 1965, p. 266). It follows from this equation that the determinant of B (or of B^{T}) is equal to $\exp(\theta i)$ for some real θ. The notation

$$\mathrm{B} = \epsilon \mathrm{C}, \quad \text{where} \quad \epsilon = \exp(\theta i/n),$$

provides a 'normalized' matrix C, *of determinant* 1, satisfying the analogous equation

$$\mathrm{C}\overline{\mathrm{C}}^{\mathrm{T}} = \mathrm{I}_n.$$

The complex affine space, with its 'unitary metric' induced by the invariant Hermitian form, is called a *unitary space*. Its isometries (that is, length-preserving transformations or 'congruent transformations') are products of unitary transformations and translations. We conclude from §8·2 that every finite group of affinities (in complex affine n-space) leaves invariant a positive definite Hermitian form which can be interpreted, in terms of a suitable basis $\mathbf{e}_1, \ldots, \mathbf{e}_n$, as the square of the length of the general vector. Transforming to an orthonormal basis (8·52) we can represent the affinities as unitary transformations.

In the familiar case when $n = 2$ (that is, in the unitary *plane*), (8·56) shows that every unitary transformation

$$u' = b_{11}u + b_{21}v, \quad v' = b_{12}u + b_{22}v \tag{8·57}$$

satisfies the relations

$$b_{11}\bar{b}_{11} + b_{12}\bar{b}_{12} = 1, \quad b_{11}\bar{b}_{21} + b_{12}\bar{b}_{22} = 0, \quad b_{11}b_{22} - b_{12}b_{21} = \exp(\theta i) = \epsilon^2.$$

These relations imply

$$b_{11} = \epsilon a, \quad b_{12} = \epsilon c, \quad b_{21} = -\epsilon\bar{c}, \quad b_{22} = \epsilon\bar{a}, \tag{8·58}$$

where $\epsilon\bar{\epsilon} = a\bar{a} + c\bar{c} = 1$. Thus the general unitary transformation in two dimensions is

$$u' = \epsilon(au - \bar{c}v), \quad v' = \epsilon(cu + \bar{a}v), \tag{8·59}$$

where $\epsilon\bar{\epsilon} = a\bar{a} + c\bar{c} = 1$.

EXERCISES

1. Any unitary transformation $u_\nu' = \Sigma b_{\mu\nu}u_\mu$ can be expressed as $u_\lambda = \Sigma \bar{b}_{\lambda\nu}u_\nu'$, and the condition $\Sigma b_{\lambda\nu}\bar{b}_{\mu\nu} = \delta_{\lambda\mu}$ is equivalent to $\Sigma b_{\lambda\mu}\bar{b}_{\lambda\nu} = \delta_{\mu\nu}$.
2. Fill in the details leading to (8·58).
3. Express (8·59) in matrix notation.

8·6 DUAL BASES

Let us now consider n new vectors $\mathbf{e}^1, \ldots, \mathbf{e}^n$, where

$$\mathbf{e}^\mu = \Sigma a^{\mu\nu}\mathbf{e}_\nu \tag{8·61}$$

and $\|a^{\mu\nu}\| = \|a_{\mu\nu}\|^{-1}$, so that

$$\Sigma a_{\lambda\mu}a^{\mu\nu} = \delta_\lambda^\nu = \begin{cases} 1 & \text{if} \quad \lambda = \nu, \\ 0 & \text{if} \quad \lambda \neq \nu, \end{cases} \tag{8·611}$$

and, of course, $a^{\mu\lambda} = \bar{a}^{\lambda\mu}$. Since

$$\Sigma a_{\lambda\mu}\mathbf{e}^\mu = \Sigma\Sigma a_{\lambda\mu}a^{\mu\nu}\mathbf{e}_\nu = \Sigma\delta_\lambda^\nu\mathbf{e}_\nu = \mathbf{e}_\lambda, \tag{8·62}$$

the n vectors $\mathbf{e}^\mu$, like $\mathbf{e}_\lambda$, span the space, and we may appropriately call them the *contravariant basis* corresponding to the 'covariant' basis $\mathbf{e}_1, \ldots, \mathbf{e}_n$. Since

$$\begin{aligned} \mathbf{e}_\lambda \cdot \mathbf{e}^\mu &= \Sigma \mathbf{e}_\lambda \cdot a^{\mu\nu}\mathbf{e}_\nu = \Sigma\bar{a}^{\mu\nu}a_{\lambda\nu} = \Sigma a_{\lambda\nu}a^{\nu\mu} \\ &= \delta_\lambda^\mu, \end{aligned} \tag{8·621}$$

each $\mathbf{e}^\mu$ is perpendicular to every $\mathbf{e}_\lambda$ except $\mathbf{e}_\mu$. In other words, $\mathbf{e}^\mu$ is normal to the $(n-1)$-flat spanned by $n-1$ of the n vectors $\mathbf{e}_\lambda$, and $\mathbf{e}_\mu$ is similarly related to the $\mathbf{e}^\lambda$. Moreover, the length of $\mathbf{e}^\mu$ is such as to make $\mathbf{e}_\mu \cdot \mathbf{e}^\mu = 1$.

In order to be able to use both bases for expressing a given vector $\mathbf{x}$, we must now denote the coefficients of $\mathbf{e}_\lambda$ by x^λ instead of x_λ. For instance, equations (8·11), (8·12), (8·31) have to be re-written as

$$\mathbf{x} = \Sigma x^\lambda \mathbf{e}_\lambda, \quad x'^\nu = \Sigma b_{\mu\nu}x^\mu,$$

$$\mathbf{x} \cdot \mathbf{y} = \mathbf{x}\mathrm{A}\bar{\mathbf{y}}^{\mathrm{T}} = \Sigma\Sigma a_{\lambda\mu}x^\lambda\bar{y}^\mu.$$

The coefficients x^λ (formerly x_λ) are called the *contravariant components* of the vector $\mathbf{x}$, in contrast to the new *covariant components* x_μ, for which

$$\mathbf{x} = \Sigma x_\mu \mathbf{e}^\mu. \tag{8·63}$$

The two kinds of component are related by the formulae

$$x^\nu = \Sigma a^{\mu\nu}x_\mu, \quad x_\nu = \Sigma a_{\mu\nu}x^\mu, \tag{8·631}$$

which are obtained by substituting (8·61) and (8·62) (in turn) in the vector identity $\Sigma x_\mu \mathbf{e}^\mu = \Sigma x^\lambda \mathbf{e}_\lambda$. The inner product of $\mathbf{x} = \Sigma x^\lambda \mathbf{e}_\lambda$ and $\mathbf{y} = \Sigma y_\mu \mathbf{e}^\mu$ is simply

$$\begin{aligned} \mathbf{x} \cdot \mathbf{y} &= \Sigma\Sigma x^\lambda \mathbf{e}_\lambda \cdot y_\mu \mathbf{e}^\mu = \Sigma\Sigma\delta_\lambda^\mu x^\lambda \bar{y}_\mu \\ &= \Sigma x^\lambda \bar{y}_\lambda. \end{aligned} \tag{8·64}$$

Being the complex conjugate of $\mathbf{y} \cdot \mathbf{x}$, $\mathbf{x} \cdot \mathbf{y}$ is also equal to $\Sigma x_\lambda \bar{y}^\lambda$. Thus the length of $\mathbf{x}$ is the square root of

$$\mathbf{x} \cdot \mathbf{x} = \Sigma x^\lambda \bar{x}_\lambda = \Sigma x_\lambda \bar{x}^\lambda.$$

The components of $\mathbf{x}$ could have been defined as its inner products with the $\mathbf{e}$'s; for

$$\mathbf{x}\cdot\mathbf{e}_\mu = \Sigma x_\lambda \mathbf{e}^\lambda\cdot\mathbf{e}_\mu = \Sigma x_\lambda \delta^\lambda_\mu$$

(8·65) $$= x_\mu,$$

and similarly $\mathbf{x}\cdot\mathbf{e}^\nu = x^\nu$. By (8·61),

$$\mathbf{e}^\mu\cdot\mathbf{e}^\nu = \Sigma a^{\mu\lambda}\mathbf{e}_\lambda\cdot\mathbf{e}^\nu = \Sigma a^{\mu\lambda}\delta^\nu_\mu$$

(8·66) $$= a^{\mu\nu}.$$

Comparing this with (8·46), we see that the reciprocity between 'covariant' and 'contravariant' is complete.

The relation between the orthonormal and oblique bases may be expressed by equations

(8·67) $$\mathbf{e}_\mu = \Sigma e_{\mu\nu}\mathbf{i}_\nu,$$

which imply

(8·671) $$e_{\mu\lambda} = \mathbf{e}_\mu\cdot\mathbf{i}_\lambda, \quad e_{\lambda\nu} = \mathbf{e}_\lambda\cdot\mathbf{i}_\nu$$

and

(8·672) $$a_{\lambda\mu} = \mathbf{e}_\lambda\cdot\mathbf{e}_\mu = \mathbf{e}_\lambda\cdot\Sigma e_{\mu\nu}\mathbf{i}_\nu = \Sigma e_{\lambda\nu}\bar{e}_{\mu\nu}.$$

There must be some coefficients $c_{\lambda\mu}$ such that $\mathbf{i}_\mu = \Sigma c_{\lambda\mu}\mathbf{e}^\lambda$. To find them, we observe that

$$e_{\lambda\mu} = \mathbf{e}_\lambda\cdot\mathbf{i}_\mu = \mathbf{e}_\lambda\cdot\Sigma c_{\nu\mu}\mathbf{e}^\nu = \Sigma\delta^\nu_\lambda \bar{c}_{\nu\mu} = \bar{c}_{\lambda\mu}.$$

Hence $c_{\lambda\mu} = \bar{e}_{\lambda\mu}$, and the expression for $\mathbf{i}_\mu$ is

(8·68) $$\mathbf{i}_\mu = \Sigma\bar{e}_{\lambda\mu}\mathbf{e}^\lambda.$$

If a vector $\mathbf{x}\,(=\mathbf{u})$ has orthogonal components u_ν, covariant components x_ν, and contravariant components x^ν, we see from (8·65) and (8·55) (with (8·67) and (8·68)) that

(8·69) $$x_\mu = \Sigma\bar{e}_{\mu\nu}u_\nu \quad\text{and}\quad u_\mu = \Sigma e_{\lambda\mu}x^\lambda,$$

in agreement with $x_\mu = \Sigma a_{\lambda\mu}x^\lambda$.

8·7 REFLECTIONS

The hyperplane, through the origin, perpendicular to a given vector $\mathbf{y} = \Sigma y_\mu\mathbf{e}^\mu = \Sigma y^\mu\mathbf{e}_\mu$ may be described as the set of all points (x) whose position vectors $\mathbf{x}$ satisfy the equation $\mathbf{x}\cdot\mathbf{y} = 0$. Accordingly, this hyperplane has the equation

$$\Sigma\bar{y}_\mu x^\mu = 0 \quad\text{or}\quad \Sigma\bar{y}^\mu x_\mu = 0.$$

The foot of the perpendicular from any point (x) to this hyperplane $\mathbf{x}\cdot\mathbf{y} = 0$ has the position vector

(8·71) $$\mathbf{x} - \frac{\mathbf{x}\cdot\mathbf{y}}{\mathbf{y}\cdot\mathbf{y}}\mathbf{y}.$$

For, if it is $\mathbf{x}+c\mathbf{y}$ it must satisfy

$$(\mathbf{x}+c\mathbf{y})\cdot\mathbf{y} = 0,$$

whence $c = -\mathbf{x}\cdot\mathbf{y}/\mathbf{y}\cdot\mathbf{y}$. The vector from this 'foot' to (x) itself is, of course,

(8·72) $$\frac{\mathbf{x}\cdot\mathbf{y}}{\mathbf{y}\cdot\mathbf{y}}\mathbf{y}.$$

A *reflection* of period p, having the hyperplane $\mathbf{x}\cdot\mathbf{y} = 0$ as its mirror, is naturally defined to be the result of multiplying this vector by a primitive pth root of unity, say $\exp(2\pi i/p)$. Thus the complex reflection transforms $\mathbf{x}$ into

(8·73) $$\mathbf{X} = \mathbf{x} + \{\exp(2\pi i/p) - 1\}\frac{\mathbf{x}\cdot\mathbf{y}}{\mathbf{y}\cdot\mathbf{y}}\mathbf{y}$$

or, if $|\mathbf{y}| = 1$, we can write simply

(8·74) $$\mathbf{X} = \mathbf{x} + \{\exp(2\pi i/p) - 1\}\,\mathbf{x}\cdot\mathbf{y}\,\mathbf{y}.$$

(We use the letter $\mathbf{X}$, rather than $\mathbf{x}'$, to avoid confusion with superscripts.)

Taking inner products with $\mathbf{e}_\lambda$ or $\mathbf{e}^\lambda$, we deduce that the reflection of period p in the hyperplane $\Sigma\bar{y}^\mu x_\mu = 0$ or $\Sigma\bar{y}_\mu x^\mu = 0$ is the transformation

(8·75) $$X_\lambda = x_\lambda + \{\exp(2\pi i/p) - 1\}y_\lambda\Sigma\bar{y}^\mu x_\mu$$

or

(8·76) $$X^\lambda = x^\lambda + \{\exp(2\pi i/p) - 1\}y^\lambda\Sigma\bar{y}_\mu x^\mu.$$

Of course this result can immediately be adapted to orthogonal coordinates, with the conclusion that the reflection of period p in the hyperplane

$$\Sigma a_\mu u_\mu = 0 \quad (\Sigma a_\mu\bar{a}_\mu = 1)$$

is

(8·77) $$u_\lambda{}' = u_\lambda + \{\exp(2\pi i/p) - 1\}\bar{a}_\lambda\Sigma a_\mu u_\mu.$$

EXERCISE

What is the analogue of (8.73) in *real* n-space?

8·8 A COMPLEX KALEIDOSCOPE

For the investigation of an n-dimensional group generated by n reflections R_ν ($\nu = 1, \ldots, n$) of periods p_ν, it is convenient to choose a coordinate system based on vectors $\mathbf{e}_\nu$, of length 1, perpendicular to the mirrors. In other words, the νth mirror is the hyperplane $\mathbf{x}\cdot\mathbf{e}_\nu = 0$ or

$$x_\nu = 0,$$

and the reflection R_ν transforms the vector $\mathbf{x}$ into

$$\mathbf{X} = \mathbf{x} + (\epsilon_\nu{}^2 - 1)\, x_\nu \mathbf{e}_\nu, \tag{8·81}$$

where

$$\epsilon_\nu = \exp(\pi i/p_\nu). \tag{8·82}$$

As a transformation of covariant coordinates, R_ν is

$$X_\lambda = x_\lambda + (\epsilon_\nu{}^2 - 1)\, a_{\nu\lambda} x_\nu, \tag{8·83}$$

and as a transformation of contravariant coordinates it is

$$\begin{aligned} X^\lambda &= x^\lambda + \delta^\lambda_\nu(\epsilon_\nu{}^2 - 1)\, x_\nu \\ &= x^\lambda + \delta^\lambda_\nu(\epsilon_\nu{}^2 - 1)\, \Sigma a_{\mu\nu} x^\mu, \end{aligned} \tag{8·84}$$

where

$$a_{\nu\nu} = 1 \quad (\nu = 1, 2, \ldots, n) \tag{8·85}$$

(because we choose $\mathbf{e}_\nu$ to be a unit vector) and the remaining coefficients (see (8·46)) are determined by the relative positions of the n mirrors (that is, by the 'shape' of the complex kaleidoscope). In other words, the transformation R_ν leaves invariant all the contravariant coordinates x^λ except x^ν, which it transforms into

$$x^\nu + (\epsilon_\nu{}^2 - 1)\, x_\nu = x^\nu + (\epsilon_\nu{}^2 - 1)\, \Sigma a_{\mu\nu} x^\mu. \tag{8·86}$$

EXERCISE

What is the analogue of (8·83) in real space?

8·9 THE TWO-DIMENSIONAL CASE

When $n = 2$, so that the mirrors are two lines in the complex affine *plane*, equations (8·84) and (8·82) show that R_1 (with $x_1 = 0$ for its mirror) is

$$\begin{aligned} X^1 &= x^1 + (\epsilon_1{}^2 - 1)(a_{11}x^1 + a_{21}x^2) = \epsilon_1{}^2 x^1 + (\epsilon_1{}^2 - 1)\, a_{21} x^2, \\ X^2 &= x^2 \qquad\qquad (\epsilon_1 = \exp(\pi i/p_1)) \end{aligned}$$

and similarly R_2 (with $x_2 = 0$ for its mirror) is

$$X^1 = x^1, \quad X^2 = (\epsilon_2{}^2 - 1)\, a_{12} x^1 + \epsilon_2{}^2 x^2 \quad (\epsilon_2 = \exp(\pi i/p_2)).$$

The corresponding matrices are

$$R_1 = \begin{bmatrix} \epsilon_1{}^2 & 0 \\ (\epsilon_1{}^2 - 1)\, a_{21} & 1 \end{bmatrix}, \quad R_2 = \begin{bmatrix} 1 & (\epsilon_2{}^2 - 1)\, a_{12} \\ 0 & \epsilon_2{}^2 \end{bmatrix}. \tag{8·91}$$

We shall find, in §9·8, that these two reflections generate a finite group if

$$a_{12} = a_{21} = -\left[\frac{\cos\left(\dfrac{\pi}{p_1} - \dfrac{\pi}{p_2}\right) + \cos\dfrac{2\pi}{q}}{2\sin\dfrac{\pi}{p_1}\sin\dfrac{\pi}{p_2}}\right]^{\frac{1}{2}},$$

where q is an integer satisfying the inequalities

$$p_\nu > 1, \quad q > 2, \quad \frac{1}{p_1} + \frac{1}{p_2} + \frac{2}{q} > 1$$

with $p_1 = p_2$ when q is odd. This group, denoted by

$$p_1[q]p_2,$$

will reappear in §11·1 as the symmetry group of the regular complex polygons

$$p_1\{q\}p_2 \quad \text{and} \quad p_2\{q\}p_1,$$

which were tentatively introduced in §4·8.

CHAPTER 9

The unitary plane, using quaternions

Let not thy left hand know what thy right hand doeth.

Matthew 6:3

We discussed quaternions in Chapters 6 and 7, not only because they provide the most natural representation for the binary polyhedral groups but also because they enable us to explore the unitary plane with scarcely any effort. Such an exploration was begun in 1961 by Donald Crowe†, who made the penetrating observation that a unitary transformation can be achieved by multiplying the general quaternion by a complex number on the left and a quaternion on the right. In §9·7 we shall obtain the symmetry groups $p_1[q]p_2$ of the regular complex polygons $p_1\{q\}p_2$, by means of a trick depending on the happy accident that the number s, defined by (2·34), is the least common multiple of p, q, r for nearly all relevant values of these integers. The phrase 'happy accident' is justified by the observation that there are two (rather uninteresting) cases in which it fails: the cases when $r = 1$ and $p = q$ is odd, and when $q = r = 2$ with p odd. What makes these groups especially appealing is the elegant conciseness of their presentations ((9·58), (9·62), (9·81)).

9·1 UNITARY GROUPS

We have seen that, in the complex affine plane with orthogonal coordinates, the general unitary transformation (preserving $u\bar{u}+v\bar{v}$) is

$$u' = \epsilon(au-\bar{c}v), \quad v' = \epsilon(cu+\bar{a}v)$$

(8·59), where ϵ, a, c are complex numbers such that

(9·11) $$\epsilon\bar{\epsilon} = a\bar{a}+c\bar{c} = 1.$$

The transformation becomes wonderfully simple and natural when we represent the point (u, v) by the quaternion

$$x = u+vj$$

and recall that

(9·111) $$(u+vj)(a+cj) = au-\bar{c}v+(cu+\bar{a}v)j$$

(see (6·14)). In fact, by comparing this expression with (8·59), we can express the unitary transformation as

(9·12) $$x' = \epsilon x\kappa,$$

where ϵ is a complex number of norm 1 and

(9·13) $$\kappa = a+cj$$

is a quaternion of norm 1. By the convention adopted in §6·3, we call this simply 'the transformation $\epsilon x\kappa$'. In particular, the 'left multiplication' ϵx transforms (u,v) into $(\epsilon u, \epsilon v)$, and the 'right multiplication' $x\kappa$ transforms (u, v) into $(au-\bar{c}v, cu+\bar{a}v)$.

Clearly, the inverse of $\epsilon x\kappa$ is $\epsilon^{-1}x\kappa^{-1}$, and the product of two such transformations, $\epsilon x\kappa$ and $\epsilon' x\kappa'$, is

$$\epsilon'(\epsilon x\kappa)\kappa' = \epsilon\epsilon' x\kappa\kappa'.$$

This neat trick works because the complex numbers ϵ are commutative although the quaternions κ are not. Since $\epsilon x\kappa = (-\epsilon)x(-\kappa)$, each unitary transformation can be expressed as $\epsilon x\kappa$ in two ways. Hence

(9·14) *The direct product of the group of all complex numbers ϵ with $\epsilon\bar{\epsilon} = 1$ and the group of all quaternions κ with $\kappa\tilde{\kappa} = 1$ is 2:1 homomorphic to the group of all two-dimensional unitary transformations.*

For a finite subgroup of the latter (continuous) group, that is, any finite group of two-dimensional unitary transformations, Du Val (1964, p. 57) proposes the symbol

(9·15) $$(\mathfrak{C}_{2m}/\mathfrak{C}_f;\ \mathfrak{R}/\mathfrak{S}),$$

where f is a divisor of $2m$, $\mathfrak{R}$ is one of the finite groups $\langle p,q,r\rangle$, and $\mathfrak{S}$ (Du Val's $\mathfrak{R}_K$) is a normal subgroup of $\mathfrak{R}$ whose quotient group is $\mathfrak{C}_{2m/f}$. This means that the whole group consists of unitary transformations $\epsilon x\kappa$, where ϵ is a power of $\exp(\pi i/m)$ and κ runs over all the elements of $\mathfrak{R}$ in such a way that each coset of $\mathfrak{C}_f$ in $\mathfrak{C}_{2m}$ is paired with the corresponding coset of $\mathfrak{S}$ in $\mathfrak{R}$ according to an isomorphism

(9·16) $$\mathfrak{C}_{2m}/\mathfrak{C}_f \cong \mathfrak{R}/\mathfrak{S}$$

between the two quotient groups $\mathfrak{C}_{2m/f}$. Since $-\epsilon x\kappa = \epsilon x(-\kappa)$, the order of $(\mathfrak{C}_{2m}/\mathfrak{C}_f;\ \mathfrak{R}/\mathfrak{S})$ is m times the order of $\mathfrak{S}$, that is, $\frac{1}{2}f$ times the order of $\mathfrak{R}$. Apart from the little complication caused by the common element -1 of $\mathfrak{C}_{2m}$ and $\mathfrak{R}$, we have here an instance of a 'subdirect product' (Hall 1959, pp. 63–4).

EXERCISES

1. Any finite group $\mathfrak{G}$ of two-dimensional unitary transformations may be described by Du Val's symbol (9·15).
2. Can $(\mathfrak{C}_{2m}/\mathfrak{C}_m;\ \mathfrak{R}/\mathfrak{S})$ be a subgroup of $(\mathfrak{C}_{4m}/\mathfrak{C}_{2m};\ \mathfrak{R}/\mathfrak{S})$?

† D. W. Crowe, The groups of regular complex polygons, *Canadian Journal of Mathematics* **13** (1961), 149–56.

9·2 A COMBINATION OF CYCLIC GROUPS

The symbol $(\mathfrak{C}_{2m}/\mathfrak{C}_f;\ \mathfrak{R}/\mathfrak{S})$ describes a unique group unless the isomorphism (9·16) can be chosen in more than one way, that is, unless $\mathfrak{C}_{2m/f}$ has an outer automorphism. Such an automorphism transforms the generator $\exp(f\pi i/m)$ of $\mathfrak{C}_{2m/f}$ into $\exp(df\pi i/m)$, where d is a number relatively prime to $2m/f$. Since we can replace d by $(2m/f)-d$, we lose nothing by assuming that $d < m/f$. When such a number $d > 1$ exists, Du Val (1964, p. 55) inserts it as a subscript.

For instance, let f, g, m, n, d be positive integers such that $f \equiv g \pmod 2$, $gm = fn$, f divides $2m$, d is relatively prime to $2m/f$, and $1 \leqslant d < m/f$. Then there is a group

$$(\mathfrak{C}_{2m}/\mathfrak{C}_f;\ \mathfrak{C}_{2n}/\mathfrak{C}_g)_d \tag{9·21}$$

consisting of the $gm = fn$ transformations

$$e^{\mu\pi i/m}x\,e^{d\nu\pi i/n},$$

where $\mu = 0, 1, \ldots, \lambda m-1$; $\nu = 0, 1, \ldots, n-1$; $\mu \equiv \nu \pmod{\lambda m/f}$, and $\lambda = 1$ or 2 according as f and g are odd or even. This description combines Du Val's types $1'$ $(\lambda = 1)$ and 1 $(\lambda = 2)$ (Du Val 1964, p. 57). It is important to notice that, when f and g are odd, d (his s) must likewise be odd.

EXERCISES

1. Express the elements of $(\mathfrak{C}_{10}/\mathfrak{C}_1;\ \mathfrak{C}_{10}/\mathfrak{C}_1)_3$ in terms of $\epsilon = \exp(\pi i/5)$.
2. Justify the statement that, if f and g are even, $(\mathfrak{C}_{2m}/\mathfrak{C}_f;\ \mathfrak{C}_{2n}/\mathfrak{C}_g)_d$ consists of
$$\pm e^{\mu\pi i/m}x\,e^{d\nu\pi i/n},$$
where $\mu = 0, 1, \ldots, m-1$; $\nu = 0, 1, \ldots, n-1$, and $\mu \equiv \nu \pmod{2m/f}$.

9·3 AN EXTENSION OF THE BINARY POLYHEDRAL GROUPS

An important special case of (9·16) occurs when $f = 2m$, so that $\mathfrak{R}$ and $\mathfrak{S}$ coincide. The group

$$(\mathfrak{C}_{2m}/\mathfrak{C}_{2m};\ \langle p,q,r\rangle/\langle p,q,r\rangle) \tag{9·31}$$

is conveniently denoted by the concise symbol

$$\langle p,q,r\rangle_m$$

(Coxeter and Moser 1972, p. 71) because it is related to

$$\langle p,q,r\rangle = \langle p,q,r\rangle_1$$

as follows. Each finite group $\langle p,q,r\rangle$ is generated (redundantly, as we have seen) by three quaternions A, B, C, which satisfy

$$A^p = B^q = C^r = ABC.$$

It is equally well generated by the three 'right multiplications' xA, xB, xC. The extension $\langle p,q,r\rangle_m$ is derived from it by adjoining the 'left multiplication' $\exp(\pi i/m)x$. Calling this new element Z, we see that $\langle p,q,r\rangle_m$ has the presentation

$$A^p = B^q = C^r = ABC = Z^m; \quad Z \rightleftarrows A, B, C \tag{9·32}$$

(implying $Z^{2m} = 1$). Since the order of $\langle p,q,r\rangle$ is $4s$, where s is given by (2·34), the order of $\langle p,q,r\rangle_m$ is $4ms$, in agreement with our general statement about the order of (9·15).

By Theorem (7·82), the group $\langle p,3,2\rangle$ $(p = 4$ or $5)$ has the representation

$$A = \begin{bmatrix} 1 & 1 \\ 2 & 3 \end{bmatrix}, \quad B = \begin{bmatrix} -1 & -3 \\ 1 & 2 \end{bmatrix}, \quad C = \begin{bmatrix} 0 & -1 \\ 1 & 0 \end{bmatrix} \tag{9·33}$$

over $GF[h+1]$ $(h = 6$ or 10, respectively). By (2·35),

$$(h+1)^2 = 4s+1.$$

Let ρ be a primitive root of the field $GF[4s+1]$, so that $\rho^{2s} = -1$. If m is a divisor of $2s$, we can derive $\langle p,3,2\rangle_m$ from $\langle p,3,2\rangle$ by working in this larger field and adjoining

$$Z = \rho^{2s/m}\,I_2. \tag{9·331}$$

Similarly, $\langle 3,3,2\rangle$ has the representation

$$A = \begin{bmatrix} -1 & -1 \\ -2 & 2 \end{bmatrix}, \quad B = \begin{bmatrix} -1 & 2 \\ 1 & 2 \end{bmatrix}, \quad C = \begin{bmatrix} 0 & 1 \\ -1 & 0 \end{bmatrix} \tag{9·34}$$

over $GF[5]$; so if $m = 2, 3, 4, 6$ or 12, we can derive $\langle 3,3,2\rangle_m$ by working in $GF[5^2]$ and adjoining

$$Z = \rho^{12/m}\,I_2, \tag{9·341}$$

where ρ is a primitive root of this field. Exceptionally, since -1 is a square in $GF[5]$, $\langle 3,3,2\rangle_2$ does not need the extended field but is representable over $GF[5]$:

$$Z = \begin{bmatrix} 2 & 0 \\ 0 & 2 \end{bmatrix} \pmod 5. \tag{9·342}$$

Since the determinant of Z is -1, $\langle 3,3,2\rangle_2$ is thus represented as a subgroup of the unimodular group modulo 5.

EXERCISES

1. If $m = 2^c n$, where n is odd, $\langle p, q, r\rangle_m$ is the direct product
$$\langle p, q, r\rangle_{2^c} \times \mathfrak{C}_n.$$
2. Which groups $\langle 3, 3, 2\rangle_m$ can be represented by matrices over $GF[3^2]$?

9·4 REFLECTIONS

If the unitary transformation (8·59) happens to be a reflection, it leaves invariant every point on the mirror, which is a certain line through (0, 0). Conversely, if the origin is not the only invariant point, the unitary transformation may reasonably be called a *reflection*, even if it is not periodic. Eliminating u and v from the equations

$$u = \epsilon(au - \bar{c}v), \quad v = \epsilon(cu + \bar{a}v), \tag{9·41}$$

we deduce $$(\epsilon a - 1)(\epsilon\bar{a} - 1) + \epsilon^2 c\bar{c} = 0$$

whence, by (9·11) and (9·13),

$$\epsilon + \bar{\epsilon} = a + \bar{a} = \kappa + \tilde{\kappa},$$

that is, $$\mathrm{Re}\,\epsilon = \mathrm{S}\kappa.$$

We have thus proved

(9·42) *The unitary transformation $\epsilon x\kappa$ is a reflection if and only if the complex number ϵ and the quaternion κ have the same real part.*

In other words, for some real angle α and some unit pure quaternion $\mathrm{P} = \xi i + \eta j + \zeta k$ (where $\xi^2 + \eta^2 + \zeta^2 = 1$), we have

$$\epsilon = \exp(\alpha i) = \cos\alpha + i\sin\alpha, \quad \kappa = \exp(\alpha\mathrm{P}) = \cos\alpha + \mathrm{P}\sin\alpha.$$

Comparison with (9·13) yields

$$a = \cos\alpha + \xi i \sin\alpha, \quad c = (\eta + \zeta i)\sin\alpha.$$

The mirror is given by either of the two consistent equations (9·41). The former can be expressed as $(a - \bar{\epsilon})u - \bar{c}v = 0$ or $(\xi + 1)iu - (\eta - \zeta i)v = 0$ or

$$(\xi + 1)u + (\zeta + \eta i)v = 0. \tag{9·43}$$

If $\alpha = \pi/p$ for some positive integer p, so that $\epsilon^p = \kappa^p = -1$, we have a reflection of period p. In particular, let us choose

$$\xi = -\cos 2\sigma, \quad \eta = 0, \quad \zeta = \sin 2\sigma,$$

so that, when P is represented on the sphere $\xi^2 + \eta^2 + \zeta^2 = 1$ with 'poles' $\pm i$, it lies on the meridian through k at an angular distance 2σ from the south pole $-i$. Then we see that $e^{\pi i/p} x e^{\pi \mathrm{P}/p}$, with

$$\mathrm{P} = -i\cos 2\sigma + k \sin 2\sigma, \tag{9·44}$$

is a reflection in the line

$$u \sin\sigma + v\cos\sigma = 0 \tag{9·45}$$

which makes an angle σ with $v = 0$.

In other words, this reflection is the transformation $\epsilon x\kappa$ where

$$\epsilon = \cos\frac{\pi}{p} + i\sin\frac{\pi}{p}$$

and $$\kappa = \cos\frac{\pi}{p} + \mathrm{P}\sin\frac{\pi}{p}$$

$$= \cos\frac{\pi}{p} - i\cos 2\sigma\sin\frac{\pi}{p} + k\sin 2\sigma \sin\frac{\pi}{p}.$$

Comparison with (9·13) yields

$$a = \cos\frac{\pi}{p} - i\cos 2\sigma\sin\frac{\pi}{p}, \quad c = i\sin 2\sigma\sin\frac{\pi}{p}.$$

Thus the coordinates u and v are transformed as follows:

$$u' = \epsilon\left\{\left(\cos\frac{\pi}{p} - i\cos 2\sigma\sin\frac{\pi}{p}\right)u + \left(i\sin 2\sigma\sin\frac{\pi}{p}\right)v\right\},$$

(9·46)

$$v' = \epsilon\left\{\left(i\sin 2\sigma\sin\frac{\pi}{p}\right)u + \left(\cos\frac{\pi}{p} + i\cos 2\sigma\sin\frac{\pi}{p}\right)v\right\}.$$

Setting $\sigma = 0$, we deduce that the reflection of period p in the line $v = 0$ is

$$u' = u, \quad v' = \epsilon^2 v. \tag{9·47}$$

EXERCISES

1. Express, in the form $\epsilon x\kappa$, the reflection of period p in the line $u + tv = 0$, where t is any complex number.
2. Express, in form $\epsilon x\kappa$, the reflections of period p in the coordinate axes $u = 0$ and $v = 0$. (In terms of u and v they are, of course,
$$u' = \epsilon^2 u, \quad v' = v \quad \text{and} \quad u' = u, \quad v' = \epsilon^2 v.)$$
3. Derive (9·45) from (8·77) (without using quaternions).

9·5 GROUPS GENERATED BY INVOLUTORY REFLECTIONS

Let us now investigate finite groups generated by two or three reflections of period 2.

When $p = 2$, the reflection $\epsilon x\kappa$ of §9·4 reduces to

$$ix\mathrm{P}.$$

For instance, if $\mathrm{P} = j\sin\theta + k\cos\theta = k\exp(\theta i)$, the mirror (9·43) becomes

$$u + \exp(\theta i)v = 0.$$

Setting $\theta = 0$ and then $\theta = 2\pi/p$, we obtain two reflections

$$ixk \quad \text{and} \quad ixk\exp(2\pi i/p)$$

whose product $x \exp(2\pi i/p)$ has period p. (Compare (6·64) with $\frac{1}{2}p$ for p.) Hence these reflections, say R_1 and R_2, satisfy the relations (1·22) and generate the ordinary dihedral group $\mathfrak{D}_p$ or $[p]$, which consists of p rotations and p reflections:

$$x \exp(2\nu\pi i/p) \quad \text{and} \quad ixk \exp(2\nu\pi i/p) \quad (\nu = 0, 1, \ldots, p-1).$$

When p is even, the value $\nu = \frac{1}{2}p$ yields $-x$; therefore

$$(9{\cdot}51) \qquad [p] \cong (\mathfrak{C}_4/\mathfrak{C}_2; \langle \tfrac{1}{2}p, 2, 2\rangle/\mathfrak{C}_p) \quad (p \text{ even}).$$

When p is odd, the value $\nu = \frac{1}{2}(p+1)$ yields $-x\exp(\pi i/p)$; therefore

$$(9{\cdot}52) \qquad [p] \cong (\mathfrak{C}_4/\mathfrak{C}_1; \langle p, 2, 2\rangle/\mathfrak{C}_p) \quad (p \text{ odd}).$$

For a group generated by *three* involutory reflections, it is natural to take these generators to be

$$(9{\cdot}53) \qquad R_\nu = ixU_\nu \quad (\nu = 1, 2, 3),$$

where the three U_ν are the pure quaternions that we discussed in §7·5, namely

$$U_1 = j, \quad U_2 = -i\cos\frac{\pi}{q} - j\cos\frac{\pi}{p} + k\sin\frac{\pi}{h}, \quad U_3 = i.$$

When $q = 3$ and $p = 4$ or 5, we see from (7·53) that

$$R_\nu{}^2 = 1, \quad (R_1R_2)^p = Z^{p+1}, \quad (R_2R_3)^3 = 1, \quad (R_1R_3)^2 = Z, \quad Z^2 = 1,$$

where Z is the transformation $-x$. The relation $(U_1U_2U_3)^{h/2} = -1$ shows that $(R_1R_2R_3)^{h/2}$ is one of the transformations $\pm ix$, but without this we already have enough relations to define

$$\langle p, 3, 2\rangle_2 \cong (\mathfrak{C}_4/\mathfrak{C}_4; \langle p, 3, 2\rangle/\langle p, 3, 2\rangle) \qquad (p = 4 \text{ or } 5),$$

of order $8s$. In fact, the above relations define a group whose order is twice that of its factor group $[p, 3]$ of order $4s$, given by setting $Z = 1$. To obtain a tidier presentation we can replace R_2 by R_2Z, so that

$$R_\nu{}^2 = 1, \quad (R_1R_2)^p = (R_2R_3)^3 = (R_1R_3)^2 = Z, \quad Z^2 = 1.$$

Since the products $A = R_1R_2$ and $B = R_2R_3$ satisfy $A^p = B^3 = (AB)^2 = Z$, the last relation is superfluous, and $\langle p, 3, 2\rangle_2$ ($p = 4$ or 5) has the presentation

$$(9{\cdot}54) \qquad R_\nu{}^2 = 1, \quad (R_1R_2)^p = (R_2R_3)^3 = (R_1R_3)^2.$$

The three R_ν are now

$$ixj, \quad ix\kappa, \quad ixi,$$

where $\kappa = jA = Bi$ in the notation of (6·63) and (7·52). What we have achieved here is a way of generating $\langle p, 3, 2\rangle_2$ by three *reflections*. In (9·32) (with $m = 2$) we found a different presentation for the same group as generated by four unitary transformations

$$xA, \quad xB, \quad xC, \quad ix,$$

which are not reflections.

Using the relations (7·56) for $\langle p, 2, 2\rangle$ (p even), we obtain

$$R_\nu{}^2 = 1, \quad (R_1R_2)^p = Z^{p+1}, \quad (R_2R_3)^2 = (R_1R_3)^2 = Z, \quad Z = 1.$$

There is no need to use the relation $(U_1U_2)^{p/2}U_3 = -1$ (which shows that $(R_1R_2)^{p/2}R_3$ is one of the transformations $\pm ix$); for since the extra relation $Z = 1$ yields the factor group $[p, 2]$, we already have enough relations to define

$$\langle p, 2, 2\rangle_2 \quad (p \text{ even}),$$

of order $8p$. For a tidier presentation we can again replace R_2 by R_2Z and then eliminate Z, so that

$$(9{\cdot}55) \qquad R_\nu{}^2 = 1, \quad (R_1R_2)^p = (R_2R_3)^2 = (R_1R_3)^2.$$

The three R_ν are now ixj, $ixj\exp(\pi i/p)$, ixi.

When ρ is odd, the same three reflections satisfy the same relations (9·55), but now, since the transformation ix does not belong, the group of order $8p$ that they generate is not $\langle p, 2, 2\rangle_2$ but

$$(\mathfrak{C}_4/\mathfrak{C}_2; \langle 2p, 2, 2\rangle/\langle p, 2, 2\rangle) \quad (p \text{ odd}).$$

Similarly, the pure quaternions (7·58), which generate $\langle 4, 3, 2\rangle$ in the manner of (7·59), yield reflections R_ν which generate

$$(\mathfrak{C}_4/\mathfrak{C}_2; \langle 4, 3, 2\rangle/\langle 3, 3, 2\rangle),$$

of order 48, and satisfy

$$R_\nu{}^2 = 1, \quad (R_1R_2)^3 = (R_2R_3)^3 = 1, \quad (R_1R_3)^2 = Z, \quad Z^2 = 1.$$

Replacing R_2 by R_2Z, we obtain the neater presentation

$$(9{\cdot}56) \qquad R_\nu{}^2 = 1, \quad (R_1R_2)^3 = (R_2R_3)^3 = (R_1R_3)^2.$$

The three R_ν are now

$$(9{\cdot}57) \qquad ix\frac{k-j}{\sqrt{2}}, \quad ix\frac{k-i}{\sqrt{2}}, \quad ix\frac{k+j}{\sqrt{2}}.$$

Comparing (9·56) and (9·54) with (7·72), we see that

$$(\mathfrak{C}_4/\mathfrak{C}_2; \langle 4, 3, 2\rangle/\langle 3, 3, 2\rangle)$$

can be represented as $GL(2, 3)$, and $\langle 5, 3, 2\rangle_2$ as the unimodular group modulo 5, as follows:

$$R_1 = \begin{bmatrix} 1 & 0 \\ 0 & -1 \end{bmatrix}, \quad R_2 = \begin{bmatrix} -1 & 0 \\ 1 & 1 \end{bmatrix}, \quad R_3 = \begin{bmatrix} 0 & 1 \\ 1 & 0 \end{bmatrix} \quad (\text{mod } 3 \text{ or } 5).$$

In the latter case we have a representation of $\langle 5,3,2\rangle_2$ over $GF[5]$, much simpler than (9·33) and (9·331) (with $2s/m = 30$), which is a representation over $GF[11^2]$. We shall find it convenient to use the symbol $GL(2,3)$ as an abbreviation for

$$(\mathfrak{C}_4/\mathfrak{C}_2;\ \langle 4,3,2\rangle/\langle 3,3,2\rangle).$$

The above results may be summarized as follows:

(9·58) *Corresponding to each group* $[p,q]$, *of order* $4s$, *generated by reflections in three planes of Euclidean 3-space, there is a group*

$$R_\nu^2 = 1, \quad (R_1R_2)^p = (R_2R_3)^q = (R_1R_3)^2$$

of order $8s$, *generated by reflections in three lines of the unitary plane. In this manner,* $[p,2]$ (p *odd*) *yields*

$$(\mathfrak{C}_4/\mathfrak{C}_2;\ \langle 2p,2,2\rangle/\langle p,2,2\rangle),$$

$[3,3]$ *yields* $(\mathfrak{C}_4/\mathfrak{C}_2;\ \langle 4,3,2\rangle/\langle 3,3,2\rangle) \cong GL(2,3)$, *and every other* $[p,q]$ *yields* $\langle p,q,2\rangle_2$.

EXERCISES

1. Comparing $A^p = B^3 = C^2 = ABC = Z^2$; $Z \rightleftarrows A, B, C$ with the presentation (9·54) for $\langle p,3,2\rangle_2$, express A, B, C, Z in terms of R_1, R_2, R_3.
2. Write down generating reflections for $\langle 4,3,2\rangle_2$ based on (7·55) instead of $j, -jA, i$.
3. The group $\langle p,2,2\rangle_2$ (p odd) is not generated by reflections.
4. In Theorem (9.58), can the expression $(R_1R_3)^2$ be replaced by $(R_3R_1)^2$?
5. Express, as quaternion transformations $\epsilon x\kappa$, the generators a and c of $GL(2,3)$ in the presentation (7·73).

9·6 OTHER GROUPS GENERATED BY THREE REFLECTIONS

We come now to a more exciting subject: reflections of arbitrary period. We saw in §9·3 that $\langle p,q,r\rangle_m$, of order $4ms$, is generated by the right multiplications xA, xB, xC along with the left multiplication $\exp(\pi i/m)\,x$, which we denoted by Z. It follows from Theorem (9·42) that, if m is a common multiple of p, q, r, this group contains the reflections

$$\exp(\pm\pi i/p)\,xA, \quad \exp(\pm\pi i/q)\,xB, \quad \exp(\pm\pi i/r)\,xC,$$

or, in the notation of (9·32),

$$Z^{\pm m/p}A, \quad Z^{\pm m/q}B, \quad Z^{\pm m/r}C.$$

If p, q, r are

$$p,2,2 \quad (p \text{ even}) \quad \text{or} \quad p,3,2 \quad (p = 3,4,5),$$

their least common multiple, s, satisfies (2·34), and the group

$$\langle p,q,r\rangle_s,$$

of order $4s^2$, is generated by the three reflections

(9·61) $$R_1 = Z^{-s/p}A, \quad R_2 = Z^{-s/q}B, \quad R_3 = Z^{-s/r}C,$$

in terms of which it has the simpler presentation

(9·62) $$R_1^p = R_2^q = R_3^r = 1, \quad R_1R_2R_3 = R_2R_3R_1 = R_3R_1R_2.$$

In fact, the relations

$$A^p = B^q = C^r = ABC = Z^s; \quad Z \rightleftarrows A, B, C$$

imply $R_1^p = Z^{-s}A^p = 1$, similarly for R_2 and R_3, and

$$R_1R_2R_3 = Z^{-(s/p)-(s/q)-(s/r)+s} = Z^{-1}.$$

Conversely, the relations

(9·621) $$R_1^p = R_2^q = R_3^r = 1, \quad R_1R_2R_3 = R_2R_3R_1 = R_3R_1R_2 = Z^{-1},$$

along with $A = Z^{s/p}R_1$. $B = Z^{s/q}R_2$, $C = Z^{s/r}R_3$, imply

$$A^p = B^q = C^r = ABC = Z^s; \quad Z \rightleftarrows A, B, C.$$

The actual instances are

(9·63) $$\langle p,2,2\rangle_p\ (p \text{ even}), \quad \langle 3,3,2\rangle_6, \quad \langle 4,3,2\rangle_{12}, \quad \langle 5,3,2\rangle_{30},$$

of orders $4p^2$, 144, 576, 3600.

When p is odd, the group (9·62) (with $q = r = 2$) is still generated by reflections, and still has order $4p^2$ (as we shall see in (10·33)); but then $\langle p,2,2\rangle_p$ is a different group, *not* generated by reflections.

EXERCISE

Represent the generating reflections for $\langle 3,3,2\rangle_6$, $\langle 4,3,2\rangle_{12}$, $\langle 5,3,2\rangle_{30}$ as matrices over $GF[4s+1]$ by the method described at the end of §9·3.

9·7 TWO-GENERATOR SUBGROUPS

Interesting subgroups of $\langle p,q,r\rangle_s$ are generated by pairs of the three R_ν. Abandoning our tacit assumption that $p \geqslant q \geqslant r$, we may consider, for instance, the subgroup generated by R_1 and R_3. This is denoted by

$$p[2q]r$$

because it has the presentation

(9·71) $$R_1^p = R_3^r = 1, \quad (R_1R_3)^q = (R_3R_1)^q.$$

In fact, since R_2 is of period q and commutes with R_3R_1,

$$(R_1R_3)^q = R_1(R_3R_1)^q R_1^{-1} = R_1(R_2R_3R_1)^q R_1^{-1} = (R_1R_2R_3)^q$$
$$= (R_2R_3R_1)^q = (R_3R_1)^q.$$

To verify the sufficiency of this presentation, we begin with the group defined by (9·71) and consider the effect of adjoining a third generator Z which commutes with R_1 and R_3 and satisfies $Z^q = (R_3R_1)^{-q}$. Defining $R_2 = (ZR_3R_1)^{-1}$, we see that the relations

(9·72) $\quad R_1^p = R_3^r = 1, \quad (R_1R_3)^q = (R_3R_1)^q = Z^{-q}, \quad Z \rightleftarrows R_1, R_3$

are equivalent to (9·621). This completes the proof that $p[2q]r$ is generated by two of the three generators of $\langle p,q,r\rangle_s$. Being a subgroup of index q, its order is

$$4s^2/q.$$

In particular, $p[2q]2$, defined by

$$R_1^p = R_3^2 = 1, \quad (R_1R_3)^q = (R_3R_1)^q,$$

is a group of order $4s^2/q$, where s is now given by (3·14). Its two conjugate elements R_1 and $R_3R_1R_3$ generate a subgroup of index 2 (order $2s^2/q$) which is denoted by

$$p[q]p$$

because, if we change the notation by writing R_2 for $R_3R_1R_3$, it has the presentation

(9·73) $\quad R_1^p = R_2^p = 1, \quad (R_1R_2)^{q/2} = (R_2R_1)^{q/2},$

where, if q is odd, $(R_1R_2)^{q/2}$ is to be construed as $R_1R_2 \ldots R_1$, the product of q R's. The bigger group $p[2q]2$ can be reconstructed from the group so presented by adjoining R_3, of period 2, which transforms R_1 into R_2 (and vice versa).

To sum up, we have obtained four groups (9·63) generated by three reflections, and the following subgroups generated by two reflections:

$p[2]p$,	$p[4]2$,	$2[q]2$,	$3[3]3$,	$3[6]2$,
$3[4]3$,	$4[3]4$,	$3[8]2$,	$4[6]2$,	$4[4]3$,
$3[5]3$,	$5[3]5$,	$3[10]2$,	$5[6]2$,	$5[4]3$,

of orders

p^2,	$2p^2$,	$2q$,	24,	48,
72,	96,	144,	192,	288,
360,	600,	720,	1200,	1800.

Strictly speaking, the first two were obtained under the assumption that p is even. But we shall see in (10·34) that $p[4]2$ is still a reflection group of order $2p^2$ when p is odd. As for $p[2]p$, this is simply the direct product of two cyclic groups of order p.

EXERCISES

1. $p[3]p$ has the alternative presentations
   (i) $R_1^p = 1, \quad R_1R_2 = R_2R_3 = R_3R_1$;
   (ii) $B^3C^2 = (BC)^p = 1$.
2. $3[3]3 \cong \langle 3, 3, 2\rangle$.
3. Represent the generating reflections for $3[3]3$ and $3[4]3$ as matrices over $GF[5^2]$. Can $3[5]3$ be represented over the same field? Can $4[3]4$ be represented over $GF[3^2]$?

9·8 THE GROUP $p_1[q]p_2$ AND ITS INVARIANT HERMITIAN FORM

The three reflections

$$\exp(-\pi i/p)\,x\mathrm{A}, \quad \exp(-\pi i/q)\,x\mathrm{B}, \quad \exp(-\pi i/r)\,x\mathrm{C},$$

which generate $\langle p,q,r\rangle_s$ (see (9·61)), involve quaternions A, B, C representing rotations through angles $2\pi/p$, $2\pi/q$, $2\pi/r$ about the vertices of a 'clockwise' spherical triangle (pqr) whose angles are π/p, π/q, π/r. Therefore the two generating reflections for $p[2q]r$ involve quaternions representing two of these three rotations. In particular, the rotations for $p[2q]2$ are through $2\pi/p$ and π about the two ends of the side ψ of the triangle $(p\,q\,2)$ in Figure 3·1B. When we pass from $p[2q]2$ to its subgroup $p[q]p$, we are, in effect, reflecting the right-angled triangle in one of its catheti to obtain an isosceles triangle whose angles are π/p, $2\pi/q$, π/p, and the rotations are through $2\pi/p$ about the ends of the 'base' (Figure 9·8A). Combining these ideas, we see that there is a group

$$p_1[q]p_2$$

defined by

(9·81) $\quad R_1^{p_1} = R_2^{p_2} = 1, \quad R_1R_2R_1 \ldots = R_2R_1R_2 \ldots$

(with q factors on each side of the last relation, and $p_1 = p_2$ if q is odd), whose generators correspond (in the above manner) to rotations through $2\pi/p_1$ and $2\pi/p_2$ about the vertices P_1 and P_2 of a spherical triangle P_1P_2Q having angles π/p_1 and π/p_2 at these two vertices and $2\pi/q$ at the remaining vertex Q, as in Figure 9·8B. By a well-known theorem of spherical trigonometry (Coxeter 1965, p. 233), the arc P_1P_2 joining these two centres of rotation has length 2σ where

(9·82) $$\cos 2\sigma = \left(\cos\frac{\pi}{p_1}\cos\frac{\pi}{p_2} + \cos\frac{2\pi}{q}\right)\Big/\sin\frac{\pi}{p_1}\sin\frac{\pi}{p_2}.$$

Since the angle-sum of the triangle P_1P_2Q is

$$\left(\frac{1}{p_1}+\frac{1}{p_2}+\frac{2}{q}\right)\pi,$$

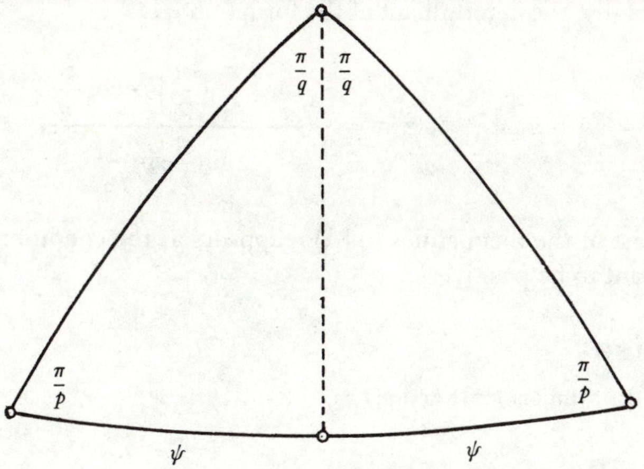

Figure 9·8A: $\left(p\,p\,\frac{q}{2}\right)$

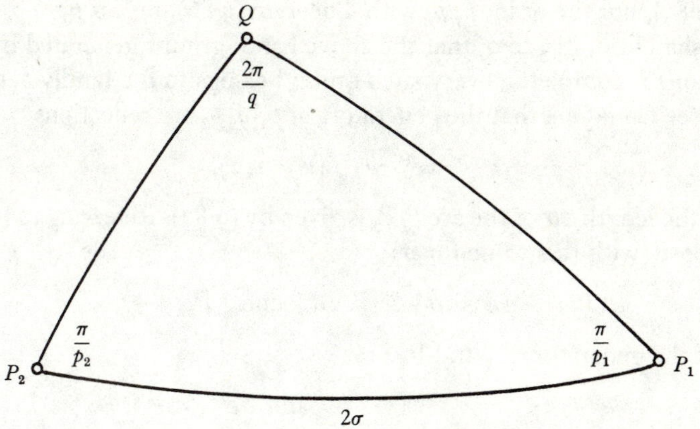

Figure 9·8B: $\left(p_1\,p_2\,\frac{q}{2}\right)$

the finite groups $p_1[q]p_2$ are given by the inequalities

$$(9{\cdot}83)\qquad p_\nu > 1,\quad q > 2,\quad \frac{1}{p_1}+\frac{1}{p_2}+\frac{2}{q} > 1.$$

The requirement that $p_1 = p_2$ when q is odd is an immediate consequence of the presentation (9·81). For, the relation

$$R_1(R_2R_1)^{(q-1)/2} = (R_2R_1)^{(q-1)/2}R_2$$

forces R_1 and R_2 to be conjugate. In fact, when q is odd, the *two* relations

$$(9{\cdot}84)\qquad R_1^p = 1,\quad R_1R_2R_1\ldots = R_2R_1R_2\ldots$$

suffice to define $p[q]p$.

For any q, the symbol $p[q]p$ may be regarded as a natural generalization of

$$2[q]2,$$

which is an embellished form of the symbol $[q]$ for the dihedral group of order $2q$, emphasizing the fact that its two generating reflections are of period 2. (See §4·8.)

We see from (9·72) that $Z^{-q}R_1 = (R_1R_3)^qR_1 = R_1(R_3R_1)^q = R_1Z^{-q}$, and similarly Z^{-q} commutes with R_3. Thus $(R_1R_3)^q$ generates the *centre* of $p[2q]r$. Since its quotient group is the ordinary polyhedral group (p, q, r), of order $2s$, the centre is of order $2s/q$: the period of $(R_1R_3)^q$ is $2s/q$, and the period of R_1R_3 itself is $2s$.

In the case of $p[q]p$, where q is odd, the centre is not generated by the element

$$R_1R_2\ldots R_1 = R_2R_1\ldots R_2$$

(which transforms R_2 into R_1, and vice versa) but by its square,

$$R_1R_2\ldots R_1R_2R_1\ldots R_2 = (R_1R_2)^q.$$

In terms of the generators of $p[2q]2$, this is $(R_1R_3)^{2q}$. Thus the order of the centre of $p[q]p$ (q odd) is s/q (where s is given by (3·14)): the period of $(R_1R_2)^q$ is s/q, and the period of R_1R_2 itself is s.

These apparently conflicting results can be reconciled by using the letter

$$h$$

to denote *the period of the product of the generators* of any reflection group, so that $h = 2s$ for $p[2q]r$, $h = s$ for $p[q]p$ and, in the case of $p_1[q]p_2$, h is given by

$$(9{\cdot}85)\qquad \frac{2}{h} = \frac{1}{p_1}+\frac{1}{p_2}+\frac{2}{q}-1.$$

The centre of $p_1[q]p_2$ has order $2h/q$ or h/q according as q is even or odd.

The expressions $4s^2/q$ and $2s^2/q$, for the orders of $p[2q]r$ and $p[q]p$, can be combined as follows:

(9·86) *The order of* $p_1[q]p_2$ *is* $2h^2/q$.

Although we have been careful to stipulate $q > 2$, we shall sometimes find it desirable to consider the group $p_1[2]p_2$ defined by

$$R_1^{p_1} = R_2^{p_2} = 1,\quad R_1R_2 = R_2R_1,$$

because this is the direct product of cyclic groups generated by two commutative reflections

$$e^{-\pi i/p_1}x\,e^{\pi i/p_1}\quad\text{and}\quad e^{\pi i/p_2}x\,e^{\pi i/p_2}$$

in perpendicular mirrors. However, the period of R_1R_2 does not agree with (9·85), nor the order $p_1 p_2$ with Theorem (9·86), unless $p_1 = p_2$.

We shall find, in § 10·9, that the above list of groups generated by *two* reflections is complete: every such group belongs to the family $p_1[q]p_2$.

We see from § 9·4 that the generators of $p_1[q]p_2$ are reflections

$$e^{i\pi/p_\nu} x e^{P_\nu \pi/p_\nu} \quad (\nu = 1, 2),$$

where the length 2σ of the arc P_1P_2 is given by (9·82). Referring to (9·44), we choose (with this value for σ)

$$P_1 = -i\cos 2\sigma + k \sin 2\sigma \quad \text{and} \quad P_2 = -i,$$

so that the generators R_1 and R_2 are

$$e^{i\pi/p_1} x e^{(-i\cos 2\sigma + k\sin 2\sigma)\pi/p_1} \quad \text{and} \quad e^{i\pi/p_2} x e^{-i\pi/p_2},$$

while the corresponding mirrors are

$$u \sin\sigma + v\cos\sigma = 0 \quad \text{and} \quad v = 0. \tag{9·87}$$

More explicitly, R_1 is the transformation (9·46) with $\exp(i\pi/p_1)$ for ϵ, p_1 for p; and R_2 is (9·47) with $\exp(i\pi/p_2)$ for ϵ.

More symmetrical expressions can be obtained by using 'oblique' coordinates, so that the mirrors for R_1 and R_2 are $x_1 = 0$ and $x_2 = 0$, as in § 8·9. Identifying these mirrors with (9·87), we set

$$x_1 = u_1 \sin\sigma + u_2\cos\sigma, \quad x_2 = -u_2$$

so that, in the notation of (8·69),

$$e_{11} = \sin\sigma, \quad e_{12} = \cos\sigma, \quad e_{21} = 0, \quad e_{22} = -1.$$

In terms of contravariant coordinates, we thus have

$$u_1 = e_{11}x^1 + e_{21}x^2 = x^1 \sin\sigma,$$

$$u_2 = e_{12}x^1 + e_{22}x^2 = x^1\cos\sigma - x^2.$$

It follows that the invariant Hermitian form is

$$u_1\bar{u}_1 + u_2\bar{u}_2 = x^1\bar{x}^1 \sin^2\sigma + (x^1\cos\sigma - x^2)(\bar{x}^1\cos\sigma - \bar{x}^2)$$

$$= x^1\bar{x}^1 - (x^1\bar{x}^2 + x^2\bar{x}^1)\cos\sigma + x^2\bar{x}^2, \tag{9·88}$$

with coefficients $\quad a_{11} = a_{22} = 1$

and, by (9·82),

$$a_{12} = a_{21} = -\cos\sigma = -\left[\frac{\cos\left(\frac{\pi}{p_1} - \frac{\pi}{p_2}\right) + \cos\frac{2\pi}{q}}{2\sin\frac{\pi}{p_1}\sin\frac{\pi}{p_2}}\right]^{\frac{1}{2}}. \tag{9·89}$$

The reflections R_1 and R_2 are given by (8·91) with this value for a_{12}.

Incidentally, the determinant of the form (9·88) is

$$1 - a_{12}a_{21} = \sin^2\sigma = -\frac{\cos\left(\frac{\pi}{p_1} + \frac{\pi}{p_2}\right) + \cos\frac{2\pi}{q}}{2\sin\frac{\pi}{p_1}\sin\frac{\pi}{p_2}}, \tag{9·891}$$

and the last of the inequalities (9·83) reappears as the condition for this determinant to be positive.

EXERCISE

Interpret the symbols $p[2]1$ and $p[1]p$.

9·9 REMARKS

The theory of linear groups in two complex variables was developed by F. Klein, L. Fuchs, C. Jordan, H. Valentiner and H. F. Blichfeldt (see Miller, Blichfeldt and Dickson 1916, p. 215). However, these early authors were less interested in the two-dimensional groups of affinities than in their central quotient groups, the one-dimensional groups of projectivities, which are isomorphic with the polyhedral groups (p, q, r). The finite groups generated by unitary reflections were completely enumerated (using a procedure quite different from ours) by Shephard and Todd.† The groups in their Tables I, II, III are, in our notation,

$$3[3]3, \quad 3[4]3, \quad 3[6]2, \quad \langle 3,3,2\rangle_6,$$

$$4[3]4, \quad 4[6]2, \quad 4[4]3, \quad \langle 4,3,2\rangle_{12},$$

$$GL(2,3), \quad \langle 4,3,2\rangle_2, \quad 3[8]2, \quad \langle 4,3,2\rangle_6,$$

$$5[3]5, \quad 5[6]2, \quad 5[4]3, \quad \langle 5,3,2\rangle_{30}, \quad 3[5]3, \quad 3[10]2, \quad \langle 5,3,2\rangle_2.$$

Crowe‡ saw the advantage of using quaternions, especially for the groups that need three generating reflections, namely

$$\langle 3,3,2\rangle_6, \quad \langle 4,3,2\rangle_{12}, \quad GL(2,3), \quad \langle 4,3,2\rangle_2,$$

$$\langle 4,3,2\rangle_6, \quad \langle 5,3,2\rangle_{30}, \quad \langle 5,3,2\rangle_2.$$

His Table I contains presentations closely resembling (9·54) and (9·62), but involving some superfluous relations. His Table II exhibits $p[2q]r$ as a subgroup of $\langle p,q,r\rangle_s$ (each case separately).

† G. C. Shephard and J. A. Todd, Finite unitary reflection groups, *Canadian Journal of Mathematics* **6** (1954), 274–304; see especially pp. 281–2.

‡ D. W. Crowe, Some two-dimensional unitary groups generated by three reflections, *Canadian Journal of Mathematics* **13** (1961), 418–26; see especially pp. 422–5.

The equation (9·111) shows that the group of unit quaternions is isomorphic to the *special unitary group* $SU(2)$. Du Val (1964, pp. vi, 39) gives a good historical account. However, the precise connection with the equation

$$(u, v)\begin{bmatrix} a & c \\ -\bar{c} & \bar{a} \end{bmatrix} = (au - \bar{c}v, cu + \bar{a}v)$$

is obscured by his preference for putting the (transposed) matrix on the left, and denoting the product of A and B by BA, so that his equations (16·3) involve a minus sign and an artificial re-ordering of w, x, y, z.

In a preliminary version of the present treatment† it was correctly remarked that *every finite unitary group generated by reflections is a subgroup of some* $\langle p, q, r\rangle_s$. However, the next sentence implies falsely that $\langle p, 2, 2\rangle_{2p}$ (with p odd) is generated by reflections. It should have been said that 'the group $\langle p, 2, 2\rangle_{2p}$, with p odd, is not generated by reflections, but its subgroup

$$(\mathfrak{C}_{4p}/\mathfrak{C}_{2p}; \langle p, 2, 2\rangle/\mathfrak{C}_{2p}),$$

generated by Z^2A, Z^pB, Z^pC, is a subgroup of index 4 in $\langle s, 2, 2\rangle_s$ with $s = 2p$'. (See the answer on page 174 to the Exercise for § 10·3.)

† Coxeter, Finite groups generated by unitary reflections, *Abhandlungen aus dem Mathematischen Seminar der Universität Hamburg* **31** (1967), 125–35; see especially p. 127. There is also a printer's error on p. 133, where the order of 'No. 24' appears as 14.4! instead of 4.4!.

CHAPTER 10

The complete enumeration of finite reflection groups in the unitary plane

> When the late Sophus Lie...was asked to name the characteristic endowment of the mathematician, his answer was the following quaternion: Phantasie, Energie, Selbstvertrauen, Selbstkritik.
>
> C. J. Keyser (1907, p. 31)

In § 10·1 we use the method of Du Val (1964) to make a complete list of the finite unitary groups in the plane. In the remaining sections of this chapter we examine each one to see whether it can be generated by reflections. The results are not new, but the method is interestingly different from that used by Shephard and Todd. It is found that their groups $G(mn, n, 2)$ (see (10·44)) are the only finite plane reflection groups not already discussed in Chapter 9.

10·1 THE FINITE UNITARY GROUPS IN THE PLANE

The discussion in §§ 7·1, 7·2, 7·3 shows that the only cases in which one finite group of quaternions has another as a normal subgroup are

$$\langle p,q,r\rangle/\mathfrak{C}_2 \cong (p,q,r),$$

$$\mathfrak{C}_{pq}/\mathfrak{C}_p \cong \mathfrak{C}_q, \quad \langle pq,2,2\rangle/\mathfrak{C}_{2p} \cong (q,2,2),$$

$$\langle pq,2,2\rangle/\mathfrak{C}_p \cong \langle q,2,2\rangle \quad (p \text{ odd}),$$

$$\langle 2p,2,2\rangle/\langle p,2,2\rangle \cong \mathfrak{C}_2, \quad \langle 4,3,2\rangle/\langle 3,3,2\rangle \cong \mathfrak{C}_2,$$

$$\langle 3,3,2\rangle/\langle 2,2,2\rangle \cong \mathfrak{C}_3, \quad \langle 4,3,2\rangle/\langle 2,2,2\rangle \cong (3,2,2).$$

Picking out the cases in which the quotient group is cyclic (including $\langle 1,2,2\rangle \cong \mathfrak{C}_4$), we deduce the following list of *finite groups of unitary transformations in the plane*:

1.	$(\mathfrak{C}_{2m}/\mathfrak{C}_f;\ \mathfrak{C}_{2n}/\mathfrak{C}_g)_d$,	of order $gm = fn$
2.	$\langle p,2,2\rangle_m$,	$4mp$
3.	$(\mathfrak{C}_{4m}/\mathfrak{C}_{2m};\ \langle p,2,2\rangle/\mathfrak{C}_{2p})$,	$4mp$
3′.	$(\mathfrak{C}_{4m}/\mathfrak{C}_m;\ \langle p,2,2\rangle/\mathfrak{C}_p)$, m and p odd,	$2mp$
4.	$(\mathfrak{C}_{4m}/\mathfrak{C}_{2m};\ \langle 2p,2,2\rangle/\langle p,2,2\rangle)$,	$8mp$
5.	$\langle 3,3,2\rangle_m$,	$24m$
6.	$(\mathfrak{C}_{6m}/\mathfrak{C}_{2m};\ \langle 3,3,2\rangle/\langle 2,2,2\rangle)$,	$24m$
7.	$\langle 4,3,2\rangle_m$,	$48m$
8.	$(\mathfrak{C}_{4m}/\mathfrak{C}_{2m};\ \langle 4,3,2\rangle/\langle 3,3,2\rangle)$,	$48m$
9.	$\langle 5,3,2\rangle_m$,	$120m$

The groups have been numbered from 1 to 9 in agreement with Du Val's table of finite groups of isometries in real Euclidean 4-space (Du Val 1964, p. 57). The rest of his table does not concern us here, because the real isometry $\tilde{a}xb$ (see Theorem (6·78)) can be interpreted as a unitary transformation only when the left multiplier a is complex (or real), that is, when the groups $\mathbf{L}$ and $\mathbf{L}_K$ in his symbol $(\mathbf{L}/\mathbf{L}_K; \mathbf{R}/\mathbf{R}_K)$ are cyclic.

In cases 2, 5, 7, 9, we have made use of the concise notation $\langle p,q,r\rangle_m$ which was introduced in (9·31).

10·2 REFLECTION GROUPS OF TYPE 1

Having established the notation $\langle p,q,r\rangle_s$ (see (9·62)) and $p_1[q]p_2$ (see (9·81)), let us begin to look systematically through the above list of finite unitary groups, to determine those that are generated by reflections.

We saw, in § 9·2, that all the tranformations in

$$(\mathfrak{C}_{2m}/\mathfrak{C}_f;\ \mathfrak{C}_{2n}/\mathfrak{C}_g)_d$$

are of the form $e^{\mu\pi i/m}x\,e^{d\nu\pi i/n}$, where $\mu \equiv \nu \pmod{\lambda m/f}$. By Theorem (9·42), such a transformation is a reflection if and only if the two exponents are equal or opposite. Hence the only group of this type generated by *one* reflection is

(10·21) $$p[2]1 \cong p[1]p \cong (\mathfrak{C}_{2p}/\mathfrak{C}_1;\ \mathfrak{C}_{2p}/\mathfrak{C}_1)_1,$$

generated by either $e^{\pi i/p}x\,e^{\pi i/p}$ or $e^{-\pi i/p}x\,e^{\pi i/p}$. The only other reflection groups of this type are generated by two commutative reflections

$$e^{\pi i/p}x\,e^{\pi i/p} \quad \text{and} \quad e^{-\pi i/q}x\,e^{\pi i/q},$$

which combine to form $e^{(aq-bp)\,\pi i/pq}x\,e^{(aq+bp)\,\pi i/pq}$.

Let f denote the greatest common divisor of p and q, and m their least common multiple: $f = (p,q), \quad m = pq/f.$

Let us also write $q' = \dfrac{q}{f} = \dfrac{m}{p}, \quad p' = \dfrac{p}{f} = \dfrac{m}{q},$

so that $p'q' = m/f$. Then the above transformation becomes

$$e^{(aq'-bp')\pi i/m}\, x\, e^{(aq'+bp')\pi i/m}.$$

In particular, when $a = p'$ and $b = q'$, it is $x \exp(2\pi i/f)$.

Since $(p', q') = 1$, we can choose a and b so that

$$aq' - bp' = 1, \quad 0 < a < p', \quad 0 \leqslant b < q'.$$

Let us define d to be the smaller of

$$aq' + bp' \quad \text{and} \quad 2p'q' - (aq' + bp').$$

Then the chosen values of a and b yield the transformation

$$e^{\pi i/m}\, x\, e^{d\pi i/m} \quad \text{or} \quad e^{\pi i/m}\, x\, e^{(2\pi i/f) - d\pi i/m},$$

and the group is the direct product

(10·22) $$p[2]q \cong \mathfrak{C}_p \times \mathfrak{C}_q \cong (\mathfrak{C}_{2m}/\mathfrak{C}_f; \mathfrak{C}_{2m}/\mathfrak{C}_f)_d.$$

Since $d = aq' + bp' \quad \text{or} \quad (q'-b)p' + (p'-a)q'$

and $(q'-b)p' - (p'-a)q' = aq' - bp' = 1,$

we may define d as the unique (odd) positive integer, less than $p'q'$, which is the sum of two consecutive integers, one divisible by p' and the other by q'. For instance, if $q' = 1$, $d = 1$.

Hence the only groups of 'type 1' that are generated by reflections are

$$(\mathfrak{C}_{2m}/\mathfrak{C}_f; \mathfrak{C}_{2m}/\mathfrak{C}_f)_d,$$

where m is a multiple of f, and d is derived by the above rule from any decomposition of m/f into relatively prime factors p' and q'. In particular, every such group with $d = 1$ is a reflection group:

(10·23) $$(\mathfrak{C}_{2m}/\mathfrak{C}_f; \mathfrak{C}_{2m}/\mathfrak{C}_f)_1 \cong m[2]f \quad (f \mid m).$$

EXERCISES

1. In the discussion just concluded, f divides m; but in §9·2, f divides $2m$. Explain the discrepancy.

2. Express, in the notation $p[2]q$, the groups

$$(\mathfrak{C}_{120}/\mathfrak{C}_2; \mathfrak{C}_{120}/\mathfrak{C}_2)_d \quad \text{for} \quad d = 1, 11, 19, 29.$$

10·3 REFLECTION GROUPS OF TYPES 2 AND 3

We see from (7·14) that the $4mp$ transformations

$$e^{\mu\pi i/m}\, x\, e^{\nu\pi i/p}, \quad e^{\mu\pi i/m}\, ix\, e^{\nu\pi i/p} j \quad (\mu = 0, \ldots, 2m-1; \nu = 0, \ldots, p-1)$$

constitute the group $\langle p, 2, 2\rangle_m$ or $(\mathfrak{C}_{4m}/\mathfrak{C}_{2m}; \langle p, 2, 2\rangle/\mathfrak{C}_{2p})$ according as m is even or odd. (In the former case we could replace μ by $\mu - \frac{1}{2}$ so as to replace $\exp(\mu\pi i/m)\, i$ by $\exp(\mu\pi i/m)$ without the extra i.) Possible generating reflections are

$$e^{\pi i/m}\, x\, e^{\pi i/m}, \quad ixj, \quad ix\, e^{\pi i/p} j.$$

The first of these belongs to the group only if m divides p. Writing mq for p, we deduce that, for all positive integers m and q, the three reflections

$$e^{\pi i/m}\, x\, e^{\pi i/m}, \quad ixj, \quad ix\, e^{\pi i/mq} j$$

(of periods m, 2, 2) generate the group

(10·31) $$\langle mq, 2, 2\rangle_m \quad (m \text{ even}) \quad \text{or} \quad (\mathfrak{C}_{4m}/\mathfrak{C}_{2m}; \langle mq, 2, 2\rangle/\mathfrak{C}_{2mq}) \quad (m \text{ odd}),$$

of order $4m^2q$.

When $m = 1$, the first generator is merely the identity, and we have the dihedral group $[2q]$, of order $4q$, generated by

$$ixj \quad \text{and} \quad ix\, e^{\pi i/q} j,$$

in agreement with (9·51) (except that there j and k were interchanged, as in (6·64)).

When $q = 1$ (and $m = p$), we have the group

(10·32) $$\langle p, 2, 2\rangle_p \quad (p \text{ even}) \quad \text{or} \quad (\mathfrak{C}_{4p}/\mathfrak{C}_{2p}; \langle p, 2, 2\rangle/\mathfrak{C}_{2p}) \quad (p \text{ odd}),$$

of order $4p^2$, generated by reflections

$$\epsilon x \epsilon, \quad ixj, \quad ix\epsilon j,$$

where $\epsilon = \exp(\pi i/p)$. Calling these generators R_1, R_2, R_3, we easily verify

(10·33) $$R_1^p = R_2^2 = R_3^2 = 1, \quad R_1R_2R_3 = R_2R_3R_1 = R_3R_1R_2.$$

These relations suffice for a presentation, since they yield $[2p]$ for the subgroup of index p generated by R_2 and R_3. On the other hand, the subgroup of index 2 generated by R_1 and R_2 is defined by

(10·34) $$R_1^p = R_2^2 = 1, \quad (R_1R_2)^2 = (R_2R_1)^2;$$

and thus we see that the order of $p[4]2$ is $2p^2$ (in agreement with theorem (9·86)) not only when p is even but also when p is odd.

EXERCISE

Is $\langle p, 2, 2\rangle_{2p}$ a reflection group?

10·4 REFLECTION GROUPS OF TYPES 3′ AND 4

Similarly, the $2mp$ transformations

$$e^{\mu\pi i/m}\, x\, e^{\nu\pi i/p}, \quad e^{\mu\pi i/m}\, ix\, e^{\nu\pi i/p} j$$

$$(\mu = 0, \ldots, 2m-1; \quad \nu = 0, \ldots, p-1; \quad \mu \equiv \nu \bmod 2)$$

constitute the group

$$(\mathfrak{C}_{4m}/\mathfrak{C}_m; \langle p, 2, 2\rangle/\mathfrak{C}_p) \quad \text{or} \quad (\mathfrak{C}_{2m}/\mathfrak{C}_m; \langle p, 2, 2\rangle/\langle \tfrac{1}{2}p, 2, 2\rangle)$$

according as m and p (both together) are odd or even. If m divides p, and p/m is odd, the group is generated by the three reflections

$$e^{\pi i/m}xe^{\pi i/m}, \quad ixj, \quad ixe^{2\pi i/p}j.$$

(If p/m were even, no product of these reflections could yield $e^{\pi i/m}ixe^{\pi i/p}j$.) Writing mq for p, we deduce that, *if q is odd*, the three reflections

$$e^{\pi i/m}xe^{\pi i/m}, \quad ixj, \quad ixe^{2\pi i/mq}j$$

(of periods m, 2, 2) generate the group

$$(10{\cdot}41) \qquad (\mathfrak{C}_{4m}/\mathfrak{C}_m; \langle mq, 2, 2\rangle/\mathfrak{C}_{mq}) \qquad (m \text{ odd})$$

or

$$(\mathfrak{C}_{2m}/\mathfrak{C}_m; \langle mq, 2, 2\rangle/\langle \tfrac{1}{2}mq, 2, 2\rangle) \qquad (m \text{ even}),$$

of order $2m^2q$.

Calling the three generators R_1, R_2, R_3, we observe that when $q = 1$, $R_2R_1^{-1}R_2R_1R_2 = R_3$, so that R_1 and R_2 suffice to generate the group. Then, since

$$R_1^m = R_2^2 = 1 \quad \text{and} \quad (R_1R_2)^2 = (R_2R_1)^2,$$

(9·71) shows that $m[4]2$ can be represented as

$$(10{\cdot}42) \quad (\mathfrak{C}_{4m}/\mathfrak{C}_m; \langle m, 2, 2\rangle/\mathfrak{C}_m) \quad \text{or} \quad (\mathfrak{C}_{2m}/\mathfrak{C}_m; \langle m, 2, 2\rangle/\langle \tfrac{1}{2}m, 2, 2\rangle)$$

according as m is odd or even[†].

On the other hand, when $m = 1$ (for any q) we have simply the dihedral group $[q]$, in agreement with (9·52).

These results may be combined with those of §10·3 to yield the following theorem:

(10·43) *Every reflection group of type 2, 3, 3′ or 4 can be generated by the three reflections*

$$e^{\pi i/m}xe^{\pi i/m}, \quad ixj, \quad ixe^{2\pi i/mn}j.$$

Its order is $2m^2n$. It is of type 2, 3, 3′, or 4 according as m and n are both even, odd and even, both odd, or even and odd.

The generators are easily seen to satisfy

$$(10{\cdot}44) \quad R_1^m = R_2^2 = R_3^2 = 1, R_2R_1R_2 = R_3R_1R_3 = R_1(R_2R_3)^n.$$

Shephard[‡] has proved that these relations suffice for a presentation of this group of order $2m^2n$, which he denotes by

$$G(mn, n, 2).$$

† D. W. Crowe, The groups of regular complex polygons, *Canadian Journal of Mathematics* **13** (1961), 155.

‡ G. C. Shephard, Abstract definitions for reflection groups, *Canadian Journal of Mathematics* **9** (1957), 273–6; see especially p. 275.

EXERCISES

1. What subgroup of $G(mn, n, 2)$ is generated by (i) R_1 and R_2? (ii) R_2 and R_3?
2. Establish the sufficiency of the relations (10·44) in the special cases (i) $m = 1$, (ii) $n = 1$, (iii) $n = 2$.

10·5 REFLECTION GROUPS OF TYPE 5

We see from (7·22) and (9·61) that $\langle 3, 3, 2\rangle_6$, of order 144, is generated by three reflections

$$\exp(\pi i/3)\,xa, \quad \exp(\pi i/3)\,xb, \quad ixi$$

($a = \frac{1}{2}(1+i+j+k)$, $b = \frac{1}{2}(1+i-j+k)$), with the presentation

$$(10{\cdot}51) \quad R_1^3 = R_2^3 = R_3^2 = 1, \quad R_1R_2R_3 = R_2R_3R_1 = R_3R_1R_2.$$

For any other reflection group $\langle 3, 3, 2\rangle_m$, m must divide 6. The value $m = 2$ is excluded because the reflections

$$ixi, \quad ixj, \quad ixk$$

do not generate the whole group $\langle 3, 3, 2\rangle_2$ but only its subgroup $\langle 2, 2, 2\rangle_2$. Thus we are left with

$$(10{\cdot}52) \qquad \langle 3, 3, 2\rangle_3 \cong 3[4]3,$$

of order 72, generated by

$$(10{\cdot}53) \qquad \exp(\pi i/3)\,xa \quad \text{and} \quad \exp(\pi i/3)\,xb,$$

that is, by $\exp(\pi i/3)\,x(1+i\pm j+k)/2$, with the presentation

$$R_1^3 = R_2^3 = 1, \quad (R_1R_2)^2 = (R_2R_1)^2.$$

10·6 REFLECTION GROUPS OF TYPE 6

If a group

$$(\mathfrak{C}_{6m}/\mathfrak{C}_{2m}; \langle 3, 3, 2\rangle/\langle 2, 2, 2\rangle)$$

is generated by reflections, it must be a subgroup of $\langle 3, 3, 2\rangle_6$; therefore $m = 1$ or 2. When $m = 2$, we have

$$3[6]2,$$

of order 48, generated by $\exp(\pi i/3)\,xa$ and ixi, with the presentation

$$(10{\cdot}61) \qquad R_1^3 = R_3^2 = 1, \quad (R_1R_3)^3 = (R_3R_1)^3;$$

and when $m = 1$ we have its subgroup

$$3[3]3,$$

of order 24, generated by R_1 and $R_3R_1R_3 = R_2$, with the presentation

$$(10{\cdot}62) \qquad R_1^3 = 1, \quad R_1R_2R_1 = R_2R_1R_2.$$

10·7 REFLECTION GROUPS OF TYPE 7

In the notation of (7·31) and (7·32), every element of $\langle 4, 3, 2\rangle$ is conjugate to a power of $A = \exp(\pi i/4)$, of $B = \frac{1}{2}(1+i+j+k)$, or of $C = 2^{-\frac{1}{2}}(i+k)$. We see from (9·61) that $\langle 4, 3, 2\rangle_{12}$, of order 576, is generated by three reflections

(10·71) $\quad R_1 = AxA, \quad R_2 = \exp(\pi i/3)\,xB, \quad R_3 = ixC,$

with the presentation

(10·72) $\quad R_1^4 = R_2^3 = R_3^2 = 1, \quad R_1R_2R_3 = R_2R_3R_1 = R_3R_1R_2.$

For any other reflection group $\langle 4, 3, 2\rangle_m$, m must divide 12. The value $m = 3$ is excluded because the reflections (10·53) do not generate the whole group $\langle 4, 3, 2\rangle_3$ but only its subgroup $3[4]3$. Thus we are left with

$$\langle 4, 3, 2\rangle_2, \quad \langle 4, 3, 2\rangle_4, \quad \langle 4, 3, 2\rangle_6,$$

of orders 96, 192, 288.

The first of these has the presentation (9·54) with $p = 4$. The second is

(10·73) $$\langle 4, 3, 2\rangle_4 \cong 4[6]2,$$

generated by AxA and ixC, that is, by R_1 and R_3. For the remaining group $\langle 4, 3, 2\rangle_6$, there is no obviously 'best' choice of three generating reflections. One possible choice, such that the axes of the corresponding rotations are coplanar, consists of the three reflections

(10·74) $$\exp(\pi i/3)\,xB, \quad ixC, \quad ixj.$$

The first and second of these, being the R_2 and R_3 of (10·71), generate the subgroup $3[8]2$ of $\langle 4, 3, 2\rangle_{12}$. In this subgroup

(10·75) $$R_2^3 = R_3^2 = 1, \quad (R_2R_3)^4 = (R_3R_2)^4,$$

(9·85) shows that R_2R_3 has period 24. We easily verify that $(R_2R_3)^{12}$ is the transformation $-x$. The remaining generator, ixj, of $\langle 4, 3, 2\rangle_6$, may be described as a new element R_4, of period 2, which transforms $3[8]2$ in such a way as to replace R_2 by $R_3R_2R_3$, and R_3 by $(R_2R_3)^{12}R_3$, so that

$$R_4^2 = 1, \quad R_4R_2R_4 = R_3R_2R_3, \quad R_4R_3R_4 = (R_2R_3)^{12}R_3.$$

Taking these relations (slightly rearranged) along with (10·75), we obtain the following presentation for $\langle 4, 3, 2\rangle_6$:

(10·76) $$R_2^3 = R_3^2 = R_4^2 = 1, \quad (R_2R_3)^4 = (R_3R_2)^4, \quad R_2 \rightleftarrows R_3R_4,$$
$$(R_2R_3)^{12} = (R_3R_4)^2.$$

A presentation for this 'No 15', with a different set of three generating reflections, was given by Crowe in his Table 1 (see the second footnote to §9·9).

EXERCISES

1. Among the generators of $\langle 4, 3, 2\rangle_{12}$, R_1 and R_2 generate $4[4]3$, of order 288. Is this $\langle 4, 3, 2\rangle_6$?
2. What subgroups of $\langle 4, 3, 2\rangle_6$ are generated by (i) R_3 and R_4; (ii) R_2 and R_4?

10·8 REFLECTION GROUPS OF TYPE 8

If a group $$(\mathfrak{C}_{4m}/\mathfrak{C}_{2m};\ \langle 4, 3, 2\rangle/\langle 3, 3, 2\rangle)$$ is generated by reflections, it must be a subgroup of $\langle 4, 3, 2\rangle_{12}$; therefore m divides 6.

When $m = 1$, so that the reflections are of period 2, we have the group $GL(2, 3)$, of order 48, defined by (9·56).

When $m = 2$, we have $4[3]4$, of order 96, generated by

$$AxA \quad \text{and} \quad AxB^{-1}AB$$

(in the notation of (7·32)), with the presentation

$$R_1^4 = 1, \quad R_1R_2R_1 = R_2R_1R_2.$$

When $m = 3$, we have $3[8]2$, of order 144, which we discussed in the preceding section.

When $m = 6$, we have $4[4]3$, of order 288, generated by AxA and $\exp(\pi i/3)\,xB$, with the presentation

$$R_1^4 = R_2^3 = 1, \quad (R_1R_2)^2 = (R_2R_1)^2.$$

EXERCISES

1. What subgroups of $(\mathfrak{C}_4/\mathfrak{C}_2;\ \langle 4, 3, 2\rangle/\langle 3, 3, 2\rangle)$ are generated by (i) R_2 and R_3; (ii) R_3 and R_1?
2. Which two-generator groups have the same orders as the three-generator groups
$$GL(2, 3), \quad \langle 4, 3, 2\rangle_2, \quad \langle 3, 3, 2\rangle_6, \quad \langle 4, 3, 2\rangle_6?$$

10·9 REFLECTION GROUPS OF TYPE 9

In the notation of (6·65), every element of $\langle 5, 3, 2\rangle$ is conjugate to a power of $A = \frac{1}{2}(\tau+\tau^{-1}i+k)$ or of $B = \frac{1}{2}(1+\tau^{-1}j+\tau k)$, or of k. We see from (9·61) that $\langle 5, 3, 2\rangle_{30}$, of order 3600, is generated by three reflections

$$\exp(\pi i/5)\,xA, \quad \exp(\pi i/3)\,xB, \quad ixk,$$

with the presentation

(10·91) $\quad R_1^5 = R_2^3 = R_3^2 = 1, \quad R_1R_2R_3 = R_2R_3R_1 = R_3R_1R_2.$

For any other reflection group $\langle 5, 3, 2\rangle_m$, m must divide 30. Thus we have the possible values

$$m = 2, 3, 5, 6, 10, 15,$$

giving groups of orders 240, 360, 600, 720, 1200, 1800.

$\langle 5,3,2\rangle_2$ has the presentation (9·54) with $p = 5$.

Omitting the generators of $\langle 5,3,2\rangle_{30}$, one by one, we obtain:

(10·92) $\langle 5,3,2\rangle_6 \cong 3[10]2$, generated by R_2 and R_3,

(10·93) $\langle 5,3,2\rangle_{10} \cong 5[6]2$, generated by R_1 and R_3,

(10·94) $\langle 5,3,2\rangle_{15} \cong 5[4]3$, generated by R_1 and R_2.

Finally, the first two of these three groups have subgroups of index 2:

(10·95) $\langle 5,3,2\rangle_3 \cong 3[5]3$,

generated by R_2 and $R_3R_2R_3$, that is, by

$$\exp(\pi i/3)\,x\mathrm{B} \quad \text{and} \quad \exp(\pi i/3)\,x(-k\mathrm{B}k),$$

and

(10·96) $\langle 5,3,2\rangle_5 \cong 5[3]5$,

generated by R_1 and $R_3R_1R_3$, that is, by

$$\exp(\pi i/5)\,x\mathrm{A} \quad \text{and} \quad \exp(\pi i/5)\,x(-k\mathrm{A}k).$$

We have now completed our verification that the only finite reflection groups in the unitary plane are the direct products $p[2]r \cong \mathfrak{C}_p \times \mathfrak{C}_r$ (including the cyclic groups $p[2]1 \cong \mathfrak{C}_p$), the group $G(pq,q,2)$ of order $2p^2q$ (see (10·44)) which is $2[q]2$ when $p = 1$ and $p[4]2$ when $q = 1$, the remaining two-generator groups $p_1[q]p_2$ $(q > 2)$ of order $2h^2/q$ (see (9·85) and Table II on page 156), and the remaining three-generator groups

$$GL(2,3), \quad \langle p,q,2\rangle_2, \quad \langle p,q,2\rangle_s, \quad \langle 4,3,2\rangle_6,$$

of orders 48, $8s$, $4s^2$, 288 (see (9·56), (9·55), (9·62), (10·76) and Table III).

EXERCISES

1. It is always true that $\langle p,q,2\rangle_p \cong p[q]p$ and $\langle p,q,r\rangle_{pr} \cong p[2q]r$?
2. Which reflection groups can be expressed as direct products?

CHAPTER 11

Regular complex polygons and Cayley diagrams

> A mathematician, like a painter or a poet, is a maker of patterns... The mathematician's patterns, like the painter's or the poet's, must be *beautiful*; the ideas, like the colours or the words, must fit together in a harmonious way. Beauty is the first test: there is no permanent place in the world for ugly mathematics.
>
> G. H. Hardy (1967, pp. 84, 85)

According to Sommerville (1929, p. 96), a polygon may be either *simple* or *complex*. It is *simple* if the number of vertices on an edge, and the number of edges through a vertex, are both 2, and *complex* if either of those numbers is greater than 2. He was evidently using the word 'complex' in its colloquial sense of 'complicated', without noticing the possible connection with complex numbers. He cited as instances the complete quadrangle and complete quadrilateral of elementary projective geometry. A more interesting example is the configuration 8_3 or $3\{3\}3$ (see §4·8) which answers a question first asked by Möbius: Can two simple quadrangles be so placed that the four vertices of each lie on the four sides of the other? In §11·1, the notion of a regular complex polygon is made precise, and such figures are completely enumerated. In §11·2, the connection with four-dimensional polytopes, already noticed in Chapter 4, is explained. In §11·3, some vertices of a complex polygon are seen to form a simple polygon that may reasonably be called a Petrie polygon. In §11·4 and §11·5 we find that, when q is even, the complex polygon $p_1\{q\}p_2$ can be interpreted as a Cayley diagram for a reflection group with a presentation such as (9·81) (with $p_1 = p_2$) or (9·62). In §11·6, some of these ideas are extended from finite polygons to apeirogons (or one-dimensional honeycombs). Finally, §11·7 employs Cayley diagrams in a different manner: to provide a general principle from which Theorem (7·42) can be deduced without the tiresome consideration of separate cases.

11·1 REGULAR COMPLEX POLYGONS

In §1·5, we derived a 'regular polygon' from the orbit of a point for the cyclic group generated by an isometry. The generalization suggested in §4·8 requires a different description, more like the definition of a regular polyhedron in §2·2. In the complex affine plane with a unitary metric, let us define a *polygon* to be a finite figure consisting of points (not all on one line) called *vertices*, and lines (not all through one point) called *edges*, satisfying the following two conditions:

(i) Every edge is incident with at least two vertices, and every vertex with at least two edges.

(ii) Any two vertices are connected by a 'chain' of successively incident edges and vertices.

The centroid of all the vertices is called the *centre* of the polygon. Similarly, the centroid of the vertices on an edge is the *centre* of the edge.

The group of all unitary transformations that preserve the incidences is called the *symmetry group* of the polygon. The figure consisting of a vertex and an incident edge is called a *flag*. The polygon is said to be *regular* if its symmetry group is *transitive on its flags*.

Consider a flag consisting of a vertex O_0 and an incident edge whose centre is O_1. By Condition (i), O_1 cannot coincide with O_0, and the line O_0O_1 cannot pass through the centre O_2 of the whole polygon; thus $O_0O_1O_2$ is a triangle. Since the symmetry group consists of unitary transformations, which are affinities, and an affinity is determined by its effect on three non-collinear points, the symmetry group of a regular polygon is *sharply* transitive on the flags: the order of the group, say g, is equal to the number of flags. Hence, if there are p_1 vertices on an edge and p_2 edges through a vertex, there are altogether g/p_1 edges and g/p_2 vertices.

If A and B are two of the vertices on an edge AB, the symmetry operation that relates the flags (A, AB) and (B, AB) leaves invariant the centre O_1 of the edge as well as the centre O_2 of the whole polygon; it is therefore a *reflection* in the line O_1O_2. There is a group of such reflections permuting all the vertices on the given edge; and since these reflections all have the same mirror, the group is cyclic, of order p_1. When the complex line AB is represented by an Argand diagram, the cyclic group of reflections appears as a cyclic group of rotations, generated by a 'primitive' rotation through $2\pi/p_1$ about a point P_1. The corresponding generator R_1 of the cyclic group of reflections is

$$e^{\pi i/p_1} x\, e^{\pi \mathrm{P}_1/p_1}$$

in the notation of §9·4. As in §4·8, we call the edge a p_1-edge; it appears in the Argand diagram as a real polygon $\{p_1\}$.

Similarly, the p_2 edges through a given vertex O_0 are permuted by a cyclic group of reflections all having the same mirror O_0O_2. If O_0 is A (or B), we may denote a 'primitive' generator of this cyclic group by R_2 (of period p_2). Since every symmetry operation yields a flag and a corresponding triangle like $O_0O_1O_2$ (right-angled at O_1), the whole symmetry group is generated by the two reflections R_1 and R_2. In other words,

(11·11) *The symmetry group of any regular polygon is generated by two reflections.*

Conversely, any two reflections that generate a finite group determine a regular polygon by Wythoff's construction, as in §1·6. If O_0O_2 and O_1O_2 are the two mirrors, with O_0O_1 perpendicular to the second one, the vertices are the images of the point O_0, and the edges are the images of the line O_0O_1. The polygon is denoted by

$$p_1\{q\}p_2,$$

where q is a certain function of the angle between the mirrors (so chosen that, in the special case when R_1 and R_2 are the usual generators for a group $p_1[q]p_2$, the same q appears in the symbol for the polygon). When the roles of the two mirrors are interchanged, the same construction yields the reciprocal polygon

$$p_2\{q\}p_1,$$

which has g/p_1 vertices, g/p_2 edges (and, of course, the same symmetry group).

Figure 9·8B indicates the classical construction for the product of rotations through $2\pi/p_1$ and $2\pi/p_2$ about two points P_1 and P_2 on the unit sphere. Great-circle arcs are drawn so as to form a clockwise spherical triangle P_1P_2Q with angles π/p_1 at P_1, π/p_2 at P_2, and (say) $2\pi/q$ at Q. The product is the rotation through $-4\pi/q$ about Q, and q is expressed in terms of the arc $P_1P_2 = 2\sigma$ by the formula (9·82). These rotations about P_1 and P_2 represent reflections R_1 and R_2 whose mirrors form an angle σ in the sense of (8·43). Thus at last we have found a precise meaning for the number q in the symbol $p_1\{q\}p_2$.

Since R_1 and R_2 generate the symmetry group of a finite polygon, q must be *rational*. In the original application of (9·83), q was an integer, and the group was $p_1[q]p_2$. In the present application, the group (being generated by two reflections) is still of the form $p_1[q']p_2$ for some integer q', but the new R_1 and R_2 may well be a different pair of reflections that serve to generate the same group (as when the dihedral group $2[5]2$ is generated by real reflections in lines forming an angle $2\pi/5$ instead of $\pi/5$).

Since the rotations about P_1 and P_2 generate a finite group, P_1P_2Q must be one of the *Schwarz triangles* that were enumerated in §2·5. Thus the problem of finding all the regular complex polygons is reduced to that of selecting, from Table I, those Schwarz triangles $(p_1\ p_2\ \tfrac{1}{2}q)$ for which p_1 and p_2 are integers. The resulting list of polygons $p_1\{q\}p_2$ appears in the third column of Table IV on pages 158–9.

For each reflection group $p_1[q]p_2$, the table shows first the 'non-starry' polygon $p_1\{q\}p_2$, then (if $p_1 \neq p_2$) its reciprocal $p_2\{q\}p_1$, and then (in certain cases) the 'star derivatives' which have the same values for p_1 and p_2 but smaller values for q and consequently greater values for σ.

In the characteristic triangle $O_0O_1O_2$ (right-angled at O_1), the side $l = O_0O_1$ is the *radius* of the edge (analogous to the half-edge of a real polygon). The other two sides, which form the angle σ, are the *circumradius* and *inradius* of the polygon:

$${}_0R = O_0O_2 = l\operatorname{cosec}\sigma, \quad {}_1R = O_1O_2 = l\cot\sigma.$$

When a star derivative is placed so as to have the same vertices as its non-starry parent, its edges contain different sets of p_1 points. Since ${}_0R$ is the same while σ is greater, l is greater and ${}_1R = {}_0R\cos\sigma$ is smaller. Dually, when the derivative is placed so as to have the same edge-lines as its parent, its vertices are different points on these same lines. Since ${}_1R$ is the same while σ is greater, l is greater and ${}_0R = {}_1R\sec\sigma$ also is greater; in other words, the parent has been 'stellated'. One of the simplest instances is $3\{3\}2$, which has the same vertices as one $3\{6\}2$ and the same edge-lines as another.

EXERCISE

The spherical tessellation of congruent triangles, beginning with the Schwarz triangle P_1P_2Q, is symmetrical by reflection in the great circle P_1P_2. Can this reflection be interpreted as a 'symmetry' of the complex polygon $p_1\{q\}p_2$?

11·2 REAL REPRESENTATIONS

When the points $(u, v) = (x_0+x_1i,\ x_2+x_3i)$ of the unitary plane are represented by the points (x_0, x_1, x_2, x_3) of a real 4-space, the g/p_1 edges of $p_1\{q\}p_2$ appear as real regular p_1-gons $2\{p_1\}2$. For, the coordinate axes can be so chosen that the p_1 vertices on an edge are

$$(l\exp(2\nu\pi i/p_1),\ v) \quad (\nu = 0, 1, \ldots, p_1-1),$$

yielding the real p_1-gon

$$\left(l\cos\frac{2\nu\pi}{p_1},\ l\sin\frac{2\nu\pi}{p_1},\ x_2,\ x_3\right).$$

Looking again at Figures 4·2A, B, D and 4·8A, B, C, D, E, we naturally ask: Can the real representation of every $p_1\{q\}p_2$ be projected onto a real plane so that the p_1-gons (representing the edges) remain regular? The answer is Yes, and the explanation is simple. Instead of going into real 4-space and then making an orthogonal projection onto a plane, we can work first in the original unitary plane (with its two complex coordinates): project the complex polygon orthogonally onto a line, and then represent the resulting set of points on the complex line by an Argand diagram. Each edge yields a set of p_1 points whose abscissae are

$$l' \exp(2\nu\pi i/p_1) \quad (\nu = 0, 1, \ldots, p_1 - 1)$$

for some $l' \leqslant l$.

We noticed, in §4·8, that the real representation of the complex polygon $p\{4\}2$ consists of the p^2 vertices and $2p$ real p-gons of the four-dimensional double prism $\{p\}\times\{p\}$ (defined in §4·5). This happens because the p^2 vertices have complex coordinates

$$(\epsilon^\mu, \epsilon^\nu) \quad (\mu, \nu = 0, 1, \ldots, p-1), \tag{11·21}$$

where $\epsilon = \exp(2\pi i/p)$ (see (4·51)). Taking the centres of the edges (or of the p-gons), we deduce that the reciprocal polygon $2\{4\}p$ has the $2p$ vertices

$$(\epsilon^\mu, 0), \quad (0, \epsilon^\nu). \tag{11·22}$$

Its real representation consists of the vertices of two p-gons in completely orthogonal planes through the origin, and the p^2 line-segments that join all the vertices of one p-gon to all the vertices of the other.

If the number of vertices of a regular complex polygon is

$$8 \quad \text{or} \quad 24 \quad \text{or} \quad 120,$$

it may happen that the real representation consists of the vertices, and some of the edges or faces, of a regular polytope

$$\{3, 3, 4\} \quad \text{or} \quad \{3, 4, 3\} \quad \text{or} \quad \{3, 3, 5\}.$$

In fact, this must happen if the symmetry group of the polygon is a subgroup of the symmetry group of the polytope. Du Val (1964, pp. 71–4†) showed that the group of direct (or 'positive') symmetries of these three polytopes (and of their reciprocals) are:

$$[4, 3, 3]^+ \cong (\langle 4, 3, 2\rangle/\langle 2, 2, 2\rangle; \langle 4, 3, 2\rangle/\langle 2, 2, 2\rangle), \text{ of order } 192,$$
$$[3, 4, 3]^+ \cong (\langle 4, 3, 2\rangle/\langle 3, 3, 2\rangle; \langle 4, 3, 2\rangle/\langle 3, 3, 2\rangle), \text{ of order } 576,$$
$$[5, 3, 3]^+ \cong (\langle 5, 3, 2\rangle/\langle 5, 3, 2\rangle; \langle 5, 3, 2\rangle/\langle 5, 3, 2\rangle), \text{ of order } 7200.$$

† See also Coxeter and Moser (1972, pp. 125, 129) and Coxeter, Abstract definitions for the symmetry groups of the regular polytopes in terms of two generators. Part II: The rotation groups, *Proceedings of the Cambridge Philosophical Society* **33** (1937), 315–24.

The relevant subgroups (see §§10·5 to 10·9) are:

$$(\mathfrak{C}_6/\mathfrak{C}_2;\ \langle 3, 3, 2\rangle/\langle 2, 2, 2\rangle) \cong 3[3]3,$$
$$(\mathfrak{C}_6/\mathfrak{C}_6;\ \langle 3, 3, 2\rangle/\langle 3, 3, 2\rangle) \cong 3[4]3,$$
$$(\mathfrak{C}_8/\mathfrak{C}_4;\ \langle 4, 3, 2\rangle/\langle 3, 3, 2\rangle) \cong 4[3]4,$$
$$(\mathfrak{C}_6/\mathfrak{C}_6;\ \langle 5, 3, 2\rangle/\langle 5, 3, 2\rangle) \cong 3[5]3,$$
$$(\mathfrak{C}_{10}/\mathfrak{C}_{10};\ \langle 5, 3, 2\rangle/\langle 5, 3, 2\rangle) \cong 5[3]5.$$

Thus the vertices of the polygons $3\{3\}3$, $3\{4\}3$ and $4\{3\}4$ belong to the polytopes $\{3, 3, 4\}$, $\{3, 4, 3\}$ and again $\{3, 4, 3\}$ (as we already saw in §4·8), while the four polygons

$$3\{5\}3, \quad 3\{\tfrac{5}{2}\}3, \quad 5\{3\}5, \quad 5\{\tfrac{5}{2}\}5$$

all have the same vertices as $\{3, 3, 5\}$. Using the results of §4·7 (see also Coxeter 1963, p. 267), we can say further that the edges of these four polygons appear as certain subsets of the two-dimensional faces of

$$\{3, 3, 5\} \text{ or } \{3, 5, \tfrac{5}{2}\}, \quad \{3, 3, \tfrac{5}{2}\} \text{ or } \{3, \tfrac{5}{2}, 5\},$$
$$\{5, \tfrac{5}{2}, 5\} \text{ or } \{5, 3, \tfrac{5}{2}\}, \quad \{5, \tfrac{5}{2}, 3\}.$$

Thus all the *self-reciprocal* polygons have regular polytopes for their real counterparts.

EXERCISE

$2\{6\}3$ has 16 vertices, $3\{6\}2$ has 24, $5\{6\}2$ has 600. Can these points be identified with the vertices of the regular polytopes $\{4, 3, 3\}$, $\{3, 4, 3\}$, $\{5, 3, 3\}$?

11·3 PETRIE POLYGONS

Analogy suggests that the *Petrie polygon* for $p_1\{q\}p_2$ should be defined as the simple h-gon formed by the orbit of the flag (O_0, O_0O_1) for the product of the two generating reflections. Thus it is a simple polygon $ABC\ldots$ whose vertices belong to the complex polygon while A and B belong to one edge, B and C to another, and so on. If $p_1\{q\}p_2$ is non-starry, the Petrie polygon is generated by R_1R_2, and h is given by (9·85). In other cases, h is still the period of the product of the matrices (8·91), where a_{12} is expressed in terms of p_1, p_2, q by the formula (9·89). Since this product

$$\begin{bmatrix} \epsilon_1{}^2 & -\epsilon_1{}^2(\epsilon_2{}^2 - 1)\cos\sigma \\ -(\epsilon_1{}^2 - 1)\cos\sigma & (\epsilon_1{}^2 - 1)(\epsilon_2{}^2 - 1)\cos^2\sigma + \epsilon_2{}^2 \end{bmatrix} \quad (\epsilon_\nu = \exp(\pi i/p_\nu))$$

has determinant $\epsilon_1{}^2\epsilon_2{}^2$ and trace

$$\epsilon_1\epsilon_2\left\{2\cos\left(\frac{\pi}{p_1} - \frac{\pi}{p_2}\right) - 4\sin\frac{\pi}{p_1}\sin\frac{\pi}{p_2}\cos^2\sigma\right\}$$
$$= -2\epsilon_1\epsilon_2\cos\frac{2\pi}{q} = 2\epsilon_1\epsilon_2\cos\left(1 - \frac{2}{q}\right)\pi,$$

its characteristic equation is

$$(\lambda-\epsilon_1\epsilon_2\epsilon)(\lambda-\epsilon_1\epsilon_2\epsilon^{-1}) = 0,$$

where $\epsilon = \exp\{(1-2/q)\pi i\}$; and its characteristic roots are

$$\exp\left[\left\{\frac{1}{p_1}+\frac{1}{p_2}\pm\left(1-\frac{2}{q}\right)\right\}\pi i\right].$$

Thus the period h is the least common multiple of the denominators of two fractions

$$\frac{1}{q}\pm\frac{1}{2}\left(1-\frac{1}{p_1}-\frac{1}{p_2}\right). \tag{11·31}$$

The various cases have been collected in Table IV.

When $p_1 = 2$ or $p_2 = 2$, this is the only kind of Petrie polygon; but when both p_1 and p_2 are greater than 2, there is another kind, having a similar description: the orbit of $R_1R_2^{-1}$ instead of R_1R_2. Using (8·91) again, we obtain the matrix

$$\begin{bmatrix} \epsilon_1^2 & -\epsilon_1^2(\epsilon_2^{-2}-1)\cos\sigma \\ -(\epsilon_1^2-1)\cos\sigma & (\epsilon_1^2-1)(\epsilon_2^{-2}-1)\cos^2\sigma+\epsilon_2^{-2} \end{bmatrix}$$

with determinant $\epsilon_1^2\epsilon_2^{-2}$ and trace

$$\epsilon_1\epsilon_2^{-1}\left\{2\cos\left(\frac{\pi}{p_1}+\frac{\pi}{p_2}\right)+4\sin\frac{\pi}{p_1}\sin\frac{\pi}{p_2}\cos^2\sigma\right\}$$

$$= 2\epsilon_1\epsilon_2^{-1}\left(2\cos\frac{\pi}{p_1}\cos\frac{\pi}{p_2}+\cos\frac{2\pi}{q}\right) = \epsilon_1\epsilon_2^{-1}(\eta+\eta^{-1}),$$

where $\eta = \exp(\pi i/t)$ and

$$\cos\frac{\pi}{t} = 2\cos\frac{\pi}{p_1}\cos\frac{\pi}{p_2}+\cos\frac{2\pi}{q} \tag{11·32}$$

(see §2·5, Ex. 4). The characteristic roots are now

$$\epsilon_1\epsilon_2^{-1}\eta^{\pm 1} = \exp\left\{\left(\frac{1}{p_1}-\frac{1}{p_2}\pm\frac{1}{t}\right)\pi i\right\}.$$

Thus the new kind of Petrie polygon is a simple h'-gon where h' is the least common multiple of the denominators of the two fractions

$$\frac{1}{2}\left\{\frac{1}{t}\pm\left(\frac{1}{p_1}-\frac{1}{p_2}\right)\right\}. \tag{11·33}$$

Table IV shows t and h' too.

Analogy with Figure 2·1B suggests that a drawing of a complex polygon should begin with a peripheral h-gon or h'-gon. Such drawings are seen in Figures 4·2A (for $3\{3\}3$, $h' = 4$), 4·2D (for $4\{3\}4$, $h = 12$), 4·8A, B, C, 11·5C, D, F, G, H. When $p_1 > 2$, we continue the drawing by inserting regular p_1-gons. When $p_1 = 2$, we insert edges so that the midpoints of the edges at a vertex are the vertices of a regular p_2-gon.

In the case of $3\{3\}3$, with its vertices numbered as in Figure 4·2A, the generating reflections may be represented as permutations

$$R_1 = (1\,3\,8)(4\,5\,7),\quad R_2 = (1\,4\,2)(5\,8\,6),$$

so that $R_1R_2R_1 = (1\,8\,5\,4)(2\,3\,6\,7) = R_2R_1R_2$, and we find

$$R_1R_2 = (1\,3\,6\,5\,7\,2)(4\,8),\quad R_1R_2^{-1} = (1\,3\,5\,7)(2\,4\,6\,8).$$

Thus the two kinds of Petrie polygon are *136572* and *1357*.

EXERCISES

1. Reconcile (9·85) with (11·31).
2. When p_1 or $p_2 = 2$, $t = \frac{1}{2}q$.
3. When $p_1 = p_2$, $h' = 2t$ (or the numerator of $2t$).
4. Obtain an explicit expression (like (9·85)) for h' when the polygon is non-starry.
5. In the above representation of $3\{3\}3$, what permutation transforms R_1 and R_2 into their inverses?

11·4 SOME USEFUL SUBGROUPS OF $p[2q]r$

As we saw in (9·71), the group $p[2q]r$, of order $4s^2/q$, has the presentation

$$R_1^p = S^r = 1,\quad (R_1S)^q = (SR_1)^q.$$

$(s^{-1}=p^{-1}+q^{-1}+r^{-1}-1.)$ The transforms of R_1 by powers of S, say

$$R_\nu = S^{\nu-1}R_1S^{1-\nu}\quad (\nu = 1, 2, \ldots, r;\ R_{\nu+r} = R_\nu),$$

generate a subgroup satisfying the relations

$$R_\nu^p = 1,\quad R_1R_2\ldots R_q = R_2\ldots R_qR_{q+1} = \ldots = R_rR_1\ldots R_{q-1}. \tag{11·41}$$

It is easy to see that this subgroup $\mathfrak{G}$ is of index r (order $4s^2/qr$), and that these relations suffice for a presentation. In fact, the group so presented has an obvious automorphism that cyclically permutes its generators R_ν. Adjoining a new element S, of period r, which transforms R_ν into $R_{\nu-1}$, we write $R_\nu = S^{\nu-1}R_1S^{1-\nu}$ and observe that the relation

$$R_1R_2\ldots R_q = R_2\ldots R_qR_{q+1}$$

yields

$$(R_1S)^q = (SR_1)^q.$$

Thus the extended group is precisely $p[2q]r$.

In the special case when q and r are relatively prime, $R_\nu \to R_{\nu-1}$ is an *inner* automorphism of $\mathfrak{G}$, since

$$(R_\nu R_{\nu+1}\ldots R_{\nu+q-1})^{-1}R_\nu(R_{\nu+1}\ldots R_{\nu+q-1}R_{\nu+q}) = R_{\nu+q}$$

and the subscripts are residues modulo r. Hence, *if r is relatively prime to q and also to the order of the centre of* $\mathfrak{G}$ (see Coxeter and Moser 1972, pp. 5–6), we have a direct product

$$(11\cdot42) \qquad p[2q]r \cong \mathfrak{C}_r \times \mathfrak{G}.$$

Since $\mathfrak{G}$ is a group generated by r (or fewer) reflections, it must occur among the groups considered in Chapter 10. In §11·5 we shall find an interesting interpretation of $p\{2q\}r$ as a Cayley diagram for $\mathfrak{G}$. Consequently it is worth while to examine each case in turn.

When $r = 2$, the relations (11·41) reduce to

$$R_1^p = R_2^p = 1, \quad R_1R_2R_1\ldots = R_2R_1R_2\ldots,$$

and $\mathfrak{G}$ is $p[q]p$. Here the only case to which (11·42) applies is

$$2[2q]2 \cong \mathfrak{C}_2 \times 2[q]2 \quad (q \text{ odd}).$$

When $q = 2$, so that we are considering $p[4]r$, the relations

$$(11\cdot43) \qquad R_\nu^p = 1, \quad R_1R_2 = R_2R_3 = \ldots = R_rR_1,$$

imply

$$R_3 = R_2^{-1}R_1R_2, \quad R_4 = R_3^{-1}R_1R_2 = R_2^{-1}R_1^{-1}R_2R_1R_2, \ldots$$
$$R_1 = R_r^{-1}R_1R_2 = R_2^{-1}R_1^{-1}\ldots R_2R_1R_2,$$

and the result of eliminating $R_3, \ldots, R_r$ is

$$R_1R_2R_1\ldots = R_2R_1R_2\ldots$$

with r factors on each side of the equation. In this case $\mathfrak{G}$ is $p[r]p$, and the instances of (11·42) are

$$p[4]3 \cong \mathfrak{C}_3 \times p[3]p \quad (p = 2,^\dagger 3, 4, 5)$$

and $3[4]5 \cong \mathfrak{C}_5 \times 3[5]3$. In the last two cases, r is no longer relatively prime to the order of the centre, but the decomposition remains valid since

$$3[4]5 \cong \langle 5, 3, 2\rangle_{15} \cong \mathfrak{C}_3 \times \mathfrak{C}_5 \times \langle 5, 3, 2\rangle.$$

In the case of $2[6]3$, the relations

$$R_\nu^2 = 1, \quad R_1R_2R_3 = R_2R_3R_1 = R_3R_1R_2$$

are of the form (9·62); thus $\mathfrak{G}$ is now $\langle 2, 2, 2\rangle_2$. (This is one of the nine non-Abelian groups of order 16. See Table 1 in Coxeter and Moser 1972, p. 134).

† Eugene Schenkman, *Group Theory* (van Nostrand, Princeton, 1965), p. 98 (III. 5. d). When the dihedral group $2[r]2$ is presented as (11·43) with $p = 2$, the generators R_ν may be represented by transpositions of consecutive symbols in a cycle of r:

$$R_1 = (1\ 2), \quad R_2 = (2\ 3), \ldots, \quad R_r = (r\ 1).$$

In the case of $2[10]3$, we have

$$R_\nu^2 = 1, \quad R_1R_2R_3R_1R_2 = R_2R_3R_1R_2R_3 = R_3R_1R_2R_3R_1,$$

so that

$$(R_1R_2R_3)^5 = R_1R_2R_3R_1R_2 \cdot R_3R_1R_2R_3R_1 \cdot R_2R_3R_1R_2R_3$$
$$= (R_2R_3R_1R_2R_3)^3 = Z, \text{ say.}$$

Now, if we define $A = (R_1R_2R_3)^2$, $B = (R_2R_3R_1R_2R_3)^{-1}Z$, $C = R_1Z$, we find

$$A^5 = Z^2, \quad B^3 = Z^{-1}Z^3 = Z^2, \quad C^2 = Z^2, \quad AB = C.$$

Comparing these relations with (9·32), we see that in this case $\mathfrak{G}$ is $\langle 5, 3, 2\rangle_2$, whose centre has order 4, and

$$2[10]3 \cong \mathfrak{C}_3 \times \langle 5, 3, 2\rangle_2 \cong \langle 5, 3, 2\rangle_6.$$

Similarly, in the case of $2[6]5$ we have

$$R_\nu^2 = 1, \quad R_1R_2R_3 = R_2R_3R_4 = R_3R_4R_5 = R_4R_5R_1 = R_5R_1R_2,$$

so that

$$(R_1R_2R_3)^5 = R_1R_2R_3 \cdot R_4R_5R_1 \cdot R_2R_3R_4 \cdot R_5R_1R_2 \cdot R_3R_4R_5$$
$$= (R_1R_2R_3R_4R_5)^3 = Z, \text{ say.}$$

Now, if we define $A = (R_1R_2R_3)^2 = R_5R_1R_2R_3R_4R_5$,

$$B = (R_1R_2R_3R_4R_5)^{-1}Z, \quad C = R_5Z,$$

we find

$$A^5 = Z^2, \quad B^3 = Z^{-1}Z^3 = Z^2, \quad C^2 = Z^2, \quad AB = C,$$

as before, so that $\mathfrak{G}$ is again $\langle 5, 3, 2\rangle_2$, and

$$2[6]5 \cong \mathfrak{C}_5 \times \langle 5, 3, 2\rangle_2 \cong \langle 5, 3, 2\rangle_{10}.$$

When $q = r+1$, we have

$$R_\nu^p = 1, \quad R_1R_2\ldots R_rR_1 = R_2\ldots R_rR_1R_2 = \ldots = R_rR_1R_2\ldots R_r,$$

whence $(R_1R_2\ldots R_rR_1)^r = (R_1R_2\ldots R_r)^{r+1}$; and $\mathfrak{G}$ has the simple presentation

$$a^r = c^{r+1}, \quad (c^{-1}a)^p = 1.$$

Similarly, when $r = q+1$, we have

$$R_\nu^p = 1, \quad R_1R_2\ldots R_q = R_2\ldots R_qR_{q+1} = \ldots = R_{q+1}R_1\ldots R_{q-1},$$

whence

$$(R_1\ldots R_qR_{q+1})^q = (R_1R_2\ldots R_q)^{q+1}$$

and

$$a^q = c^{q+1}, \quad (c^{-1}a)^p = 1.$$

To some extent these results duplicate others already obtained (see Coxeter and Moser 1972, p. 76). But (7·73) shows that, in the case of $2[8]3$ or $2[6]4$, $\mathfrak{G}$ is $GL(2, 3)$. Although

$$2[8]3 \cong \mathfrak{C}_3 \times GL(2, 3),$$

there is no such decomposition for $2[6]4 \cong \langle 4, 3, 2\rangle_4$.

These and other 'reflection subgroups' of reflection groups are listed in Table III on page 157 (compare Table 4 of Coxeter and Moser 1972, p. 136). Each group appears in one row and one column, and each entry is the index of the column-group as a subgroup of the row-group.

EXERCISES

1. What subgroup of $p[q]p$ (q odd) is generated by R_1 and all its conjugates?
2. What subgroup of $R_1^3 = 1$, $R_1R_2R_1R_2R_1 = R_2R_1R_2R_1R_2$ is generated by R_1 and $R_2^{-1}R_1R_2$? (P. McMullen.)
3. What subgroup of $R_1^4 = 1$, $R_1R_2R_1 = R_2R_1R_2$ is generated by R_1 and R_2^2?
4. What subgroup of $R_1^3 = R_2^4 = 1$, $(R_1R_2)^2 = (R_2R_1)^2$ is generated by R_1 and R_2^2?

11·5 CAYLEY DIAGRAMS FOR REFLECTION GROUPS

Consider a non-starry polygon $p\{2q\}r$ and its group $p[2q]r$. Since R_1 cyclically permutes the p vertices on an edge, the conjugate reflection

$$R_\nu = S^{\nu-1}R_1S^{1-\nu} \quad (\nu = 2, \ldots, r)$$

cyclically permutes the p vertices on another of the r edges that radiate from the initial vertex. Since these p-edges appear, in the real representation, as simple p-gons (or just edges, if $p = 2$), it is natural to use one of r different colours for each of the r representative p-gons (or edges). Since the r reflections R_ν generate a subgroup of $p[2q]r$, the colouring can be continued consistently over the whole of the real representation. We conclude that this arrangement of $4s^2/pq$ coloured p-gons (or edges), meeting by r's at $4s^2/qr$ points, provides a Cayley diagram (with r colours) for the group (11·41). Although the arrangement of points and p-gons appears first in a Euclidean 4-space, there is a conspicuous advantage (beyond the obvious advantage of making it visible) to be gained by projecting it onto a suitable plane (see, for instance, Figure 11·5A): there is no need to mark arrows along the edges of the regular p-gons ($p > 2$) provided these edges are understood to be directed in a positive sense round each p-gon. This happy state of affairs becomes obvious when we first project the complex polygon onto a complex line and then represent the resulting one-dimensional figure on the Argand plane. (See the beginning of page 105.)

The results of §11·4 show that, in this sense,

$p\{2q\}2$ and	$p\{4\}q$	are Cayley diagrams for $p[q]p$,
	$2\{6\}3$	is a Cayley diagram for $\langle 2, 2, 2\rangle_2$,
$2\{10\}3$ and	$2\{6\}5$	are Cayley diagrams for $\langle 5, 3, 2\rangle_2$,
$2\{8\}3$ and	$2\{6\}4$	are Cayley diagrams for $GL(2, 3)$.

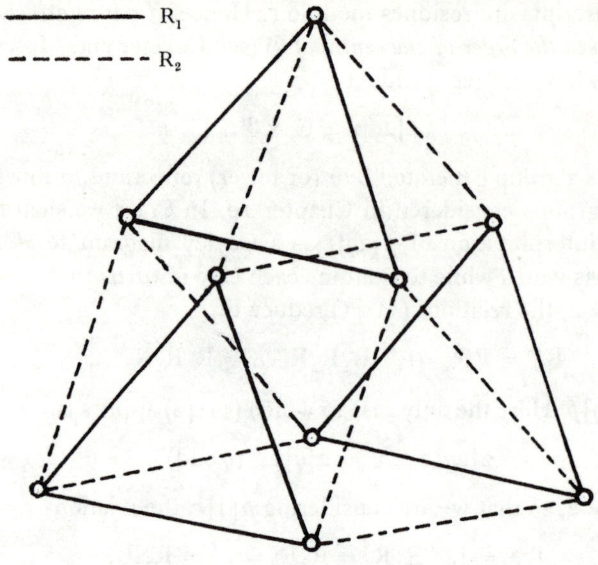

Figure 11·5A: $3\{4\}2$, a Cayley diagram for $3[2]3 \cong \mathfrak{C}_3 \times \mathfrak{C}_3$

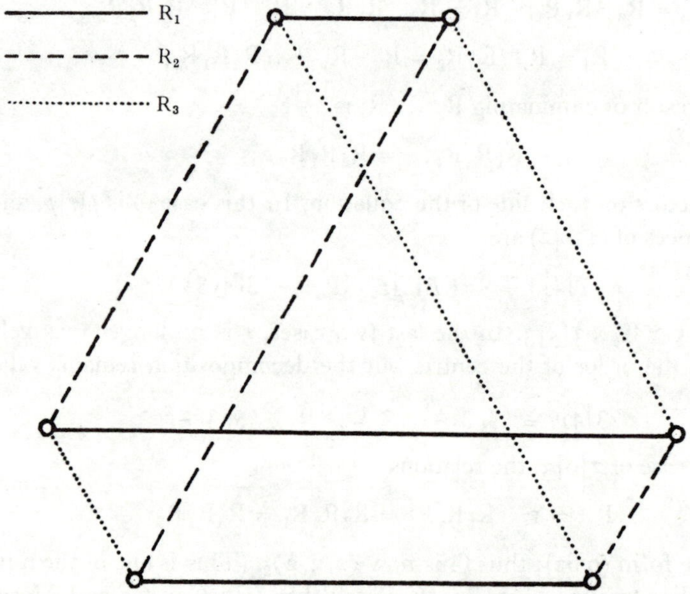

Figure 11·5B: $2\{4\}3$, a Cayley diagram for $2[3]2 \cong \mathfrak{D}_3$ in the form $R_1^2 = 1$, $R_1R_2 = R_2R_3 = R_3R_1$

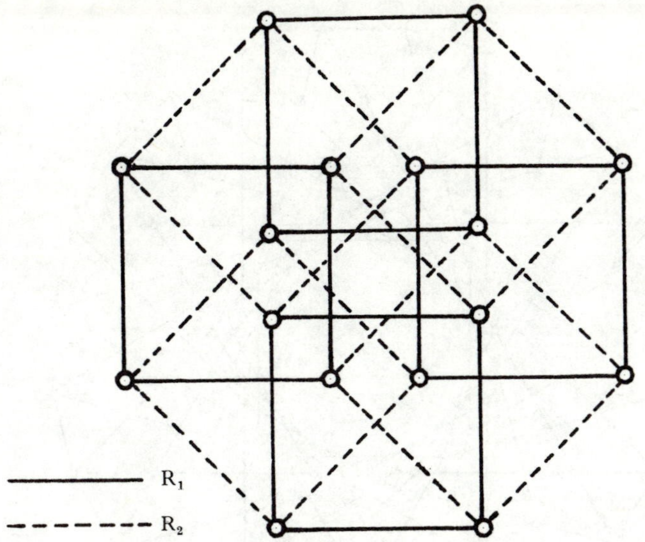

Figure 11·5C: 4{4}2, a Cayley diagram for 4[2]4 $\cong \mathfrak{C}_4 \times \mathfrak{C}_4$

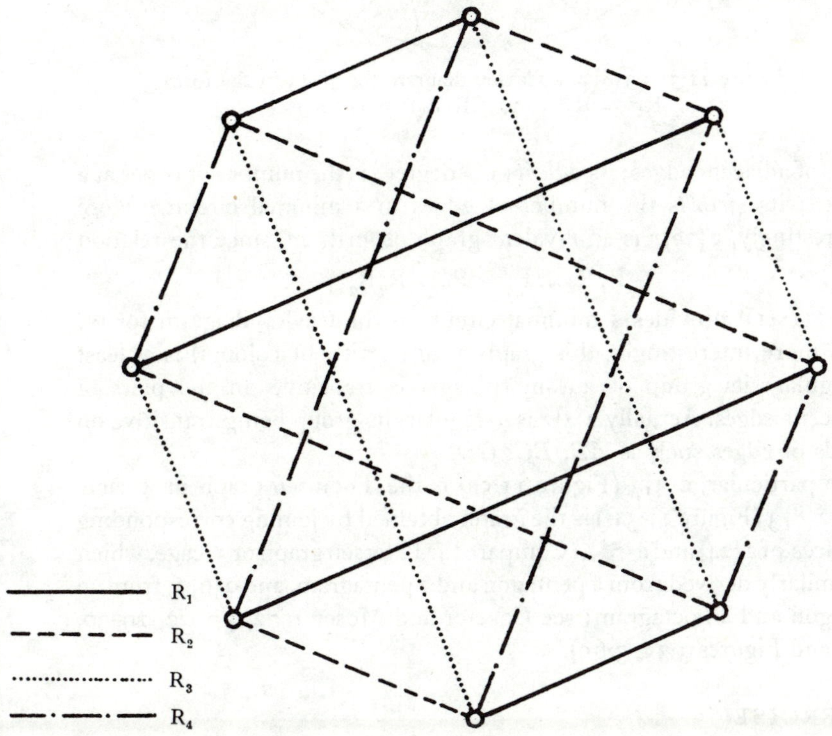

Figure 11·5D: 2{4}4, a Cayley diagram for 2[4]2 $\cong \mathfrak{D}_4$ in the form $R_\nu^2 = 1$, $R_1R_2 = R_2R_3 = R_3R_4 = R_4R_1$

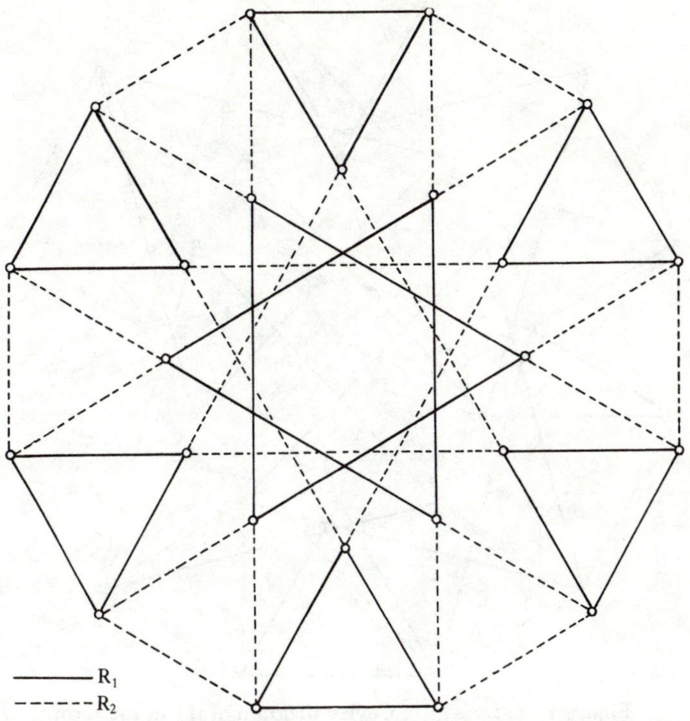

Figure 11·5E: 3{6}2, a Cayley diagram for 3[3]3 in the form $R_1^3 = 1, \quad R_1R_2R_1 = R_2R_1R_2$

Since $h = h' = 6$ for 3{4}2, it would perhaps have been more natural to let the six outermost vertices in Figure 11·5A form a *regular* hexagon; but then the remaining three would all coincide in the middle. The same change, from a regular $2p$-gon to an equilateral $2p$-gon (whose sides touch a circle) is appropriate for p{4}2 whenever p is odd (see §4·8).

The six vertices of 2{4}3 (see Figure 11·5B) correspond to the centres of the six equilateral triangles (in Figure 11·5A) which represent the edges of the reciprocal polygon 3{4}2. Similarly, the eight vertices of 2{4}4 (Figure 11·5D) correspond to the centres of the eight squares (four of each 'colour' in Figure 11·5C) which represent the edges of 4{4}2; and the midpoints of the edges of 2{4}4 correspond to the vertices of 4{4}2.

Figure 11·5E is easy to draw, because, apart from the essential variations of colour, it is just a dodecagon {12} with an inscribed dodecagram $\{\frac{12}{5}\}$. Since four of the equilateral triangles are concentric, the corresponding projection of the reciprocal 2{6}3 has four of its vertices coincident in the middle. For a more satisfactory way to draw the Cayley diagram

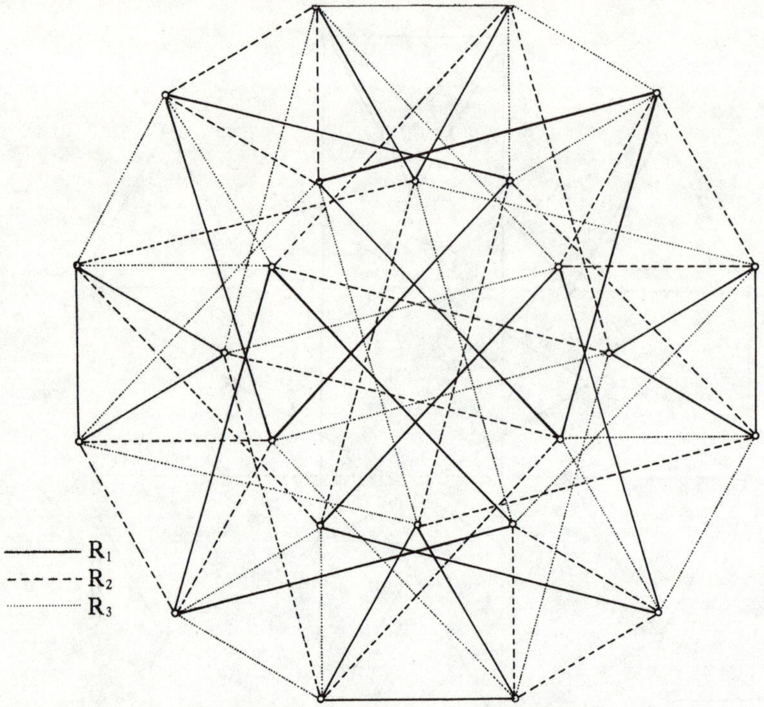

Figure 11·5F: $3\{4\}3$, a Cayley diagram $3[3]3$ in the form
$R_1^3 = 1, \quad R_1R_2 = R_2R_3 = R_3R_1$

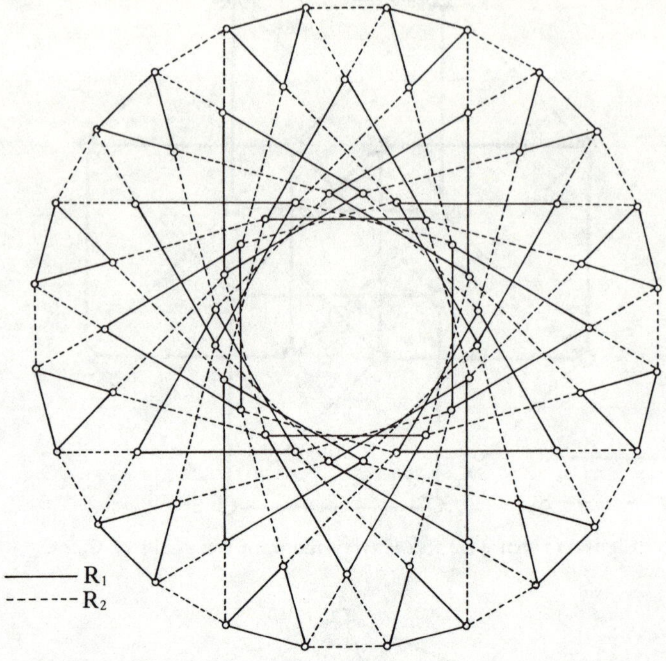

Figure 11·5G: $3\{8\}2$, a Cayley diagram for $3[4]3$ in the form
$R_1^3 = R_2^3 = 1, \quad (R_1R_2)^2 = (R_2R_1)^2$

for $\langle 2, 2, 2\rangle_2$, see Coxeter and Moser (1972, pp. 22 and 30, Figure 3·3C and 3·61).

Such coincidences have been artfully avoided in Figure 11·5F (compare Figure 4·8A) by locating the twelve 'inner' vertices of $3\{4\}3$ at the vertices of a hexgon and the mid-points of its edges, and then drawing the appropriate equilateral triangles (with the result that the twelve 'outer' vertices form an equiangular dodecagon inscribed in a circle).

Figure 11·5G looks complicated until we recognize it as a $\{24\}$ with three inscribed octagrams $\{\frac{8}{3}\}$. However, we must carefully notice that the edges of the octagrams involve both colours (twice) and have blank segments in the middle. Figure 11·5H can be derived from this by using the centres of all the 48 equilateral triangles; or, if Figure 11·5H is drawn first, we can use the midpoints of all its 72 edges to obtain Figure 11·5G.

In the language of graph-theory, the real representation of $p\{2q\}r$ with $p > 2$ is a $2r$-valent graph of girth min $(p, 2q)$, having $4s^2/qr$ vertices and $4s^2/q$ edges, usually 1-regular. (A graph is 1-*regular* if its group of automorphisms is transitive on the vertices and edges but not on the pairs of adjacent edges; its *valency* or 'degree' is the number of edges at a vertex; its *girth* is the number of edges in a minimal circuit.) More interestingly, $2\{2q\}r$ is an r-valent graph of girth $2q$, since the relation

$$R_1R_2 \ldots R_q = R_2 \ldots R_qR_{q+1}$$

(see (11·41)) provides a minimal circuit in the Cayley diagram for $\mathfrak{G}$. Still more interestingly, this graph $2\{2q\}r$ (without colours) is at least 2-regular: its group of automorphisms is transitive on the pairs of adjacent edges. Actually $2\{4\}r$ is 3-regular, its group being transitive on triads of edges such as AB, BC, CD.

In particular, $2\{4\}3$ (Figure 11·5B) is the Thomsen graph or 4-cage, and $2\{8\}3$ (Figure 11·5H) is the graph obtained by joining corresponding vertices of a $\{24\}$ and a $\{\frac{24}{5}\}$. (Compare the Petersen graph or 5-cage, which is similarly derived from a pentagon and a pentagram, and $2\{6\}3$, from an octagon and an octagram; see Coxeter and Moser 1972, pp. 22, 29–30, 116 and Figures 3·3C, 3·6C).

EXERCISE

How can the real representation of $3\{3\}3$ (Figure 4·2A) be coloured so as to serve as a Cayley diagram for the quaternion group $\langle 2, 2, 2\rangle$?

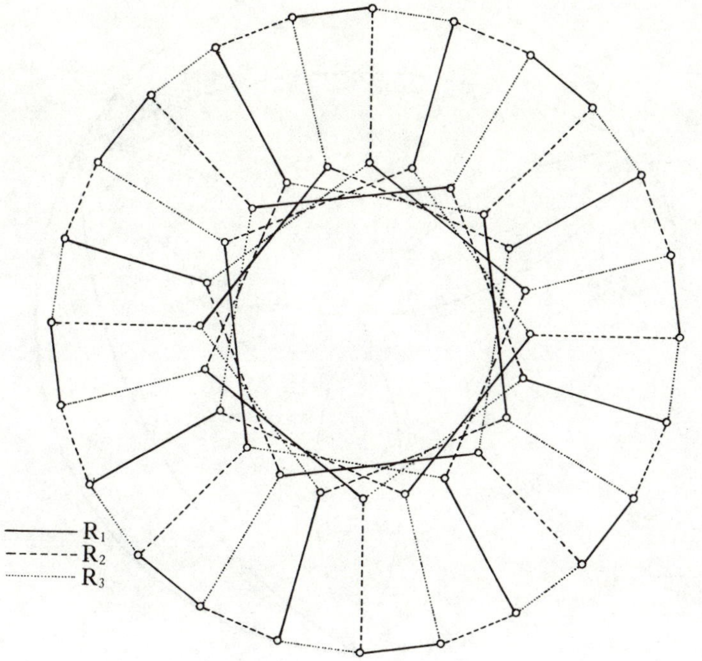

Figure 11·5H: $2\{8\}3$, a Cayley diagram for $GL(2, 3)$ in the form
$R_1^2 = 1, \quad R_1R_2R_3R_1 = R_2R_3R_1R_2 = R_3R_1R_2R_3$

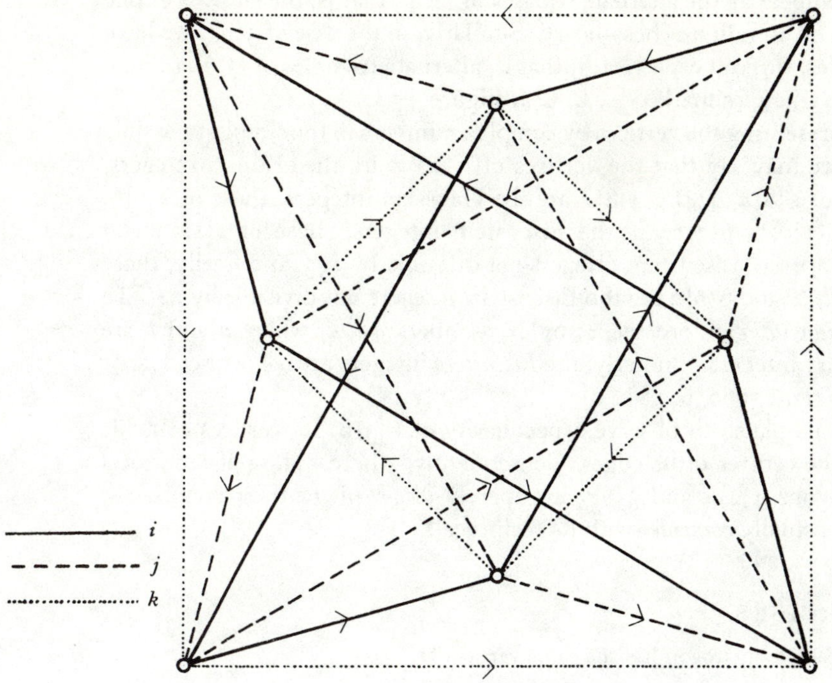

Figure 11·5I: A Cayley diagram for the quaternion group $\langle 2, 2, 2\rangle$

11·6 APEIROGONS

The real apeirogon $\{\infty\}$ (§1·6), which can be regarded as a degenerate polygon $\{q\}$, is the special case $2\{\infty\}2$ of the complex apeirogon $p_1\{q\}p_2$ whose vertices are the orbit of the origin with respect to the group generated by unitary reflections of periods p_1 and p_2 about points $u = l$ and $u = 0$ of the complex line. As transformations of the complex abscissa u, these reflections R_1 and R_2 are evidently

$$\exp(2\pi i/p_1)(u-l)+l \quad \text{and} \quad \exp(2\pi i/p_2)\,u.$$

As transformations of the Argand plane, they are rotations through $2\pi/p_1$ and $2\pi/p_2$ about two vertices $(l, 0)$ and $(0, 0)$ of a triangle whose angles are π/p_1, π/p_2, $2\pi/q$, where

(11·61) $$\frac{1}{p_1}+\frac{1}{p_2}+\frac{2}{q} = 1$$

(compare (9·83) and Figure 9·8B).

Since the possible periods of rotations that can generate an infinite discrete group are 2, 3, 4, 6, and since 4 never occurs with 3 or 6 (see Coxeter 1963, pp. 62–3), the only values for p_1, p_2 (with $p_1 \geqslant p_2$) are

$$2, 2;\quad 3, 2;\quad 3, 3;\quad 4, 2;\quad 4, 4;\quad 6, 2;\quad 6, 3;\quad 6, 6.$$

Thus there are exactly twelve regular apeirogons:

(11·62) $$\begin{matrix} 2\{\infty\}2, & 3\{12\}2, & 2\{12\}3, & 3\{6\}3, & 4\{8\}2, & 2\{8\}4, \\ 4\{4\}4, & 6\{6\}2, & 2\{6\}6, & 6\{4\}3, & 3\{4\}6, & 6\{3\}6. \end{matrix}$$

As McMullen remarks, 'it is a little difficult to draw complex figures on real paper', so let us be content to describe the corresponding figures in the Argand plane. By considering the rotations that represent R_1 and R_2, we see that the vertices and edges of $p_1\{4\}p_2$ appear as the vertices and faces of the real tessellation $\{p_1, p_2\}$. The only apeirogon $p_1\{q\}p_2$ with q odd is $6\{3\}6$; here the vertices belong to $\{3, 6\}$ again, but the hexagons representing the edges are those that enclose sets of six triangles. When $p_1 = 2$, the edges appear as edges (or digons), namely the edges of $\{\frac{1}{2}q, p_2\}$. In the case of $p_1\{q\}2$ the edges appear as p_1-gons, two at each vertex, these two being interchanged by a half-turn; so we have the vertices and p_1-gons of the tessellation $\begin{Bmatrix} p_1 \\ \frac{1}{2}q \end{Bmatrix}$. This is an unambiguous description of $3\{12\}2$ or $6\{6\}2$, but in the case of $4\{8\}2$ it means that the edges appear as the alternate squares of $\{4, 4\}$, that is, the squares of one colour on an infinite chess-board. Similarly, in the case of $3\{6\}3$ we have triangles, three at each vertex, that is, alternate triangles of $\{3, 6\}$. (These are the triangles marked A, B, C in Figure 3·3A.)

Representing the vertices by complex numbers in the simplest possible way, we may say that the vertices of $2\{\infty\}2$ are the ordinary integers, those of $2\{8\}4$, $4\{4\}4$, $4\{8\}2$ are the Gaussian integers, those of $2\{6\}6$, $3\{4\}6$, $3\{6\}3$, $6\{3\}6$ are the Eisenstein integers, those of $2\{12\}3$ and $6\{4\}3$ are the Eisenstein integers not divisible by $i\sqrt{3}$, and finally, those of $3\{12\}2$ and $6\{6\}2$ are the Eisenstein integers not divisible by 2. (The *Gaussian* integers are the complex numbers $a+bi$, where a and b are ordinary integers. Similarly, the *Eisenstein* integers are $a+b\omega$; see Hardy and Wright 1960, p. 188.)

Let us pause to observe a peculiarity of $6\{3\}6$: its vertices coincide with the centres of its edges. Consequently, whereas the self-reciprocal apeirogons $4\{4\}4$ and $3\{6\}3$ are merely *congruent* to their reciprocals, $6\{3\}6$ actually *coincides* with its reciprocal!

EXERCISES

1. Which apeirogon has the same vertices as

   (i) two reciprocal $4\{4\}4$'s,
   (ii) two reciprocal $3\{6\}3$'s,

while each edge joins a vertex of one to a neighbouring vertex of the other?

2. Are there any starry apeirogons?
3. For a non-starry finite polygon $p_1\{q\}p_2$, the symmetry group has the presentation (9·81). Does this remain valid for an apeirogon?
4. The real representation of a finite polygon $2\{q\}p$ is a p-valent graph of girth q. Does this remain valid when $2\{q\}p$ is an apeirogon?

11·7 A GENERAL TREATMENT FOR THE BINARY POLYHEDRAL GROUPS

To set the stage for a general proof of Theorem (7·42), let us look once more at Figure 3·3A, which is a Cayley diagram for the infinite group (3, 3, 3) with the presentation

$$A^3 = B^3 = C^3 = ABC = 1.$$

Figure 11·7A shows the analogous diagram for the finite group

$$(2, 2, 2) \cong \mathfrak{D}_2,$$

of order 4:

$$A^2 = B^2 = C^2 = ABC = 1.$$

Here we have abandoned the usual convention whereby an involutory element was represented by a single undirected edge, and we regard the

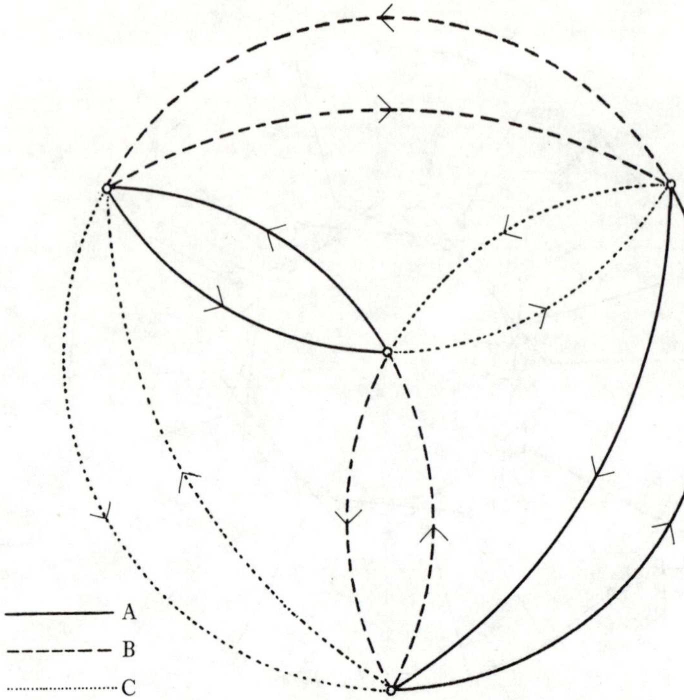

Figure 11·7A: A Cayley diagram for (2, 2, 2)

Cayley diagram as a two-dimensional complex (or 'map') with oriented cells filling and covering the inversive plane, which is topologically a 2-sphere. The cells represent the relations in the presentation, and are as follows: two digons for each of the relations $A^2 = 1$, $B^2 = 1$, $C^2 = 1$, and four triangles for the relation $ABC = 1$. One of the four triangles appears as the peripheral circuit, which may be regarded as surrounding the 'point at infinity' of the inversive plane. More obviously, it bounds the finite part of the diagram and may be decomposed into the nine circuits that bound the remaining cells: six digons and three triangles. Since the plane is simply-connected, the peripheral circuit, which is the 'product' of these nine elementary circuits, can be gradually shrunk down to the lowest vertex in the figure by eliminating first the three outer digons, then the three triangles, then the inner digons C^2, A^2, and finally the inner digon B^2, The corresponding reduction of the word ABC is

$$ABC = A^2A^{-1}B^2B^{-1}C^2C^{-1} = A^{-1}B^{-1}C^{-1}$$
$$= (BCA)^{-1}BC(CAB)^{-1}CA(ABC)^{-1}AB$$
$$= BC\ CA\ AB = BC^2A^2B = B^2 = 1.$$

Figure 11·7A may also be interpreted as a 'coset diagram' (Coxeter and Moser 1972, p. 31) for the quaternion group $\langle 2, 2, 2\rangle$, enabling us to see that the relations

$$A^2 = B^2 = C^2 = ABC = Z$$

imply $Z^2 = 1$. For, the same reduction of the peripheral circuit now yields

$$\begin{aligned} Z = ABC &= A^2A^{-1}B^2B^{-1}C^2C^{-1} = ZA^{-1}ZB^{-1}ZC^{-1} = A^{-1}B^{-1}C^{-1}Z^3 \\ &= Z^{-1}BCZ^{-1}CAZ^{-1}ABZ^3 \\ &= BC\,CA\,AB = BZ\,ZB = B^2Z^2 = Z^3. \end{aligned}$$

Clearly, this kind of procedure can be applied to any finite group $\langle p, q, r\rangle$ with $p, q, r \geqslant 2$. More generally,

(11·71) *Suppose a given group has a cyclic centre generated by Z, and a presentation*

$$f_i(A, B, \ldots) = Z^{\nu_i} \quad (i = 1, 2, \ldots).$$

Each relation corresponds to a set of oriented polygons π_i in the Cayley diagram for the central quotient group

$$f_i(A, B, \ldots) = 1 \quad (i = 1, 2, \ldots).$$

If these polygons fit together so as to form a tessellation covering the topological sphere (or the inversive plane), let n_i, with an appropriate sign, be the number of polygons of type π_i. Then the period of Z is $\Sigma n_i \nu_i$.

To prove this, we consider the Cayley diagram and compare the results of shrinking the peripheral circuit (or any other circuit) to a single point (or 'point-circuit') in two distinct ways, corresponding to the two discs into which the circuit decomposes the 2-sphere. More symmetrically, we may begin with any point-circuit (such as the point at infinity) and 'shrink' it to another point-circuit by contracting over *all* the oriented polygons in the Cayley diagram. The details are omitted because the above example makes the procedure sufficiently clear.

If p, q, r are all greater than 1 and satisfy (2·33), the group (p, q, r), defined by (3·32), is a finite rotation group whose order $2s$ is given by (2·34). Its Cayley diagram (see the remark after Theorem (3·33)) consists of $2s/p$ p-gons representing $A^p = 1$, $2s/q$ q-gons representing $B^q = 1$, $2s/r$ r-gons representing $C^r = 1$, all positively oriented, and $2s$ negatively oriented triangles representing $ABC = 1$. According to Theorem (11·71) (with $\nu_i = 1$ for $i = 1, 2, 3, 4$), the relations

$$A^p = B^q = C^r = ABC = Z$$

imply that Z has period $\dfrac{2s}{p}+\dfrac{2s}{q}+\dfrac{2s}{r}-2s = 2.$

Thus Theorem (7·41) has been proved in its full generality, and Theorem (7·42) also, except for the trivial necessity of giving separate consideration to the cyclic group $\langle p, p, 1\rangle$.

It is interesting to observe that each finite group $\langle p, q, r\rangle$ remains finite when any or all of p, q, r are replaced by $-p$, $-q$, $-r$, respectively, though the order of the centre increases. (See Coxeter and Moser 1972, p. 70.) Since the relation $A^{-p} = Z$ can be replaced by $A^p = Z^{-1}$, Theorem (11·71) is still relevant though now $\nu_i = \pm 1$. Thus, if

$$p, q, r > 1 \quad \text{and} \quad p^{-1}+q^{-1}+r^{-1} > 1,$$

the relations $\qquad A^{\pm p} = B^{\pm q} = C^{\pm r} = ABC = Z$

(with any distribution of signs) imply $Z^{2s/s'} = 1$, where

$$s = (p^{-1}+q^{-1}+r^{-1}-1)^{-1}$$

and

(11·72) $$s' = |\pm p^{-1}\pm q^{-1}\pm r^{-1}-1|^{-1}.$$

In other words,

(11·73) *The order of any finite group $\langle \pm p, \pm q, \pm r\rangle$ (with $p, q, r > 1$) is $4s^2/s'$.*

EXERCISES

1. Apply Theorem (11·71) to
   (i) $A^2 = B^2 = (AB)^p$ (see §7·5, Ex. 2),
   (ii) $A^p = B^q = Z$, $(AB)^r = 1$,
   (iii) $a^p = b^q = Z$, $(a^{-1}b)^r = 1$ (see, e.g., (7·73)).
2. Make a list of the values of s/s' for the various groups $\langle \pm p, \pm q, \pm r\rangle$.
3. In which cases do we have
$$\langle \pm p, \pm q, \pm r\rangle \cong \mathfrak{C}_{s/s'} \times \langle p, q, r\rangle?$$
4. Applying Theorem (11·71) to $\langle p, q, 2\rangle$ and the rotation group of the spherical tessellation or Platonic solid $\{p, q\}$, express n_i in terms of the numbers N_0, N_1, N_2, and deduce that the period of Z is $N_0-N_1+N_2$. (See §2·2, Ex. 1.)

11·8 REMARKS

The regular complex polygons were first enumerated by Shephard, in 1951. His pioneering work was complicated; but the different procedure in §11·1 provides a check, showing that his list (of three infinite families plus forty-seven separate items) is perfectly correct. For the non-starry polygons he used a symbol $p_1(g)p_2$, which was later replaced by $p_1\{q\}p_2$ (see Coxeter and Moser 1972, p. 80). For the twenty-nine star polygons he used no symbol, but he identified them by means of the number

$$P = \cos\frac{\pi}{p_1}+i\cos 2\sigma\sin\frac{\pi}{p_1}$$

(compare (9·46)). He listed them in three columns[†] as follows:

$2\{m/k\}2$	$3\{\frac{10}{3}\}2$	$2\{\frac{10}{3}\}5$
$3\{3\}2$	$3\{5\}2$	$5\{\frac{5}{2}\}3$
$2\{3\}3$	$2\{\frac{5}{2}\}3$	$5\{\frac{10}{3}\}3$
$3\{\frac{8}{3}\}2$	$2\{5\}3$	$5\{3\}3$
$2\{\frac{8}{3}\}3$	$2\{\frac{10}{3}\}3$	$3\{\frac{5}{2}\}5$
$4\{3\}2$	$5\{3\}2$	$3\{\frac{10}{3}\}5$
$2\{3\}4$	$5\{5\}2$	$3\{3\}5$
$4\{\frac{8}{3}\}3$	$5\{\frac{10}{3}\}2$	$3\{\frac{5}{2}\}3$
$3\{\frac{8}{3}\}4$	$2\{3\}5$	$5\{\frac{5}{2}\}5$
$3\{\frac{5}{2}\}2$	$2\{5\}5$	

The treatment in §11·1 owes much to McMullen, who likes to emphasize the starry nature of $3\{3\}2$ and $5\{\frac{5}{2}\}3$ (for instance) by writing these symbols as $3\{\frac{6}{2}\}2$ and $5\{\frac{10}{4}\}3$. (Moreover, when this has been done, the even numerator is the smallest integer such that an equation

$$R_1R_2R_1\ldots = R_2R_1R_2\ldots$$

holds with this number of R's on each side.) It was he also who suggested the treatment of apeirogons in §11·6.

We see from (11·22) that the vertices and edges of $2\{4\}p$ form the 'complete bipartite graph' usually denoted by $K_{p,\,p}$.

† G. C. Shephard, Regular complex polytopes, *Proceedings of the London Mathematical Society* (3), **2** (1952), 82–97; see especially p. 92.

§11·4 and §11·5 were inspired by R. M. Foster, who has tabulated hundreds of trivalent 'symmetrical' graphs (s-regular for $1 \leqslant s \leqslant 5$). He noticed that in many cases (though certainly not all) the graph can be described as the Cayley diagram for a group having three involutory generators which are cyclically permuted by an outer automorphism. Naturally, our polygons $2\{2q\}3$ ($q = 2, 3, 4, 5$) occur in his list among the graphs having $48q/(6-q)^2$ vertices (and girth $2q$). $2\{6\}3$ and $2\{8\}3$ are $G(8, 3)$ and $G(24, 5)$ in the notation of Roberto Frucht, J. E. Graver and M. E. Watkins.[†]

The essential idea of §11·7 occurred to John Conway in 1968 in a flash of intuition, a few minutes after the end of my lecture in Cambridge, where I regretted that Theorem (7·42) had been proved only by separate consideration of the various cases. His first written account was highly condensed and remained obscure till Shephard elucidated it three years later.[‡]

EXERCISE

Can the Petersen graph serve as a Cayley diagram?

† The groups of the generalized Petersen graphs, *Proceedings of the Cambridge Philosophical Society* **70** (1971). 211–18.

‡ J. H. Conway, H. S. M. Coxeter and G. C. Shephard, The centre of a finitely generated group, *Tensor* **25** (1972), 405–18; **26** (1972), 477.

CHAPTER 12

Regular complex polytopes defined and described

> Surely with as good reason as had Archimedes to have the cylinder, cone and sphere engraved on his tombstone might our distinguished countrymen [Arthur Cayley and George Salmon] leave testamentary directions for the cubic eikosiheptagram to be engraved on theirs. Spirit of the Universe! whither are we drifting, and when, where, and how is all this to end?
>
> J. J. Sylvester†

In §12·1 the conditions for a configuration to be a regular polytope are carefully stated, and it is proved that the symmetry group of such a polytope is generated by a sequence of unitary reflections, of periods

$$p_1, p_2, \ldots, p_n,$$

such that any two non-consecutive reflections are commutative. These ideas lead to the generalized Schläfli symbol (12·12). In §12·2, the invariant Hermitian form is expressed in terms of the p's and q's; then the generalized cube and the generalized 'octahedron' are described. The next four sections develop the sequence of polytopes

$$3\{3\}3, \quad 3\{3\}3\{3\}3, \quad 3\{3\}3\{3\}3\{3\}3$$

(having 8, 27, 240 vertices) which culminates in the four-dimensional honeycomb $3\{3\}3\{3\}3\{3\}3\{3\}3$. These are described in considerable detail because of their connection with the celebrated configuration of 27 lines on the general cubic surface in projective 3-space. It will be seen that the binary tetrahedral group $3[3]3$ is a subgroup of index 9 in the Hessian group of order 216, which is a factor group of the symmetry group $3[3]3[3]3$ of the 'Hessian polyhedron' $3\{3\}3\{3\}3$; this group of order 648 is a subgroup of index 40 in the simple group of order 25920 which, in turn, is a factor group of the symmetry group $3[3]3[3]3[3]3$ of the 'Witting polytope' $3\{3\}3\{3\}3\{3\}3$. Finally, in §12·7 and §12·8, the honeycomb of Witting polytopes is seen to be one of a large family of complex honeycombs, interrelated in remarkable ways.

12·1 DEFINITIONS

Extending the ideas of §11·1 to unitary n-space, let us consider a finite non-empty set of subspaces, μ-flats ($0 \leqslant \mu < n$), with a relation of incidence, defined by saying that a μ-flat and a ν-flat are *incident* if one of them is a proper subspace of the other. (This implies $\mu \neq \nu$.) A subset of this set of flats is said to be *connected* if any two of its flats can be joined by a 'chain' of successively incident flats. If $\lambda < \nu - 1$, any incident λ-flat and ν-flat determine a *medial figure* consisting of all the μ-flats ($\lambda < \mu < \nu$) that are incident with both. Here we are assuming $\lambda \geqslant 0$ and $\nu < n$; but it is sometimes convenient to allow the values of λ and ν to include -1 and n, meaning that there is one (-1)-flat, the empty set, and one n-flat. the whole space. Then we say that each μ-flat for $0 \leqslant \mu \leqslant n-1$ is incident with both these 'trivial' flats; in other words, the whole set, without its trivial flats, is *their* medial figure.

Such a set of flats is called a *polytope* if it satisfies the following two conditions:

(i) Every medial figure includes at least two μ-flats for each μ

$$(-1 \leqslant \lambda < \mu < \nu \leqslant n).$$

(ii) Every medial figure with $\lambda < \nu - 2$ is connected.

It is to be understood that these conditions include their extreme situations: we allow $\lambda = -1$ or $\nu = n$ or both.

Any polytope has a *reciprocal*, in which each μ-flat is replaced by an $(n-\mu-1)$-flat and all incidences are dualized.

Any polytope has a group of automorphisms, permuting the μ-flats for each μ and preserving the incidences. (Of course, the order of this group is not necessarily greater than 1.) For our present purposes we shall find it desirable to recognize the fact that we are working in a *unitary* space and to consider especially those automorphisms that are achieved by unitary transformations. These 'symmetry operations' form a subgroup called the *symmetry group* of the polytope.

In an n-dimensional polytope Π_n, a μ-flat and all its incident λ-flats ($\lambda < \mu$) form a sub-polytope Π_μ, called a μ-dimensional *element* of Π_n. In particular there are *vertices* Π_0, *edges* Π_1, *faces* Π_2, ..., and *cells* Π_{n-1}. Each Π_μ ($\mu \geqslant 0$) has a *centre* O_μ, which is the centroid of its vertices. In particular, Π_n itself has a centre O_n, which is invariant for the symmetry group. The figure consisting of one element of each dimension number, all mutually incident, is called a *flag*.

The polytope is said to be *regular* if its symmetry group is transitive on

† *Proceedings of the London Mathematical Society* **2** (1867), 155.

its flags. As in §11·1, this transitivity is sharp: the order, g, of the symmetry group is equal to the number of flags.

By condition (i), the medial figure of two incident elements

$$\Pi_{\mu\pm 1} \quad (0 \leqslant \mu \leqslant n-1)$$

consists of at least two elements, say $p_{\mu+1}$ elements. Let Π_μ and Π_μ' be two of them. The symmetry operation that relates the two flags

$$(\Pi_0, \ldots, \Pi_{\mu-1}, \Pi_\mu, \Pi_{\mu+1}, \ldots, \Pi_{n-1})$$

and

$$(\Pi_0, \ldots, \Pi_{\mu-1}, \Pi_\mu', \Pi_{\mu+1}, \ldots, \Pi_{n-1})$$

leaves invariant the centres $O_0, \ldots, O_{\mu-1}, O_{\mu+1}, \ldots, O_n$, and thus leaves invariant every point of the hyperplane spanned by these n points: it is a *reflection* in this hyperplane. Such reflections, permuting the elements of the medial figure, form a subgroup of the symmetry group. Since these reflections all have the same mirror, the subgroup is cyclic, of order $p_{\mu+1}$, and is generated by a 'primitive' reflection $R_{\mu+1}$. In particular, R_1 permutes the p_1 vertices on an edge Π_1, and its mirror is the hyperplane $O_1 O_2 \ldots O_n$. Since the n lines $O_0O_1, O_1O_2, \ldots, O_{n-1}O_n$ are mutually orthogonal, the simplex $O_0 O_1 \ldots O_n$ is a (complex) orthoscheme (see §5·7). Its cells (except $O_0 O_1 \ldots O_{n-1}$) lie in the mirrors for the reflections $R_1, R_2, \ldots, R_n$, which generate the symmetry group[†]. Since all non-adjacent pairs of these mirrors are orthogonal, the reflections satisfy the relations

$$R_\mu^{\,p_\mu} = 1, \quad R_\mu \rightleftarrows R_\nu \quad (\mu < \nu - 1). \tag{12·11}$$

Conversely, any n reflections that satisfy such relations, and generate a finite group, can be used to determine a regular polytope by Wythoff's construction: its vertices are obtained by applying the group to a point O_0 that lies on all the mirrors except $O_1 O_2 \ldots O_n$.

Similarly, the symmetry group of the medial figure of two incident elements $\Pi_{\mu-1}$ and Π_ν ($\mu < \nu$) arises from the orthoscheme $O_\mu \ldots O_\nu$ and is generated by the $\nu - \mu$ reflections $R_{\mu+1}, \ldots, R_\nu$. Applying these reflections to the point O_μ, we obtain a *medial polytope*, that is, a $(\nu-\mu)$-dimensional polytope whose λ-dimensional elements have the same relations of incidence as the $\Pi_{\mu+\lambda}$'s in the medial figure. Three special cases are important.

First, when $\nu - \mu = 2$ we have a medial *polygon* whose group is generated by reflections $R_{\nu-1}$ and R_ν of periods $p_{\nu-1}$ and p_ν. If this polygon is

$$p_{\nu-1}\{q_{\nu-1}\}p_\nu$$

† For a more complete account, see Peter McMullen, Combinatorially regular polytopes, *Mathematika* **14** (1967), 142–50; especially p. 146, Proposition (3.9).

(see §11·1), we can use the q's so defined to set up a symbol

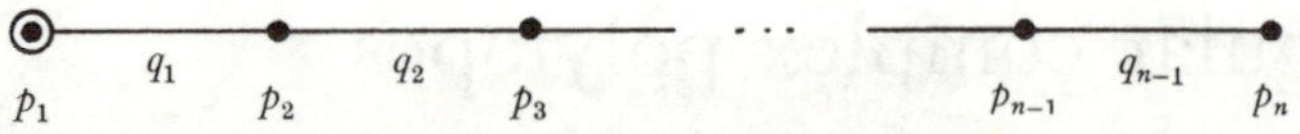

or, for easier printing,

$$p_1\{q_1\}p_2\{q_2\} \cdots p_{n-1}\{q_{n-1}\}p_n, \tag{12·12}$$

which specifies the polytope Π_n completely. In fact, the generating kaleidoscope involves reflections of periods $p_1, p_2, \ldots, p_n$ in a sequence of mirrors such that all non-adjacent pairs are orthogonal while the angle σ_μ between the mirrors for R_μ and $R_{\mu+1}$ is given by (9·82) in the form

$$\cos 2\sigma_\mu = \left(\cos\frac{\pi}{p_\mu}\cos\frac{\pi}{p_{\mu+1}} + \cos\frac{2\pi}{q_\mu}\right)\bigg/\sin\frac{\pi}{p_\mu}\sin\frac{\pi}{p_{\mu+1}}. \tag{12·13}$$

The inequalities

$$p_\mu > 1, \quad q_\mu > 2, \quad \frac{1}{p_\mu} + \frac{1}{p_{\mu+1}} + \frac{2}{q_\mu} > 1 \tag{12·131}$$

provide a necessary condition for the existence of such a polytope. The ring round the first node of the graph indicates that the vertices of the polytope are the images of a point lying on all the mirrors *except* the first. We shall sometimes have occasion to mention a 'truncated' polytope (usually not regular) indicated by ringing the tth node (instead of the first); its vertices are the images of a point lying on all the mirrors except the tth. (See Coxeter 1963, p. 200.)

The second important case of a medial polytope is when $\mu = 0$, so that we are using only the first ν of the n reflections. We thus obtain the ν-dimensional element

$$\Pi_\nu = p_1\{q_1\}p_2\{q_2\} \ldots p_{\nu-1}\{q_{\nu-1}\}p_\nu \tag{12·14}$$

of Π_n.

The third case is when $\mu = 1$ and $\nu = n$. We now have an $(n-1)$-dimensional polytope whose vertices constitute the orbit of O_1 for the group generated by $R_2, \ldots, R_n$. By comparison with §2·1, we naturally call it the *vertex figure* of Π_n, and we can describe Π_n inductively as the regular polytope whose cell and vertex figure are

$$p_1\{q_1\}p_2\{q_2\} \cdots p_{n-2}\{q_{n-2}\}p_{n-1} \tag{12·15}$$

and

$$p_2\{q_2\} \cdots p_{n-2}\{q_{n-2}\}\, p_{n-1}\{q_{n-1}\}p_n.$$

The symbol $\{q_1, q_2, \ldots, q_{n-1}\}$ for a real polytope (see §5·7) may be regarded as a convenient abbreviation for

$$2\{q_1\}2\{q_2\} \ldots 2\{q_{n-1}\}2.$$

By taking the generating reflections R_μ in the reverse order, we see that the reciprocal of Π_n is

$$p_n\{q_{n-1}\}p_{n-1} \cdots \{q_2\}p_2\{q_1\}p_1.$$

Since a regular polytope is a special kind of *configuration* (Veblen and Young 1910, p. 38; Coxeter, 1963, p. 12), many of its numerical properties can be exhibited in a matrix

$$\begin{bmatrix} N_0 & N_{01} & \cdots & N_{0,n-1} \\ N_{10} & N_1 & \cdots & N_{1,n-1} \\ \cdots & \cdots & \cdots & \cdots \\ N_{n-1,0} & N_{n-1,1} & \cdots & N_{n-1} \end{bmatrix}$$

where N_μ (or $N_{\mu\mu}$, or $N_{n\mu}$) is the number of Π_μ's while $N_{\mu\nu}$ ($\mu \neq \nu$) is the number of Π_ν's incident with each Π_μ, so that

(12·16) $$N_\mu N_{\mu\nu} = N_\nu N_{\nu\mu}.$$

(Notice how these four numbers are situated in the matrix at the corners of a square.) For instance, the numerical matrix for the polygon $p_1\{q\}p_2$ is

$$\begin{bmatrix} N_0 & N_{01} \\ N_{10} & N_1 \end{bmatrix} = \begin{bmatrix} g/p_2 & p_2 \\ p_1 & g/p_1 \end{bmatrix}$$

where g is given by (9·85) and (9·86).

The order of symmetry group, being equal to the number of flags, is given by any one of the n formulae

$$\begin{aligned} g &= N_0 N_{01} N_{12} \ldots N_{n-2,n-1} \\ &= N_{10} N_1 N_{12} \ldots N_{n-2,n-1} \\ &= \quad \cdots \qquad \cdots \qquad \cdots \\ &= N_{10} N_{21} N_{32} \ldots N_{n-1,n-2}\, N_{n-1} \end{aligned}$$

which can be traced as 'broken lines' in the matrix (Coxeter 1963, pp. 130–1). The matrix for the reciprocal polytope is derived by applying the half-turn that replaces N_μ by $N_{n-\mu-1}$.

A polytope is said to be *r-symmetric* if its symmetry operations include a dilatation

(12·17) $$x_\lambda' = \exp(2\pi i/r)\, x_\lambda \quad (\lambda = 1, \ldots, n)$$

of period $r > 1$. For instance, it is 2-symmetric if it is symmetric by a central inversion (§1·1), that is, if its vertices occur in antipodal pairs; thus all regular real polytopes are 2-symmetric except the odd polygons and the simplexes ($n \geqslant 2$) (Coxeter 1963, p. 226).

Defining a *diameter* of a polytope to be the line joining the centre to any vertex, we notice that the vertices of an r-symmetric regular polytope occur in sets of r on its N_0/r diameters, such sets being permuted by a group $\mathfrak{C}_r$ of dilatations. This $\mathfrak{C}_r$ (generated by (12·17)) is the *centre* of the symmetry group; its quotient group, the *central quotient group*, may be regarded as a *collineation group* in the complex projective $(n-1)$-space whose points represent the lines through the origin in the unitary n-space.

Since r depends only on the symmetry group, reciprocal polytopes are r-symmetric together, with the same r.

The enumeration of regular polygons in §11·1 shows that, apart from $2\{q\}2$ with an odd numerator for q, every regular polygon is r-symmetric for some r, and every regular star polygon has the same r as the non-starry polygon that has the same vertices. In fact, if the symmetry group is $p_1[q]p_2$ (so that q is an integer), we have

(12·18) $$r = (2, q)\, h/q,$$

where h is given by (9·85) and $(2, q)$ means 1 or 2 according as q is odd or even. Moreover, the collineation group is polyhedral, namely

$$(p_1, q, 2) \quad \text{or} \quad (p_1, \tfrac{1}{2}q, p_2)$$

according as q is odd or even.

If $q_1, \ldots, q_{n-1}$ are integers, and $p_\mu = p_{\mu+1}$ whenever q_μ is odd, the symmetry group of the polytope (12·12) (or of its reciprocal) is denoted by

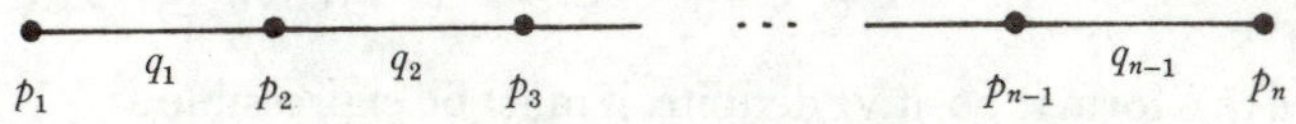

or
$$p_1[q_1]p_2[q_2] \cdots p_{n-1}[q_{n-1}]p_n$$

(or $p_n[q_{n-1}]p_{n-1} \cdots [q_2]p_2[q_1]\,p_1$). It satisfies the relations (12·11) and

(12·19) $$\mathrm{R}_\mu \mathrm{R}_{\mu+1} \mathrm{R}_\mu \ldots = \mathrm{R}_{\mu+1} \mathrm{R}_\mu \mathrm{R}_{\mu+1} \ldots$$

(with q_μ factors on each side, as in (9·81)). Till Chapter 13, we shall not attempt to decide whether these relations suffice for a presentation.

EXERCISES

1. What does condition (i) say when $\lambda = -1$ and $\nu = n$?
2. What does condition (ii) say when $\lambda = -1$ and $\nu = 2$? When $\lambda = n-3$ and $\nu = n$?
3. Instead of condition (ii) could we have written simply: Every medial figure with $\lambda = \nu - 3$ is connected?
4. Write out the numerical matrix for the real 24-cell $2\{3\}2\{4\}2\{3\}2$.
5. For a convex polytope $2\{q_1\}2 \ldots \{q_{n-1}\}2$, express q_μ in terms of the N's.
6. For a complex polytope $p_1\{q_1\}p_2 \ldots \{q_{n-1}\}p_n$, express p_μ in terms of the N's.

12·2 HERMITIAN FORMS

We have seen that a regular complex polytope

$$p_1\{q_1\}p_2\{q_2\} \cdots p_{n-1}\{q_{n-1}\}p_n$$

can be derived by Wythoff's construction from reflections of periods p_ν in mirrors forming an orthoscheme whose acute angles σ_μ are given by

(12·13). As in §8·8, we use an oblique coordinate system based on unit vectors $\mathbf{e}_\nu$ perpendicular to these mirrors, so that the inner products of pairs of these vectors are

$$a_{\nu\nu} = \mathbf{e}_\nu \cdot \mathbf{e}_\nu = 1 \quad (\nu = 1, 2, \ldots, n)$$

$$a_{\mu\nu} = \mathbf{e}_\mu \cdot \mathbf{e}_\nu = 0 \quad (\mu < \nu - 1)$$

and

$$(12{\cdot}21) \quad a_{\mu,\mu+1} = \mathbf{e}_\mu \cdot \mathbf{e}_{\mu+1} = -\cos\sigma_\mu = -\left(\frac{\cos\left(\dfrac{\pi}{p_\mu} - \dfrac{\pi}{p_{\mu+1}}\right) + \cos\dfrac{2\pi}{q_\mu}}{2\sin\dfrac{\pi}{p_\mu}\sin\dfrac{\pi}{p_{\mu+1}}}\right)^{\frac{1}{2}}$$

(see (9·89)), and the invariant Hermitian form is

$$\mathbf{x}\cdot\mathbf{x} = \Sigma\Sigma\, a_{\mu\nu} x^\mu \bar{x}^\nu$$

$$(12{\cdot}22) \quad = x^1\bar{x}^1 - (x^1\bar{x}^2 + x^2\bar{x}^1)\cos\sigma_1 + x^2\bar{x}^2 - (x^2\bar{x}^3 + x^3\bar{x}^2)\cos\sigma_2 + \ldots - (x^{n-1}\bar{x}^n + x^n\bar{x}^{n-1})\cos\sigma_{n-1} + x^n\bar{x}^n.$$

Since this form is positive definite, it must be expressible as

$$u_1\bar{u}_1 + u_2\bar{u}_2 + \ldots + u_n\bar{u}_n,$$

where the u's are linear combinations of the x's. (In practice they can be found by inspection, as in the answers to Exx. 2 and 3 of §8·2.) These expressions

$$u_\mu = \Sigma\, e_{\lambda\mu} x^\lambda$$

can then be 'transposed' to yield $x_\nu = \Sigma\,\bar{e}_{\nu\mu} u_\mu$, as in (8·69). Now R_ν appears as a reflection of period p_ν in the hyperplane $x_\nu = 0$ or

$$\Sigma\,\bar{e}_{\nu\mu} u_\mu = 0,$$

and we can use (8·77) to express R_ν as

$$u_\lambda' = u_\lambda + \{\exp(2\pi i/p_\nu) - 1\} e_{\nu\lambda} \Sigma\,\bar{e}_{\nu\mu} u_\mu.$$

To obtain one vertex of the polytope (on all the mirrors except the first), we solve the $n-1$ equations $x_\lambda = 0$ or

$$\Sigma\,\bar{e}_{\lambda\mu} u_\mu = 0 \quad (\lambda = 2, \ldots, n).$$

To obtain the remaining vertices we apply the reflections R_ν repeatedly, for various values of ν. By including the equation

$$\Sigma\,\bar{e}_{1\mu} u_\mu = 0$$

instead of $\Sigma\,\bar{e}_{n\mu} u_\mu = 0$ (so as to take a point on all the mirrors except the *last*), we obtain the reciprocal polytope.

For instance, generalizing the $p\{4\}2$ whose p^2 vertices are given by (11·21), let us consider Shephard's 'generalized orthotope'

or

$$(12{\cdot}23) \qquad \gamma_n^p = p\{4\}2\{3\}2\ldots\{3\}2.$$

(In the graph we naturally omit the mark q_ν whenever it is 3, and p_ν whenever it is 2.) We now have

$$a_{12} = -\sqrt{\tfrac{1}{2}}, \quad a_{23} = a_{34} = \ldots = -\tfrac{1}{2},$$

so the form is

$$x^1\bar{x}^1 - \sqrt{\tfrac{1}{2}}(x^1\bar{x}^2 + x^2\bar{x}^1) + x^2\bar{x}^2 - \tfrac{1}{2}(x^2\bar{x}^3 + x^3\bar{x}^2) + \ldots - \tfrac{1}{2}(x^{n-1}\bar{x}^n + x^n\bar{x}^{n-1}) + x^n\bar{x}^n$$

$$= u_1\bar{u}_1 + u_2\bar{u}_2 + \ldots + u_{n-1}\bar{u}_{n-1} + u_n\bar{u}_n,$$

where $u_1 = x^1 - \dfrac{x^2}{\sqrt{2}}, \quad u_2 = \dfrac{x^2 - x^3}{\sqrt{2}}, \ldots, u_{n-1} = \dfrac{x^{n-1} - x^n}{\sqrt{2}}, \quad u_n = \dfrac{x^n}{\sqrt{2}}.$

Transposing the matrix of coefficients, we obtain the mirrors

$$u_1 = 0, \quad \frac{-u_1 + u_2}{\sqrt{2}} = 0, \quad \frac{-u_2 + u_3}{\sqrt{2}} = 0, \quad \ldots, \quad \frac{-u_{n-1} + u_n}{\sqrt{2}} = 0.$$

Thus R_1 multiplies u_1 by $\exp(2\pi i/p)$ (leaving the rest of the coordinates unchanged), R_2 is the transposition (1 2) (interchanging u_1 and u_2), R_3 is (2 3), …, and R_n is $(n-1 \;\; n)$. The group

$$p[4]2[3]2\ldots[3]2,$$

generated by these n reflections, is clearly the *wreath product*

$$\mathfrak{C}_p \wr \mathfrak{S}_n$$

of the cyclic group generated by R_1 and the symmetric group generated by the $n-1$ transpositions. In other words, it is the semidirect product of $\mathfrak{C}_p{}^n$ and $\mathfrak{S}_n$; thus its order is $p^n n!$. Solving the $n-1$ equations

$$u_1 = u_2, \quad u_2 = u_3, \ldots, u_{n-1} = u_n,$$

we obtain an initial vertex that may conveniently be taken to be (1, 1, …, 1). Applying the group, we obtain the p^n points whose coordinates range independently over the powers of $\exp(2\pi i/p)$; these are the vertices of the *generalized n-cube* γ_n^p. Fixing one of the coordinates, we obtain one of the pn cells γ_{n-1}^p. Fixing $n-1$ of the coordinates, we obtain one of the np^{n-1} edges γ_1^p. In fact, the edges fall into n sets of p^{n-1} parallel edges, one for each of the n complex dimensions, and the polytope γ_n^p can be described as the *Cartesian product* of n γ_1^p's. (See Coxeter 1963, p. 124 for the case when $p = 2$.)

Applying the same group to the point (0, …, 0, 1), we obtain the pn

points for which one coordinate is a power of $\exp(2\pi i/p)$ while all the others are zero; these are the vertices of the reciprocal *generalized cross polytope*

$$(12\cdot24) \qquad \beta_n^p = 2\{3\}2\ldots\{3\}2\{4\}p,$$

whose cells consist of p^n regular simplexes

$$\alpha_{n-1} = 2\{3\}2\ldots\{3\}2 = \{3,\ldots,3\}.$$

Clearly, both γ_n^p and β_n^p are p-symmetric. The one-dimensional regular polytope

p

which has p vertices, may be denoted by either γ_1^p or β_1^p. It is represented in the Argand diagram by an ordinary regular p-gon.

EXERCISES

1. Compute N_μ and $N_{\mu\nu}$ for the generalized n-cube, γ_n^p.
2. Does the Euler–Schälfli formula $\sum_{-1}^{n}(-1)^\mu N_\mu = 0$ remain valid for γ_n^p?
3. What n-dimensional polytope has for vertices the points

$$(\epsilon^\mu, \epsilon^{2\mu}, \ldots, \epsilon^{n\mu}) \quad (\epsilon = \exp\{2\pi i/(n+1)\}; \quad \mu = 0, 1, \ldots, n)?$$

12·3 THE HESSIAN POLYHEDRON

In the case of the *Hessian polyhedron*

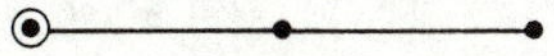

3 3 3

or $3\{3\}3\{3\}3$, the Hermitian form is

$$x^1\bar{x}^1 - \sqrt{\tfrac{1}{3}}(x^1\bar{x}^2 + x^2\bar{x}^1) + x^2\bar{x}^2 - \sqrt{\tfrac{1}{3}}(x^2\bar{x}^3 + x^3\bar{x}^2) + x^3\bar{x}^3 = u_1\bar{u}_1 + u_2\bar{u}_2 + u_3\bar{u}_3$$

where $$u_1 = -x^1 + \frac{x^2}{\sqrt{3}}, \quad u_2 = \frac{x^2}{\sqrt{3}}, \quad u_3 = \frac{x^2}{\sqrt{3}} - x^3.$$

Thus the three mirrors are

$$(12\cdot31) \qquad u_1 = 0, \quad \frac{u_1+u_2+u_3}{\sqrt{3}} = 0, \quad u_3 = 0;$$

R_1 and R_3 multiply u_1 and u_3 (respectively) by

$$\omega = \exp(2\pi i/3)$$

and R_2 is

$$(12\cdot32) \qquad u_\lambda' = u_\lambda - \frac{i\omega^2}{\sqrt{3}}(u_1+u_2+u_3).$$

For the initial vertex (on the second and third mirrors) we may take $(1, -1, 0)$. The reflection R_1 yields $(\omega, -1, 0)$ and then $(\omega^2, -1, 0)$; R_2 transforms these points into $(0, \omega^2, -\omega)$ and $(-\omega, 0, 1)$. Continuing, we find that $3\{3\}3\{3\}3$ has just the 27 vertices†

$$(12\cdot33) \qquad (0, \omega^\mu, -\omega^\nu), \quad (-\omega^\nu, 0, \omega^\mu), \quad (\omega^\mu, -\omega^\nu, 0),$$

where μ and ν take (independently) the three values 0, 1, 2 or (equally well) 1, 2, 3. Making the latter choice, we may conveniently use the concise symbols

$$(12\cdot331) \qquad 0\mu\nu, \quad \nu0\mu, \quad \mu\nu0$$

for these 27 points. For instance, the points

$$(1, -1, 0), \quad (\omega, -1, 0), \quad (\omega^2, -1, 0), \quad (0, \omega^2, -\omega), \quad (-\omega, 0, 1)$$

appear as 330, 130, 230, 021, 103.

This notation is sufficiently concise to allow us to represent the generating reflections as permutations:

$$\begin{aligned}
&R_1 = (101\ 201\ 301)(102\ 202\ 302)(103\ 203\ 303)\\
&\qquad (110\ 210\ 310)(120\ 220\ 320)(130\ 230\ 330),\\
(12\cdot34)\quad &R_2 = (012\ 230\ 103)(013\ 102\ 320)(021\ 203\ 130)\\
&\qquad (023\ 310\ 201)(031\ 120\ 302)(032\ 301\ 210),\\
&R_3 = (011\ 012\ 013)(021\ 022\ 023)(031\ 032\ 033)\\
&\qquad (101\ 102\ 103)(201\ 202\ 203)(301\ 302\ 303).
\end{aligned}$$

Since the symbol $3\{3\}3\{3\}3$ is a palindrome, this polyhedron is self-reciprocal (like the tetrahedron $2\{3\}2\{3\}2$). Having 27 vertices, it also has 27 faces $3\{3\}3$, one of which, derived from $(1, -1, 0)$ by applying R_1 and R_2, has the 8 vertices

$$(12\cdot35) \qquad 012,\ 021,\ \lambda03,\ \lambda30 \quad (\lambda = 1, 2, 3)$$

(see Figure 12·3A). Since the vertex figure is another $3\{3\}3$, there are 8 edges at each vertex, and 72 edges altogether (appearing in Figure 12·3B as equilateral triangles, 8 of which are seen also in Figure 12·3A):

	011, 202, 330, etc.	6
	032, 301, 210, etc.	6
(12·36)	023, 302, 230, etc.	6
	011, 021, 031, etc.	18
	011, 120, 203, etc.	36

(Here 'etc.' means that we may permute the digits 1, 2, 3, or the three positions, or both.)

† Coxeter, The polytope 2_{21}, whose 27 vertices correspond to the lines on the general cubic surface, *American Journal of Mathematics* **62** (1940), 457–86; see especially p. 469.

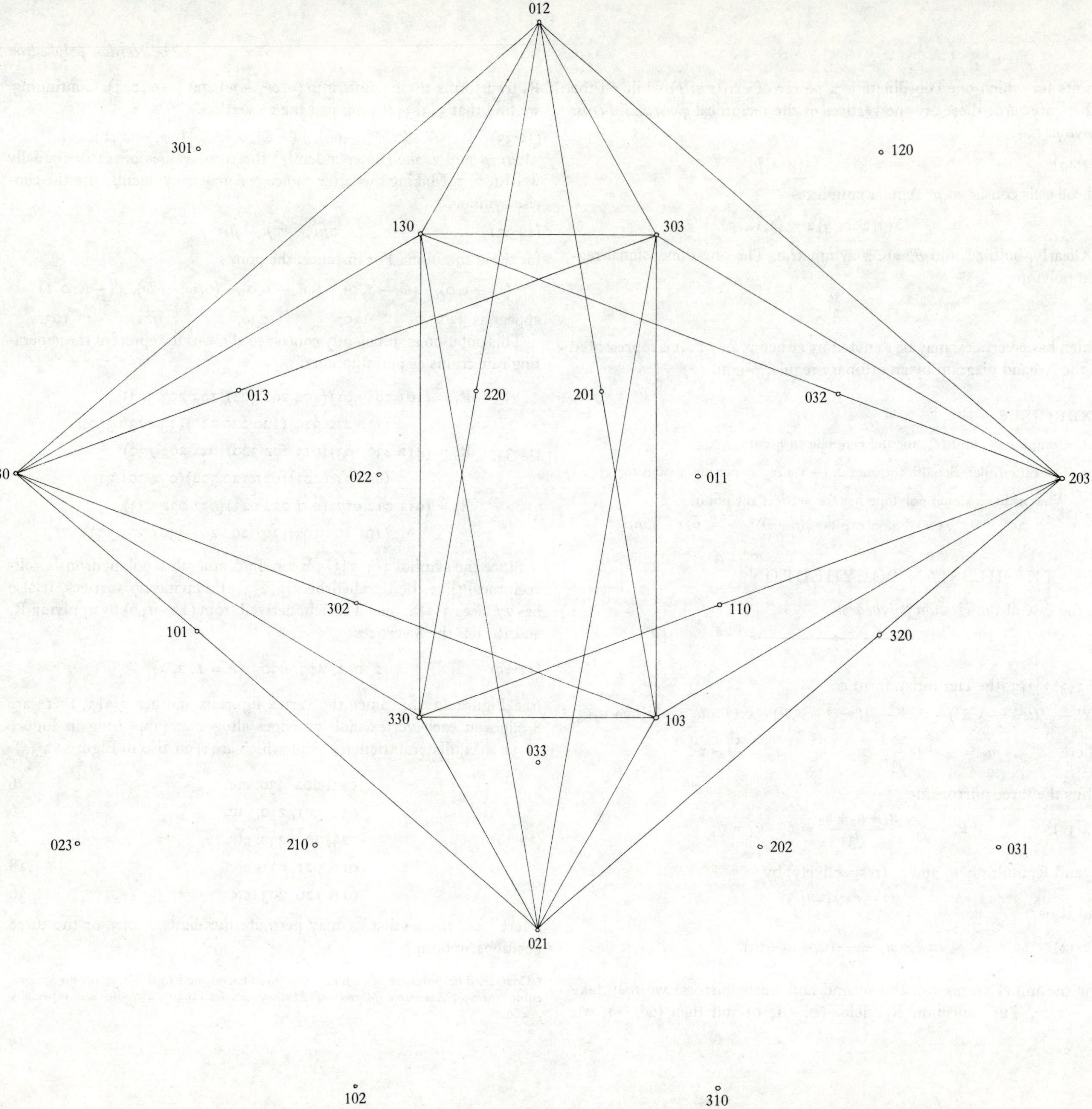

Figure 12·3A: The twenty-seven vertices of $3\{3\}3\{3\}3$, and one of its twenty-seven faces $3\{3\}3$

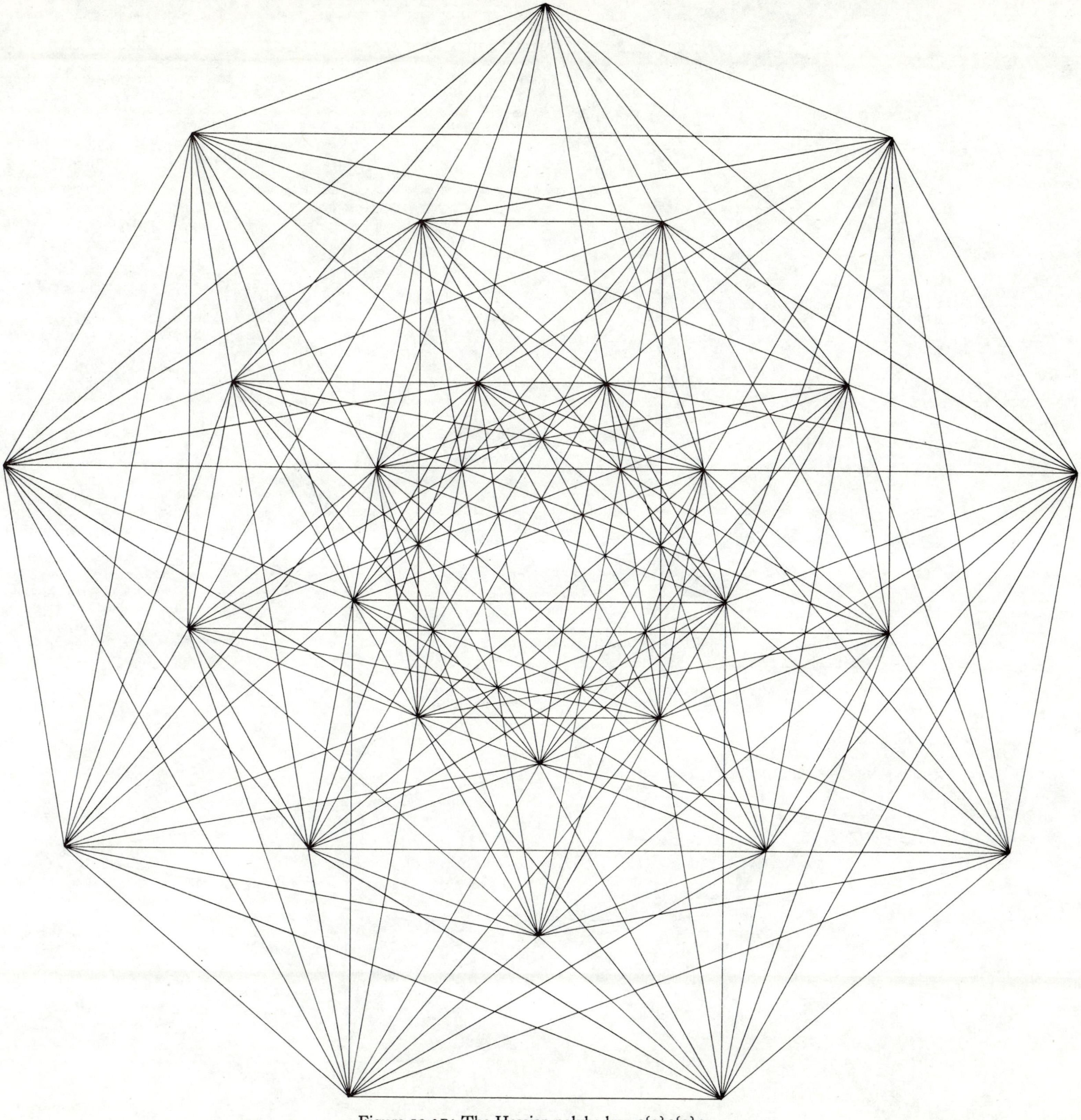

Figure 12·3B: The Hessian polyhedron $_3\{3\}_3\{3\}_3$

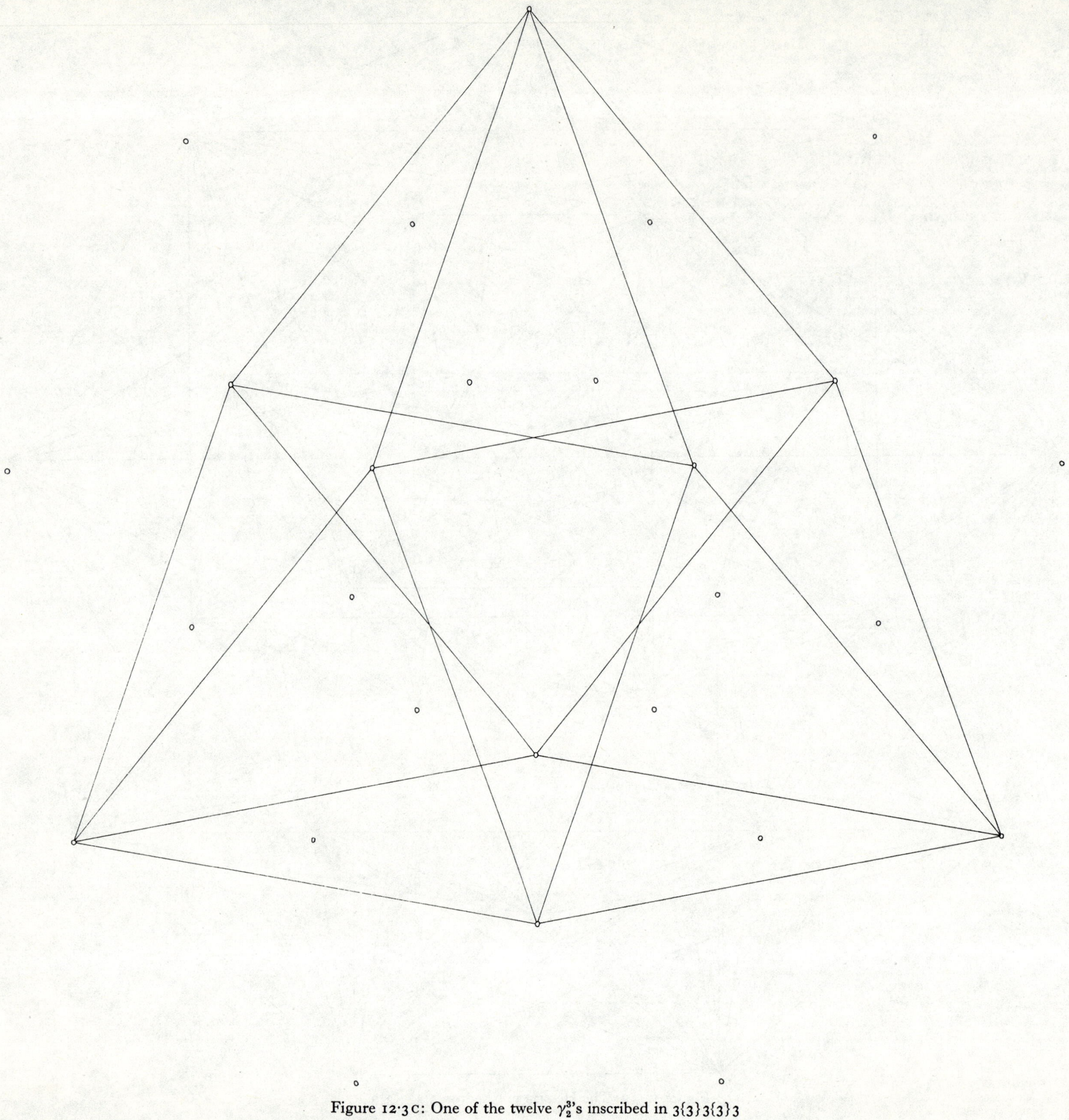

Figure 12·3c: One of the twelve γ_2^3's inscribed in $3\{3\}3\{3\}3$

Two vertices are naturally said to be *adjacent* if they belong to the same edge; their symbols agree in an even number of places, that is, in just two places or nowhere. The 72 edges occur in 36 pairs of *opposites*, such that any two vertices belonging to opposite edges are *non-adjacent*; their symbols agree in just one place. In the above list, the first two edges are opposite to each other, the third is opposite to

032, 203, 320,

derived by interchanging 2 and 3. The opposites of the fourth and fifth are

101, 201, 301 and 101, 210, 023,

derived by interchanging the first and second digits.

Although the edges occur in pairs of opposites, the vertices obviously do not (as 27 is an odd number). In fact, since (12·17) with $r = 3$ is a symmetry operation, $3\{3\}3\{3\}3$ is not 2-symmetric but 3-symmetric: the 27 vertices lie by threes on nine diameters. In the notation of (12·331), the three vertices on any diameter are derived from one another by cyclically permuting the digits 1, 2, 3 wherever they occur; for instance,

one diameter contains 011, 022, 033,
and another, 012, 023, 031.

Similarly, the 72 edges lie by sixes on twelve *planes of symmetry*

$$(12·37)\quad u_1 = 0,\quad u_2 = 0,\quad u_3 = 0,\quad \omega^\lambda u_1 + \omega^\mu u_2 + \omega^\nu u_3 = 0$$
$$(\lambda, \mu, \nu = 0, 1, 2 \text{ independently}).$$

These planes include the mirrors (12·31) for the generating reflections R_1, R_2, R_3; in fact, they are the mirrors for all the reflections that occur in $3[3]3[3]3$. For instance, the plane $u_1 = 0$ (which is the mirror for R_1) contains the six edges that appear as rows and columns in the matrix

$$\begin{bmatrix} 011 & 012 & 013 \\ 021 & 022 & 023 \\ 031 & 032 & 033 \end{bmatrix}.$$

The nine vertices that lie on these six edges have coordinates

$$(0, \omega^\mu, -\omega^\nu);$$

hence, by (11·21), they belong to a $3\{4\}2$ or γ_2^3. Since

$$R_1 R_2 R_1 = R_2 R_1 R_2 \quad\text{and}\quad R_2 R_3 R_2 = R_3 R_2 R_3,$$

the generators R_1, R_2, R_3 are mutually conjugate; and of course every reflection in the group is conjugate to some $R_\nu^{\pm 1}$. Thus *the section of* $3\{3\}3\{3\}3$ *by any plane of symmetry is a polygon* $3\{4\}2$ (see Figure 12·3c).

The nine diameters and twelve planes of symmetry are related in a remarkable way: every two diameters determine a plane of symmetry which contains a third diameter. This can be seen by arranging the diameters γ_1^3 as the elements of a 3×3 matrix

$$(12·38)\quad \begin{bmatrix} (011, 022, 033) & (101, 202, 303) & (110, 220, 330) \\ (023, 031, 012) & (302, 103, 201) & (230, 310, 120) \\ (032, 013, 021) & (203, 301, 102) & (320, 130, 210) \end{bmatrix}$$

and observing the twelve ways in which three diameters contain the nine vertices of a γ_2^3: the three rows, three columns, and six 'diagonals' (or 'determinant terms'). For instance, the first column yields the γ_2^3 that was exhibited above.

$3\{3\}3\{3\}3$ is called the *Hessian polyhedron* because the incidences of the diameters and planes of symmetry are the same as the incidences of the points and lines of the *Hessian configuration*

$$\begin{bmatrix} 9 & 4 \\ 3 & 12 \end{bmatrix}$$

or $(9_4, 12_3)$ (Miller, Blichfeldt and Dickson 1916, p. 335): nine points lying by threes on twelve lines, with four of the lines through each point. To make this transition from complex affine 3-space to the complex projective plane, we merely have to interpret the affine coordinates (12·33) for the 27 points (12·331) as homogeneous coordinates

$$(12·39)\quad \begin{matrix} (0, 1, -1) & (-1, 0, 1) & (1, -1, 0) \\ (0, 1, -\omega) & (-\omega, 0, 1) & (1, -\omega, 0) \\ (0, 1, -\omega^2) & (-\omega^2, 0, 1) & (1, -\omega^2, 0) \end{matrix}$$

for the nine points given by $x^3 + y^3 + z^3 = xyz = 0$. These are the nine points of inflexion of the plane cubic curve

$$x^3 + y^3 + z^3 - 3axyz = 0 \quad (a \neq 0, 1)$$

whose Hessian is $x^3 + y^3 + z^3 + (a - 4a^{-2})xyz = 0$. This configuration serves to justify Sylvester's use of the word 'real' in his famous problem of 1893 (see Coxeter 1969, pp. 65, 181):

> Prove that it is not possible to arrange any number of real points so that a right line through every two of them shall pass through a third, unless they all lie in the same right line.

More simply, we may remark that the Hessian polyhedron yields a Hessian configuration when we consider the section of the diagonals and planes of symmetry by a plane of general position.

The notation (12·331), as an abbreviation for the coordinates (12·33), seems so obvious that one might be tempted to forget its original use by Beniamino Segre (1942, pp. 3–4) as a notation for the 27 lines on a general cubic surface in complex projective 3-space. In that notation, two of the lines intersect if their symbols agree in just one place, but two of the lines are skew if their symbols agree in two places or nowhere.

Thus the 27 lines on the cubic surface represent the 27 vertices of $3\{3\}3\{3\}3$ in such a way that two of the lines are intersecting or skew according as the corresponding vertices are non-adjacent or adjacent. As the agreement seems so perfect, one naturally asks why the group of automorphisms of the 27 lines has order 51 840 whereas the order of $3[3]3[3]3$ is only 648. The explanation is that, among the pairs of non-adjacent vertices of $3\{3\}3\{3\}3$, we have included pairs that belong to a diameter. In fact, there are altogether $9+36$ triads of mutually non-adjacent vertices: 9 lying on diameters, and 36, such as

$$120, \; 210, \; 330,$$

forming equilateral triangles, three inscribed in each of the twelve γ_2^{3}'s. (See the three concentric equilateral triangles not drawn in Figure 12·3 C.) However, on the cubic surface there is no corresponding distinction: the 45 triads of intersecting lines are all alike; their planes are just the 45 tritangent planes (Segre 1942, p. 5). The nine diameters correspond to one of the forty ways in which the 27 lines can be distributed into nine triads, each triad forming a tritangent plane (Baker 1946, p. 37). The representation of the nine diameters by the points of the configuration $(9_4, 12_3)$ corresponds to the fact that there are 12 ways of distributing these 9 tritangent planes into 3 Steiner trihedra (Segre 1942, pp. 12–14) so that each plane belongs to 4 of the trihedra.

Thus the group of automorphisms of the polyhedron (namely, the group $3[3]3[3]3$ augmented by the involutory antiprojective collineation that replaces each coordinate u_ν by its complex conjugate $\bar{u}_\nu$ and thus transforms each R_ν into its inverse) is a subgroup of index 40 in the group of automorphisms of the 27 lines.

When the three complex coordinates u_ν in unitary 3-space are replaced by six Cartesian coordinates in Euclidean 6-space, the polyhedron $3\{3\}3\{3\}3$ yields the uniform polytope 2_{21} (Coxeter 1963, pp. 202–3, 211) whose 27 vertices *faithfully* represent the 27 lines. This polytope is of special interest as a 'two-distance set' of 27 points: if the edges are of length 1, all the other pairs of vertices are at distance $\sqrt{2}$. (As before, the edges represent pairs of skew lines; the 'other pairs' represent pairs of intersecting lines.)

Figure 12·3 B was first published† as a drawing of 2_{21}, but since the number of real edges at a vertex of 2_{21} is just twice the number of complex edges at a vertex of $3\{3\}3\{3\}3$, the same figure serves just as well for the polyhedron. In fact, it serves still better for the polyhedron, as the 72 complex edges appear as equilateral triangles‡ which represent a subset of the 720 triangular faces of the real polytope; the rest of these 720 faces are foreshortened so as to appear as isosceles or scalene triangles or mere triads of collinear points.

† See p. 462 of the paper cited on page 119.

EXERCISES

1. Find the circumradius ${}_0R$, midradius ${}_1R$ and inradius ${}_2R$ of $3\{3\}3\{3\}3$ in terms of the radius of an edge as unit of measurement.

2. What happens to the Hessian polyhedron and the cubic surface when the coordinates are regarded as elements of the Galois field $GF[2^2]$ with moduli 2 and $\omega^2+\omega+1$?

3. Justify the above statement that 2_{21} is the real six-dimensional counterpart of the complex polyhedron $3\{3\}3\{3\}3$.

4. The polygon $3\{3\}3$ forms, with its centre and diameters, a Hessian configuration.

12·4 OTHER COMPLEX POLYHEDRA

A polyhedron closely related to $3\{3\}3\{3\}3$ (for reasons that will soon become clear) is

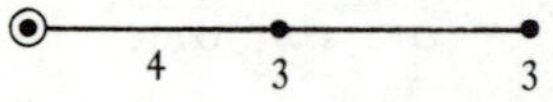

or

$$2\{4\}3\{3\}3,$$

for which the Hermitian form is

$$x^1\bar{x}^1 - \sqrt{\tfrac{1}{2}}(x^1\bar{x}^2 + x^2\bar{x}^1) + x^2\bar{x}^2 - \sqrt{\tfrac{1}{3}}(x^2\bar{x}^3 + x^3\bar{x}^2) + x^3\bar{x}^3$$
$$= u_1\bar{u}_1 + u_2\bar{u}_2 + u_3\bar{u}_3,$$

where $u_1 = -\dfrac{i\omega^2 x^1}{\sqrt{2}} + \dfrac{x^2}{\sqrt{3}}, \quad u_2 = \dfrac{i\omega x^1}{\sqrt{2}} + \dfrac{x^2}{\sqrt{3}}, \quad u_3 = \dfrac{x^2}{\sqrt{3}} - x^3.$

The three mirrors are now

(12·41) $$\frac{u_1 - \omega u_2}{\sqrt{2}} = 0, \quad \frac{u_1+u_2+u_3}{\sqrt{3}} = 0, \quad u_3 = 0;$$

R_2 and R_3 (of period 3) are the same as in § 12·3, but R_1 (of period 2) is

$$u_1' = \omega u_2, \quad u_2' = \omega^2 u_1, \quad u_3' = u_3,$$

transforming $(0, \omega^\mu, -\omega^\nu)$, $(-\omega^\nu, 0, \omega^\mu)$, $(\omega^\mu, -\omega^\nu, 0)$ into

$$(\omega^{\mu+1}, 0, -\omega^\nu), \quad (0, -\omega^{\nu-1}, \omega^\mu), \quad (-\omega^{\nu+1}, \omega^{\mu-1}, 0).$$

‡ For a more perspicuous view, entirely separating the 72 equilateral triangles, see Coxeter, The equianharmonic surface and the Hessian polyhedron, *Annali di Matematica* (4), **97** (1973), in press.

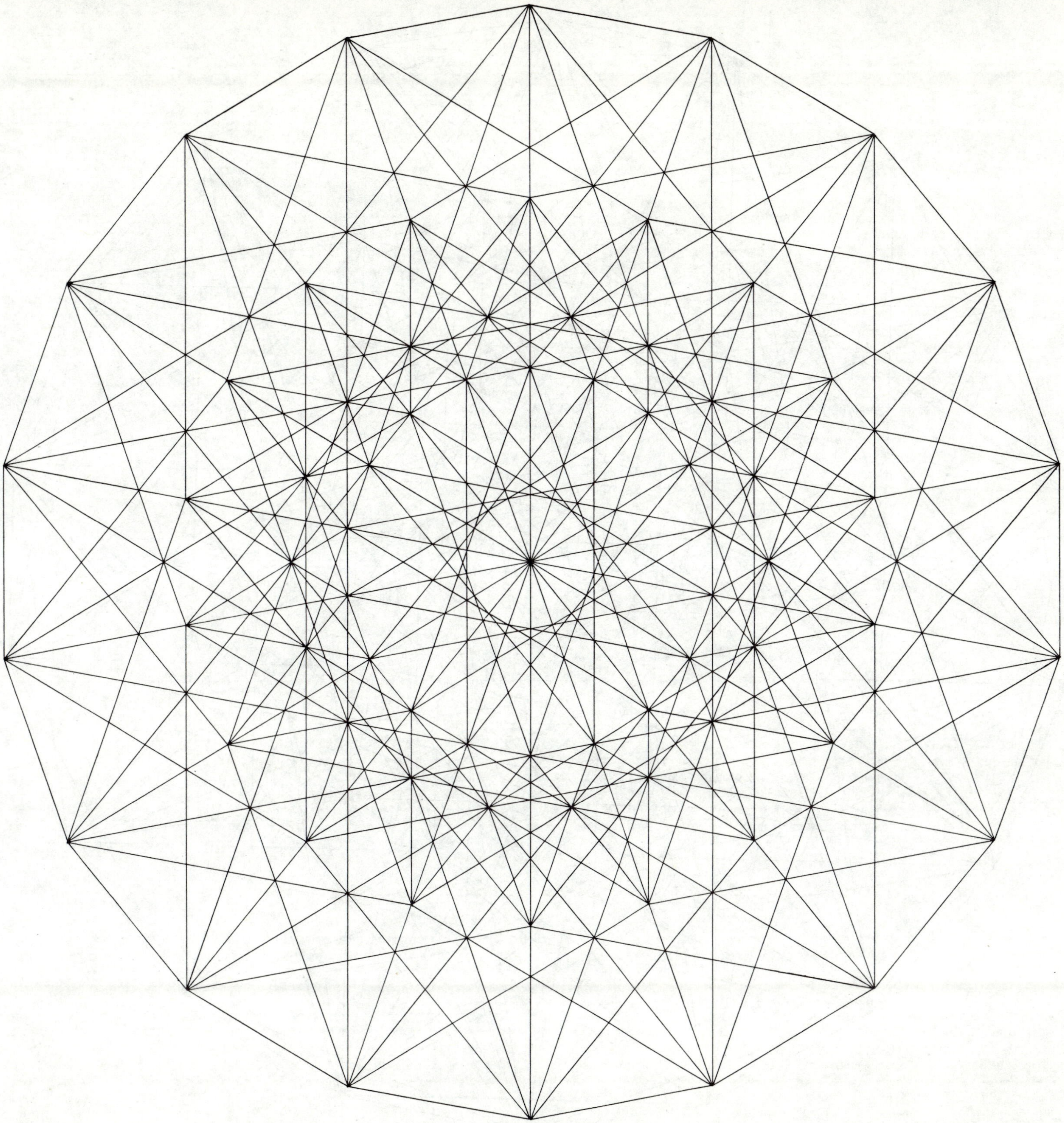

Figure 12·4A: The polyhedron $2\{4\}3\{3\}3$

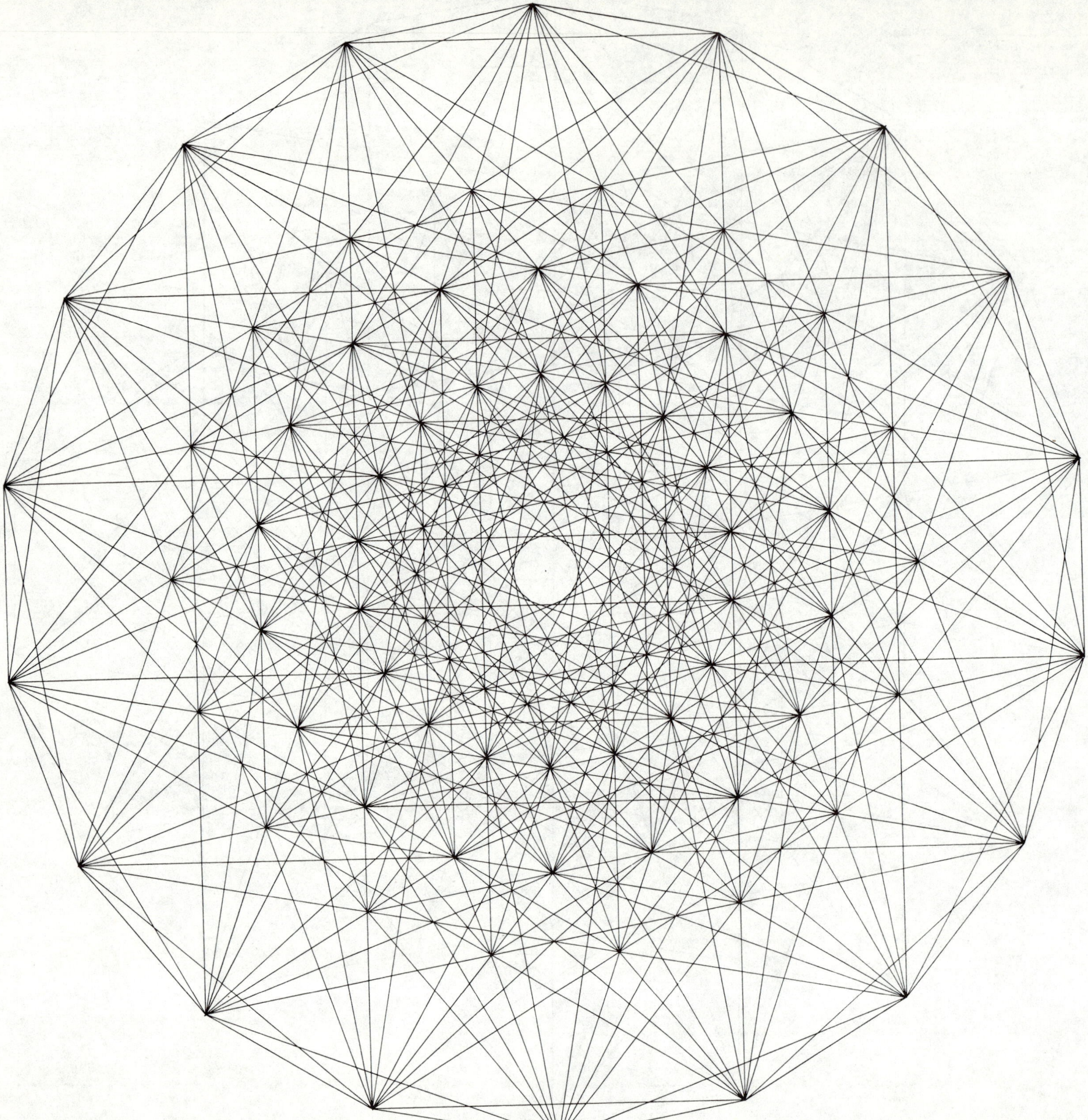

Figure 12·4B: The polyhedron $3\{3\}3\{4\}2$

(Notice the new positions for the minus signs.) Thus $2\{4\}3\{3\}3$ has the 54 vertices

$$(0, \pm\omega^{\mu}, \mp\omega^{\nu}),\quad (\mp\omega^{\nu}, 0, \pm\omega^{\mu}),\quad (\pm\omega^{\mu}, \mp\omega^{\nu}, 0),$$

namely the vertices of two reciprocal $3\{3\}3\{3\}3$'s of the same size, each derivable from the other by the central inversion

$$(12\cdot42)\qquad u_1' = -u_1,\quad u_2' = -u_2,\quad u_3' = -u_3.$$

There are 216 edges, each joining a vertex of one $3\{3\}3\{3\}3$ to one of the 8 nearest vertices of the other, for instance, joining $(0, -1, 1)$ to one of the 8 points (12·35).

In Figure 12·4A (drawn by McMullen) the 54 vertices appear as 3 rings of 18. Just outside the innermost ring of vertices we notice a ring of crossing points which are not vertices; their conspicuous appearance is an unfortunate consequence of this particular projection. The true vertices lie in sets of 6 on 9 lines through the centre of the drawing; but of course the centre itself is not a vertex.

In the plane $u_3 = -1$ we find the 6 vertices

$$(0, \omega^{\mu}, -1) \quad\text{and}\quad (\omega^{\nu}, 0, -1)$$

of one of the 72 faces $2\{4\}3$; another one, derived from this by applying R_2, lies in the plane

$$\omega(u_1+u_2)+\omega^2 u_3 = i\sqrt{3}.$$

Since the vertex figure $3\{3\}3$ has 8 vertices, the polyhedron has 8 edges at each vertex, 216 edges altogether. These are counted 3 times among the 72×9 edges of the 72 faces.

$2\{4\}3\{3\}3$ has the same nine diameters as $3\{3\}3\{3\}3$, but each contains six (instead of three) vertices, forming a γ_1^6. These nine γ_1^6's appear in Figure 12·4A as regular hexagons, three inscribed in each of the concentric $\{18\}$'s ('rings of 18'), *not* as the '9 lines through the centre of the drawing'.

On the line of intersection of the mirrors for R_1 and R_2, we find $(\omega, 1, \omega^2)$ as a typical vertex of the reciprocal polyhedron

$$3\{3\}3\{4\}2,$$

which has 72 vertices, 216 edges, and 54 faces $3\{3\}3$. Applying R_3, and then the other reflections again, we soon locate the whole set of

$$27+27+18$$

vertices $(\omega^{\lambda}, \omega^{\mu}, \omega^{\nu})$, $(-\omega^{\lambda}, -\omega^{\mu}, -\omega^{\nu})$ and $(\pm i\omega^{\lambda}\sqrt{3}, 0, 0)$, permuted. (Here again, λ, μ, ν run independently over the values 0, 1, 2.) In Figure 12·4B, these 72 points appear in 4 rings of 18. We recognize them as being the centres of the 72 edges (12·36) of $3\{3\}3\{3\}3$.

It is interesting to observe how remarkably the three complex polyhedra

$$2\{4\}3\{3\}3,\quad 3\{3\}3\{3\}3,\quad 3\{3\}3\{4\}2$$

resemble the real cube, tetrahedron and octahedron

$$2\{4\}2\{3\}2,\quad 2\{3\}2\{3\}2,\quad 2\{3\}2\{4\}2.$$

In both sets of three, the vertices of the first belong to two reciprocal specimens of the second, while the vertices of the third are the centres of the edges of the second.

Notice that $2\{4\}3\{3\}3$ and $3\{3\}3\{4\}2$ are 6-symmetric. In the case of $3\{3\}3\{4\}2$, the six vertices $(\pm i\omega^{\lambda}\sqrt{3}, 0, 0)$ form a $\gamma_1^6(=\beta_1^6)$ on one of the twelve diameters, which appear in Figure 12·4B as $\{6\}$'s inscribed in the four $\{18\}$'s.

These diameters distribute themselves into four triads of mutually orthogonal diameters, namely, lines joining the origin to the following triads of points:

$$(12\cdot43)\qquad \begin{array}{lll} (1, 0, 0), & (0, 1, 0), & (0, 0, 1);\\ (1, 1, 1), & (\omega, \omega^2, 1), & (\omega^2, \omega, 1);\\ (\omega, 1, 1), & (\omega^2, \omega^2, 1), & (1, \omega, 1);\\ (\omega^2, 1, 1), & (1, \omega^2, 1), & (\omega, \omega, 1). \end{array}$$

In other words, there is a *compound* consisting of four β_3^6's whose 4×18 vertices are just the 72 vertices of $3\{3\}3\{4\}2$.

Figure 12·4C shows a typical face. (Compare Figure 4.2A on page 30.)

The twelve *planes of symmetry* (12·37) of $3\{3\}3\{3\}3$, being mirrors for reflections of period 3 (such as R_2 and R_3), play the same role for

$$2\{4\}3\{3\}3 \quad\text{and}\quad 3\{3\}3\{4\}2;$$

but these two reciprocal polyhedra have nine additional planes of symmetry which are mirrors for reflections of period 2 (such as R_1). These are the nine planes†

$$u_{\mu}-\omega^{\lambda}u_{\nu} = 0 \quad (\mu \neq \nu).$$

Although the plane $u_2 = u_3$ (for instance) contains no vertices of $2\{4\}3\{3\}3$, it contains 24 vertices

$$(\omega^{\lambda}, \omega^{\mu}, \omega^{\mu}),\quad (-\omega^{\lambda}, -\omega^{\mu}, -\omega^{\mu}),\quad (\pm i\omega^{\lambda}\sqrt{3}, 0, 0)$$

of $3\{3\}3\{4\}2$. These belong to a $3\{4\}3$ (Figure 4·8A and Figure 12·4D).

† G. C. Shephard and J. A. Todd, Finite unitary reflection groups, *Canadian Journal of Mathematics* 6 (1954), 274–304; see especially p. 297.

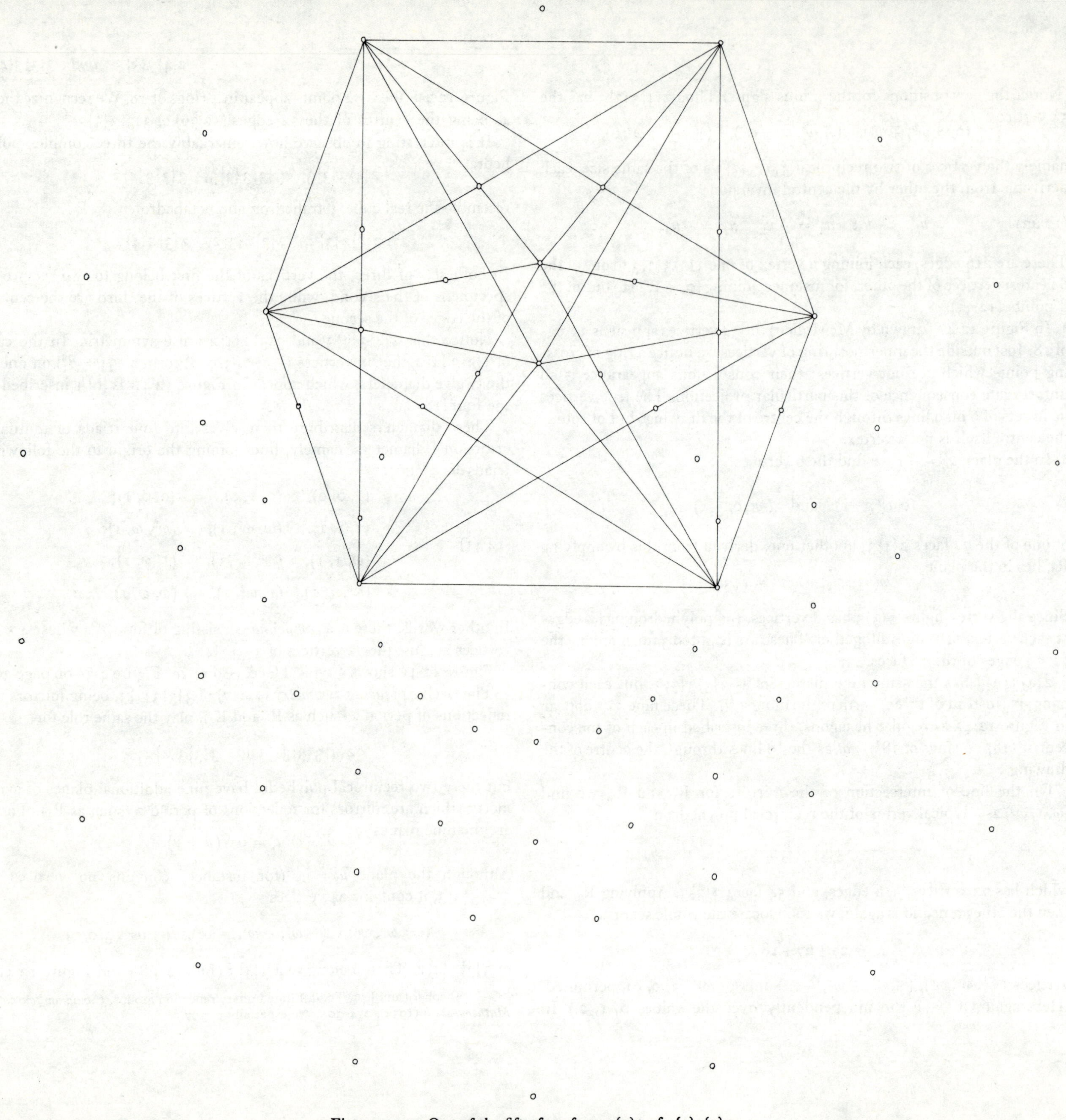

Figure 12·4c: One of the fifty-four faces 3{3}3 of 3{3}3{4}2

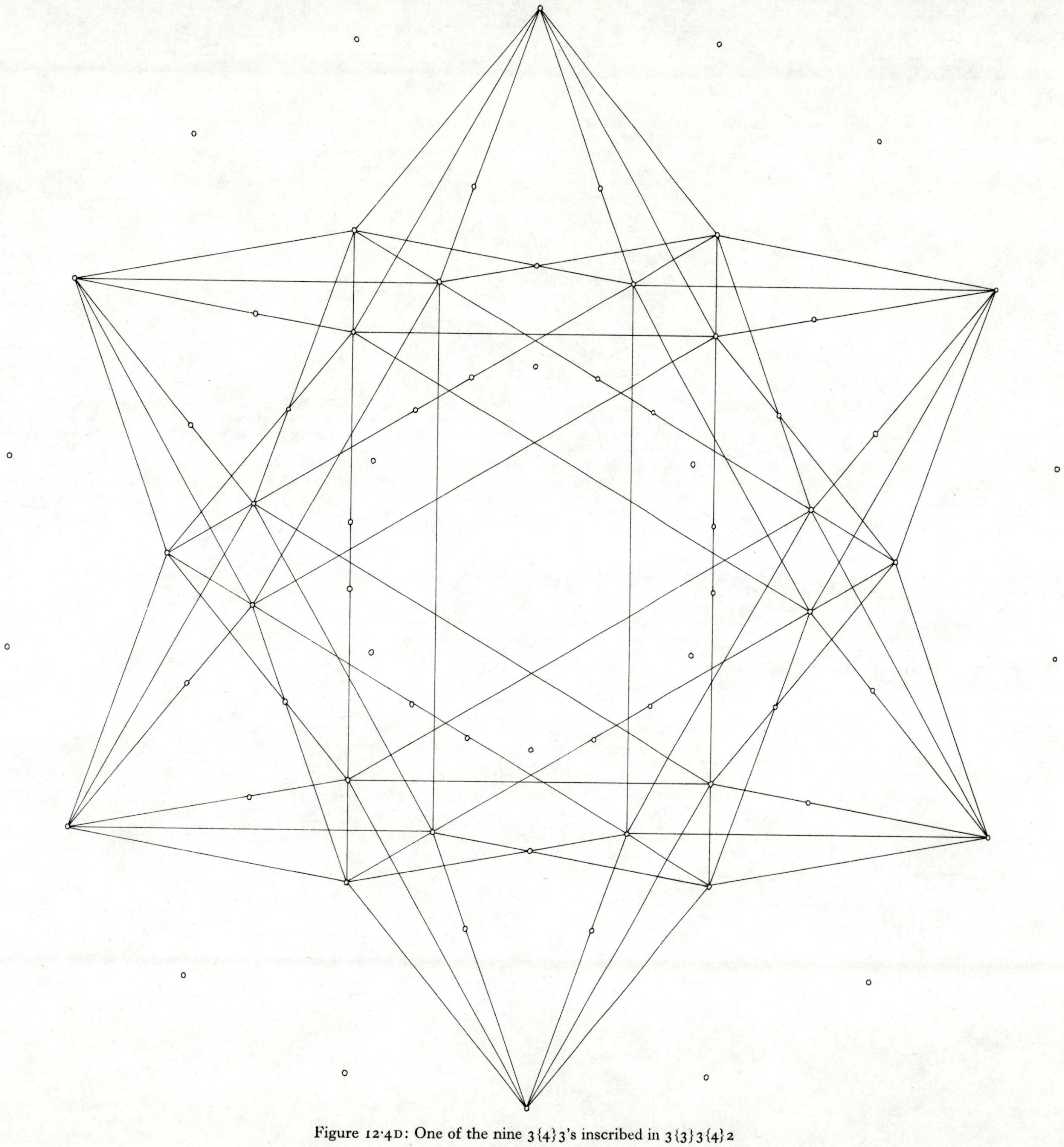

Figure 12·4D: One of the nine $3\{4\}3$'s inscribed in $3\{3\}3\{4\}2$

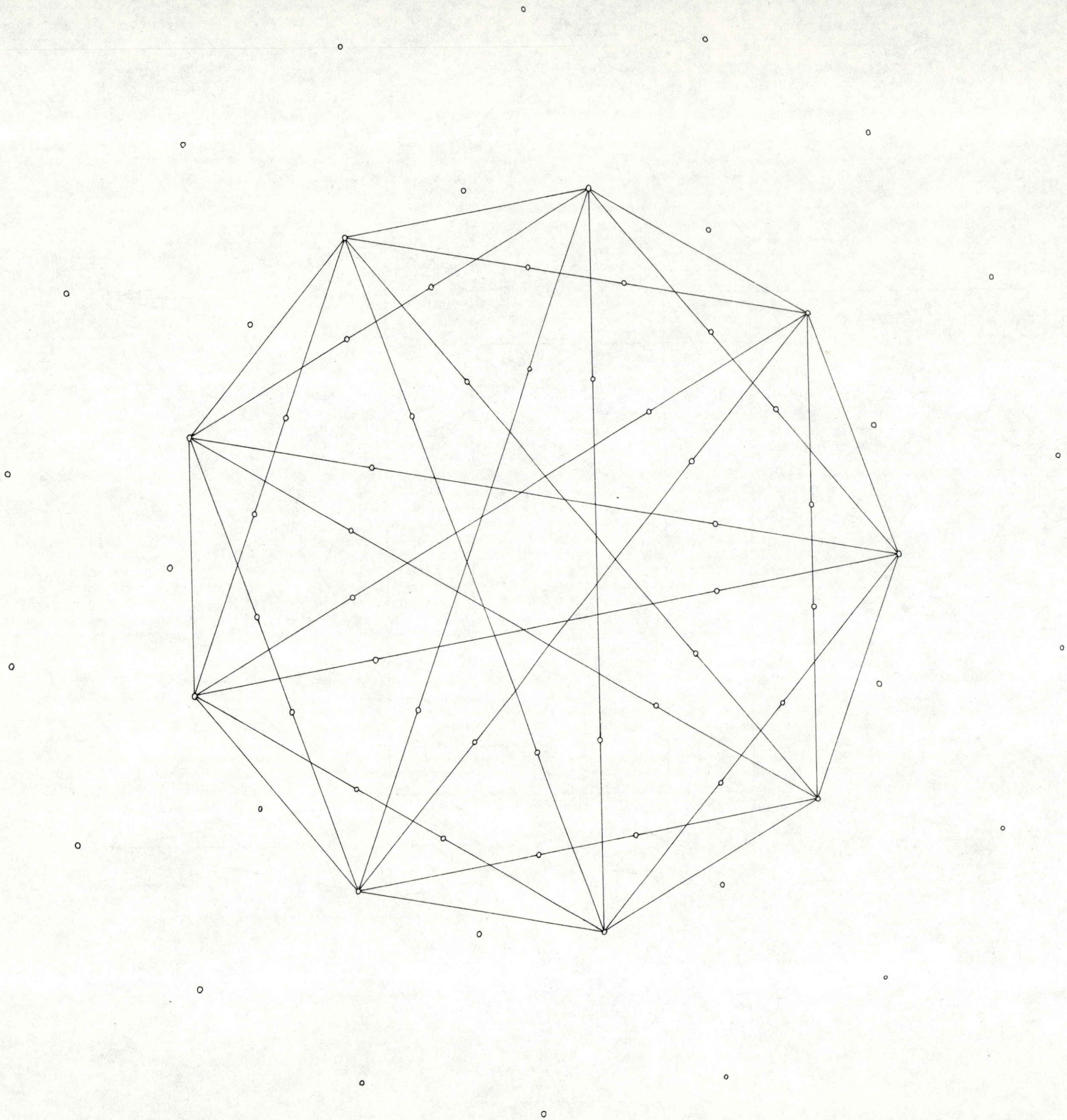

Figure 12·4E: One of the eight β_3^{3}'s inscribed in $3\{3\}3\{4\}2$

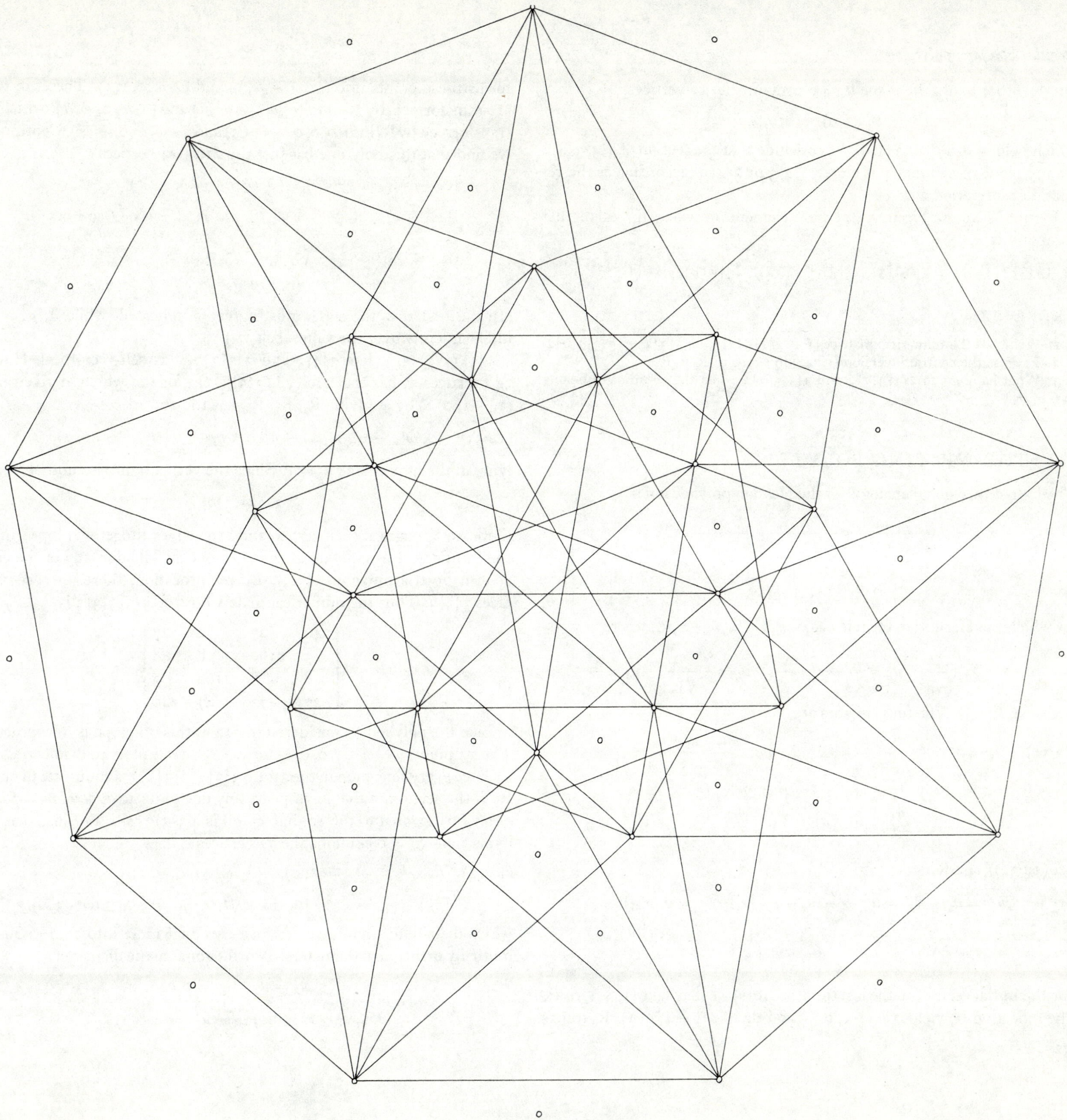

Figure 12·4F: One of the eight γ_3^3's inscribed in $3\{3\}3\{4\}2$

On the other hand, the plane $u_1 = 0$ contains the 12 vertices

$$(0, \pm i\omega^\lambda\surd 3, 0), \quad (0, 0, \pm i\omega^\lambda\surd 3),$$

which belong to a β_2^6 or $2\{4\}6$. We conclude that the section of $3\{3\}3\{4\}2$ by any plane of symmetry is either $3\{4\}3$ or $2\{4\}6$ (according as the reflection is of period 2 or 3).

Figure 12·4E and Figure 12·4F show elements of two reciprocal regular compounds:

$$(3\{3\}3\{4\}2)[8\beta_3^3]2(2\{4\}3\{3\}3), \quad 2(3\{3\}3\{4\}2)[8\gamma_3^3](2\{4\}3\{3\}3).$$

EXERCISES

1. Write out the numerical matrices for $2\{4\}3\{3\}3$, $3\{3\}3\{3\}3$, $3\{3\}3\{4\}2$.
2. Express the central inversion (12·42) in terms of $R_1R_2R_3$.
3. What happens to $2\{4\}3\{3\}3$ and $3\{3\}3\{4\}2$ when the coordinates belong to $GF[2^2]$?

12·5 THE WITTING POLYTOPE

The four-dimensional analogue of the Hessian polyhedron is

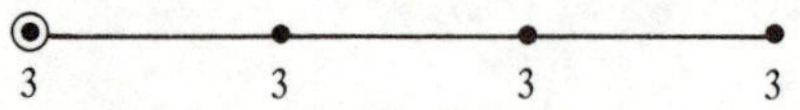

or

$$3\{3\}3\{3\}3\{3\}3,$$

for which the Hermitian form is $u_1\bar{u}_1 + u_2\bar{u}_2 + u_3\bar{u}_3 + u_4\bar{u}_4$, where

$$u_1 = -x^1 + \frac{x^2}{\surd 3}, \quad u_2 = \frac{x^2}{\surd 3} - x^3 + \frac{x^4}{\surd 3}, \quad u_3 = \frac{x^2 - x^4}{\surd 3}, \quad u_4 = \frac{x^4}{\surd 3}$$

(see §8·2, Ex. 3). The four mirrors are

$$(12{\cdot}51) \quad u_1 = 0, \quad \frac{u_1 + u_2 + u_3}{\surd 3} = 0, \quad u_3 = 0, \quad \frac{u_2 - u_3 + u_4}{\surd 3} = 0.$$

Thus R_1 and R_3 multiply u_1 and u_3 (respectively) by ω; R_2 is

$$u_\lambda' = u_\lambda - \frac{i\omega^2}{\surd 3}(u_1 + u_2 + u_3) \quad (\lambda = 1, 2, 3), \quad u_4' = u_4$$

(see (12·32)), and R_4 is

$$u_1' = u_1, \quad u_\mu' = u_\mu - \frac{i\omega^2}{\surd 3}(u_2 - u_3 + u_4) \quad (\mu = 2 \text{ or } 4),$$

$$u_3' = u_3 + \frac{i\omega^2}{\surd 3}(u_2 - u_3 + u_4).$$

For the initial vertex (on the last three mirrors) we may take $(1, -1, 0, 1)$. The reflection R_1 yields $(\omega, -1, 0, 1)$ and then $(\omega^2, -1, 0, 1)$; R_2 transforms these points into $(0, \omega^2, -\omega, 1)$ and $(-\omega, 0, 1, 1)$. These, in turn, are transformed by R_3 into $(0, \omega^2, -\omega^2, 1)$ and $(-\omega, 0, \omega, 1)$, which are transformed by R_4 into $(0, 0, 0, -i\omega\surd 3)$ and $(-\omega, \omega, 0, -\omega^2)$. Continuing, we find that this polytope has just the $216 + 24$ vertices†

$$(12{\cdot}52) \quad \begin{array}{l} (0, \pm\omega^\mu, \mp\omega^\nu, \pm\omega^\lambda), \quad (\mp\omega^\nu, 0, \pm\omega^\mu, \pm\omega^\lambda), \\ \qquad (\pm\omega^\mu, \mp\omega^\nu, 0, \pm\omega^\lambda), \quad (\mp\omega^\lambda, \mp\omega^\mu, \mp\omega^\nu, 0), \\ (\pm i\omega^\lambda\surd 3, 0, 0, 0), \quad (0, \pm i\omega^\lambda\surd 3, 0, 0), \\ \qquad (0, 0, \pm i\omega^\lambda\surd 3, 0), \quad (0, 0, 0, \pm i\omega^\lambda\surd 3). \end{array}$$

(In the first row, the signs must agree as indicated, while λ, μ, ν run independently over the values 0, 1, 2.)

$3\{3\}3\{3\}3\{3\}3$, like $3\{3\}3$ and $3\{3\}3\{3\}3$, is self-reciprocal. Having 240 vertices, it has also 240 cells $3\{3\}3\{3\}3$, one of which, derived from $(1, -1, 0, 1)$ by applying R_1, R_2, R_3, has the 27 vertices

$$(12{\cdot}53) \quad (0, \omega^\mu, -\omega^\nu, 1), \quad (-\omega^\nu, 0, \omega^\mu, 1), \quad (\omega^\mu, -\omega^\nu, 0, 1),$$

lying in the hyperplane $u_4 = 1$. Since the vertex figure is another

$$3\{3\}3\{3\}3,$$

there are 27 edges at each vertex, and 2160 edges altogether (appearing in the frontispiece as equilateral triangles not inscribed in any of the eight concentric triacontagons, $\{30\}$). By reciprocation, there are also 2160 faces $3\{3\}3$. Thus the numerical matrix for $3\{3\}3\{3\}3\{3\}3$ is

$$\begin{bmatrix} 240 & 27 & 72 & 27 \\ 3 & 2160 & 8 & 8 \\ 8 & 8 & 2160 & 3 \\ 27 & 72 & 27 & 240 \end{bmatrix}.$$

Like the polyhedra considered in §12·4, this polytope is 6-symmetric. The six points $(\pm i\omega^\lambda\surd 3, 0, 0, 0)$ form a β_1^6 on one of its 40 diameters.

Since every reflection in the group $3[3]3[3]3[3]3$ is conjugate to one of $R_\nu^{\pm 1}$, the section of the polytope by any one of its 40 *hyperplanes of symmetry* (orthogonal to the 40 diameters) is a $3\{3\}3\{4\}2$. For instance, the hyperplane $u_4 = 0$ contains the 72 vertices

$$(12{\cdot}54) \quad (\mp\omega^\lambda, \mp\omega^\mu, \mp\omega^\nu, 0), \quad (\pm i\omega^\lambda\surd 3, 0, 0, 0),$$
$$(0, \pm i\omega^\lambda\surd 3, 0, 0), \quad (0, 0, \pm i\omega^\lambda\surd 3, 0),$$

which distribute themselves (as we saw in §12·4) into four triads of mutually orthogonal diameters, all orthogonal to the diameter

$$u_1 = u_2 = u_3 = 0.$$

† See p. 480 of the paper cited on page 119.

Thus each of the 40 diameters belongs, in four ways, to a tetrad of mutually orthogonal diameters; and there are altogether 40 such tetrads. In other words, there is a *compound* consisting of 40 β_4^6's whose 40×24 vertices are just the 240 vertices of $3\{3\}3\{3\}3\{3\}3$, each counted four times.

Since any two orthogonal diameters belong to just one tetrad, and each of the 40 tetrads provides six such pairs of diameters, there are altogether 240 *diametral planes* spanned by pairs of orthogonal diameters. The section of the polytope by such a plane is a β_2^6 or $2\{4\}6$. This kind of diametral plane may alternatively be described as the intersection of two orthogonal hyperplanes of symmetry.

On the other hand, the section by the common plane of two non-orthogonal hyperplanes of symmetry is a $3\{4\}3$. For instance, the plane $u_1 = u_2 + u_3 = 0$ contains the 24 vertices

$$(0, \pm\omega^\mu, \mp\omega^\mu, \pm\omega^\lambda), \quad (0, 0, 0, \pm i\omega^\lambda\sqrt{3}).$$

Since there are 9 such planes in each hyperplane of symmetry, and 4 hyperplanes of symmetry through each plane (for instance,

$$u_1 = 0 \quad \text{and} \quad \omega^\lambda u_1 + u_2 + u_3 = 0$$

through the plane just considered), there are altogether 90 such $3\{4\}3$'s.

In other words, the polytope has 240 diametral planes of the first kind, and 90 of the second kind.

$3\{3\}3\{3\}3\{3\}3$ may reasonably be called the *Witting polytope* because the incidences of its diameters, diametral planes and hyperplanes of symmetry are the same as the incidences of the points, lines and planes of the Witting configuration†

$$\begin{bmatrix} 40 & 12 & 12 \\ 2 & 240 & 2 \\ 12 & 12 & 40 \end{bmatrix} \quad \text{or} \quad \begin{bmatrix} 40 & 9 & 12 \\ 4 & 90 & 4 \\ 12 & 9 & 40 \end{bmatrix}$$

in complex projective 3-space. The 40 points and 40 planes of this configuration have the same homogeneous coordinates (derived from (12·52))

$$(12\cdot55) \qquad \begin{array}{l} (0, 1, -\omega^\nu, \omega^\lambda), \quad (-\omega^\nu, 0, 1, \omega^\lambda), \\ \qquad\qquad (\omega^\mu, -\omega^\nu, 0, 1), \quad (1, \omega^\mu, \omega^\nu, 0), \\ (1, 0, 0, 0), \quad (0, 1, 0, 0), \quad (0, 0, 1, 0), \quad (0, 0, 0, 1). \end{array}$$

The last four of these points (and planes) form one of Witting's 40 tetrahedra (arising from our 40 tetrads of mutually orthogonal diameters).

The Witting configuration and its connection with the cubic surface become much clearer when the complex projective space is replaced by the finite space $PG(3, 2^2)$, which consists of 85 points, 357 lines, and 85 planes. In the field $GF[2^2]$, regarded as a quadratic extension of $GF[2]$, the *conjugate* of any element is its square, and so the Hermitian form $\Sigma u\bar{u}$ is the same as the cubic form Σu^3.

Let us first consider, for a moment, the finite *plane* $PG(2, 2^2)$ which consists of 21 points and 21 lines. The ternary cubic (or Hermitian) form Σu^3 splits the 21 into $9+12$. The nine points (12·39) (with the minus signs deleted, as now $-1 = 1$) constitute the cubic curve

$$(12\cdot56) \qquad u_1^3 + u_2^3 + u_3^3 = 0,$$

while the twelve points (12·43) satisfy

$$u_1^3 + u_2^3 + u_3^3 = 1.$$

Dually, the 21 lines of this finite plane consists of the 9 tangents and 12 secants of the 'curve'.

Similarly in the finite space $PG(3, 2^2)$, the quaternary form Σu^3 yields the splitting

$$85 = 45 + 40, \quad 357 = 27 + 90 + 240.$$

The 45 points

$$(0, 0, \omega^\nu, 1) \quad \text{permuted, and} \quad (\omega^\lambda, \omega^\mu, \omega^\nu, 1)$$

constitute the cubic surface

$$(12\cdot57) \qquad u_1^3 + u_2^3 + u_3^3 + u_4^3 = 0,$$

while the 40 points (12·55) (with the minus signs deleted) satisfy

$$u_1^3 + u_2^3 + u_3^3 + u_4^3 = 1.$$

The 27 lines whose Plücker coordinates

$$(p_{14}, p_{24}, p_{34}, p_{23}, p_{31}, p_{12})$$

are

$$(0, \omega^\mu, \omega^\nu, 0, \omega^{-\mu}, \omega^{-\nu}), \quad (\omega^\nu, 0, \omega^\mu, \omega^{-\nu}, 0, \omega^{-\mu}), \quad (\omega^\mu, \omega^\nu, 0, \omega^{-\mu}, \omega^{-\nu}, 0)$$

(compare (12·32)) lie entirely on the cubic surface; for instance, the line $(\omega^\mu, \omega^\nu, 0, \omega^{-\mu}, \omega^{-\nu}, 0)$, which we naturally denote by $\mu\nu 0$, joins the points

$$(\omega^\mu, \omega^\nu, 0, 0), \quad (0, 0, \omega^{-\mu-\nu}, 1)$$

and therefore contains also $(\omega^\mu, \omega^\nu, \omega^{\lambda-\mu-\nu}, \omega^\lambda)$ $(\lambda = 0, 1, 2)$. These 27 lines lie by threes on 45 planes whose tangential coordinates are the same as the ordinary coordinates for the 45 points; for instance, the plane

$$u_1 + u_2 + u_3 + u_4 = 0$$

contains the three lines

$$(0, 1, 1, 0, 1, 1), \quad (1, 0, 1, 1, 0, 1), \quad (1, 1, 0, 1, 1, 0),$$

† Alexander Witting, Ueber Jacobi'sche Functionen k^{ter} Ordnung zweier Variabler, *Mathematische Annalen* **29** (1887), 157–70; see especially p. 169.

for which Segre's symbols are 033, 303, 330. Unlike the lines in a tritangent plane of the real or complex cubic surface (which form a triangle), these three lines are concurrent, all passing through the *Eckardt point* (1, 1, 1, 1). In fact, all the 45 points on the cubic surface in $PG(3, 2^2)$ are Eckardt points. The remaining two lines through (1, 1, 1, 1) in the plane $\Sigma u = 0$ join (1, 1, 1, 1) to $(0, 1, \omega, \omega^2)$ and $(0, 1, \omega^2, \omega)$. The join of (1, 1, 1, 1) and $(0, 1, \omega, \omega^2)$ contains also

$$(1, 0, \omega^2, \omega), \quad (1, \omega^2, \omega, 0), \quad (1, \omega, 0, \omega^2)$$

and thus has (1, 1, 1, 1) as its only common point with the surface. Hence the 90 lines of this type (representing the 90 diametral planes of the second kind) are *tangents* of the surface, and the above-mentioned 45 planes are *tangent planes*. (The name 'tritangent plane' is no longer appropriate, as there is only one point of contact.) The remaining 240 lines in the space (representing the 240 diametral planes of the first kind) each contain two of the 40 points and three of the 45; thus they are *secants* of the surface. Similarly, the remaining 40 planes (representing the 40 hyperplanes of symmetry) are *secant planes*, each containing a cubic curve such as (12·56).

The mirrors (12·51) are now

$$u_1 = 0, \quad u_1+u_2+u_3 = 0, \quad u_3 = 0, \quad u_2+u_3+u_4 = 0,$$

and the reflections in them are *homologies* of period 3: R_1 and R_3 (as before) multiply u_1 and u_3 (respectively) by ω; R_2 is

$$u_\lambda' = u_\lambda + \omega^2(u_1+u_2+u_3) \quad (\lambda = 1, 2, 3), \quad u_4' = u_4,$$

and R_4 is

$$u_1' = u_1, \quad u_\lambda' = u_\lambda' + \omega^2(u_2+u_3+u_4) \quad (\mu = 2, 3, 4).$$

Thus R_1 transforms the line 011, which joins the points

$$(0, \omega, \omega, 0) \quad \text{and} \quad (\omega, 0, 0, 1),$$

into the join of $(0, \omega, \omega, 0)$ and $(\omega^2, 0, 0, 1)$, which is

$$(0, \omega, \omega, 0, 1, 1) \quad \text{or} \quad (0, \omega^2, \omega^2, 0, \omega, \omega) \quad \text{or} \quad 022.$$

Similarly, R_2 transforms 011 into the join of $(0, \omega, \omega, 0)$ and $(\omega^2, 1, 1, 1)$, which is again 022. Proceeding in this manner with all four generators and all 27 lines, we obtain the permutations

$$\begin{aligned} R_1 = {} & (011\ 022\ 033) \quad (023\ 031\ 012) \quad (032\ 013\ 021) \\ & \cdot(101\ 302\ 203) \quad (103\ 301\ 202) \quad (102\ 303\ 201) \\ & \cdot(110\ 320\ 320) \quad (120\ 330\ 210) \quad (130\ 310\ 220), \\ R_2 = {} & (011\ 022\ 033) \quad (101\ 202\ 303) \quad (110\ 220\ 330) \\ & \cdot(023\ 120\ 103) \quad (302\ 012\ 310) \quad (230\ 201\ 031) \\ & \cdot(032\ 102\ 130) \quad (203\ 210\ 013) \quad (320\ 021\ 301), \\ R_3 = {} & (011\ 023\ 032) \quad (021\ 033\ 012) \quad (031\ 013\ 022) \\ & \cdot(101\ 203\ 302) \quad (301\ 103\ 202) \quad (201\ 303\ 102) \\ & \cdot(110\ 220\ 330) \quad (320\ 130\ 210) \quad (230\ 310\ 120), \\ R_4 = {} & (011\ 033\ 022) \quad (120\ 210\ 330) \quad (303\ 102\ 201) \\ & \cdot(101\ 130\ 031) \quad (310\ 012\ 302) \quad (023\ 203\ 220) \\ & \cdot(110\ 103\ 013) \quad (230\ 202\ 032) \quad (320\ 301\ 021). \end{aligned} \tag{12·58}$$

Since the Witting polytope has 240 cells $3\{3\}3\{3\}3$, each of which has a symmetry group of order 648, the order of the group $3[3]3[3]3[3]3$ is 155 520. Since the polytope is 6-symmetric, the central quotient group

$$3[3]3[3]3[3]3/\mathfrak{C}_6 \tag{12·59}$$

has order 25920. When the 40 diameters are represented by the 40 points of the Witting configuration, this quotient group (generated by the permutations (12·58)) appears as the group of projective collineations that preserve the configuration. The order of the collineation group can be increased to 51 840 by adjoining the antiprojective collineation that replaces each coordinate by its *complex conjugate* (or, in the finite geometry, by its *square*). Since the 27 lines have the same group of automorphisms in $PG(3, 2^2)$ as in the complex space, this group of order 51 840 is $[3^{2,2,1}]$ (see Coxeter 1963, pp. 200–2), and the subgroup (12·59) of projective collineations (generated by the permutations (12·57)) is $[3^{2,2,1}]^+$ (see Coxeter and Moser 1972, p. 125), the simple group of order 25920.

The real counterpart of $3\{3\}3\{3\}3\{3\}3$ is the 8-dimensional uniform polytope 4_{21} (Coxeter 1968, p. 32). Each of its 240 vertices belongs to 56 edges (corresponding to the 56 vertices of its 7-dimensional vertex figure, 3_{21}). Thus a projection of 4_{21} can be derived from the frontispiece by inserting the edges of certain polygons; namely forty regular hexagons, five inscribed in each of the eight concentric triacontagons.

EXERCISES

1. Find the radii ${}_0R$, ${}_1R$, ${}_2R$, ${}_3R$ of $3\{3\}3\{3\}3\{3\}3$.
2. Describe a significant difference between the triads of lines permuted in (12·34) and those permuted in (12·58).
3. Express the antiprojective collineation $u_\lambda' = \bar{u}_\lambda$ as a permutation of the 27 lines $0\mu\nu$, $\nu 0 \mu$, $\mu\nu 0$.

12·6 THE HONEYCOMB OF WITTING POLYTOPES

Pushing the analogy one stage further, we obtain the symbol

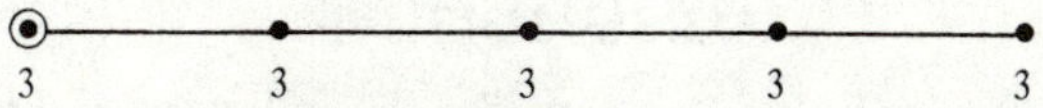

or

$$3\{3\}3\{3\}3\{3\}3\{3\}3,$$

for which the Hermitian form is

$$x^1\bar{x}^1 - \sqrt{\tfrac{1}{3}}(x^1\bar{x}^2 + x^2\bar{x}^1) + x^2\bar{x}^2 - \sqrt{\tfrac{1}{3}}(x^2\bar{x}^3 + x^3\bar{x}^2) + x^3\bar{x}^3$$
$$- \sqrt{\tfrac{1}{3}}(x^3\bar{x}^4 + x^4\bar{x}^3) + x^4\bar{x}^4 - \sqrt{\tfrac{1}{3}}(x^4\bar{x}^5 + x^5\bar{x}^4) + x^5\bar{x}^5$$
$$= u_1\bar{u}_1 + u_2\bar{u}_2 + u_3\bar{u}_3 + u_4\bar{u}_4,$$

where $u_1 = -x^1 + \dfrac{x^2}{\sqrt{3}}, \quad u_2 = \dfrac{x^2}{\sqrt{3}} - x^3 + \dfrac{x^4}{\sqrt{3}}, \quad u_3 = \dfrac{x^2 - x^4}{\sqrt{3}}, \quad u_4 = \dfrac{x^4}{\sqrt{3}} - x^5.$

Since there are only four u's instead of five, the form is semidefinite, and we have an *apeirotope* or *honeycomb*, whose symmetry group is generated by reflections of period 3 in five mirrors

$$(12\cdot61) \quad u_1 = 1, \quad \frac{u_1 + u_2 + u_3}{\sqrt{3}} = 0, \quad u_2 = 0, \quad \frac{u_2 - u_3 + u_4}{\sqrt{3}} = 0, \quad u_4 = 0,$$

forming a simplex in the unitary 4-space. (Compare (12·51). We write $u_1 = 1$, rather than $u_1 = 0$, so as to make the mirrors form a four-dimensional simplex instead of being concurrent.) The initial vertex (on the last four mirrors) is simply the origin. The reflection R_1, namely

$$u_1' = \omega(u_1 - 1) + 1, \quad u_\mu' = u_\mu \quad (\mu = 2, 3, 4),$$

yields $(1 - \omega, 0, 0, 0)$ or $(i\omega^2\sqrt{3}, 0, 0, 0)$, and then $(\omega - \omega^2, 0, 0, 0)$ or $(i\sqrt{3}, 0, 0, 0)$; R_2 (see (12·32)) transforms these points into

$$(1, \omega, \omega, 0) \quad \text{and} \quad (\omega, \omega^2, \omega^2, 0)$$

which, in their turn, are transformed by R_3 into

$$(1, \omega^2, \omega, 0) \quad \text{and} \quad (\omega, 1, \omega^2, 0),$$

and these by R_4 (see §12·5) into

$$(1, 0, -1, -\omega^2) \quad \text{and} \quad (\omega, 0, -\omega, -1).$$

Continuing, we find that the vertices of $3\{3\}3\{3\}3\{3\}3\{3\}3$ form a *lattice* consisting of all the points whose four coordinates are Eisenstein integers satisfying the congruences

$$(12\cdot62) \qquad u_1 + u_2 + u_3 \equiv u_2 - u_3 + u_4 \equiv 0 \pmod{i\sqrt{3}}.$$

10

Since 3 is a multiple of $i\sqrt{3}$, namely $3 = -(i\sqrt{3})^2$, the modulus

$$i\sqrt{3} = 1 + 2\omega$$

can be replaced by $1 - \omega$, and the congruences (12·62) imply

$$u_3 - u_2 \equiv u_1 - u_3 \equiv u_2 - u_1 \equiv u_4 \pmod{i\sqrt{3}}.$$

Since $1 \equiv \omega \equiv \omega^2 \pmod{i\sqrt{3}}$, the lattice is invariant for multiplication of any coordinate by ω (in particular, for R_3 and R_5). It is also invariant for R_1, which adds $(1 - u_1)(1 - \omega)$ to u_1; for R_2, which adds the Eisenstein integer

$$\omega^2(u_1 + u_2 + u_3)/i\sqrt{3}$$

to each of u_1, u_2, u_3; and for R_4, which adds

$$\omega^2(u_2 - u_3 + u_4)/i\sqrt{3}$$

to each of u_2, $-u_3$, u_4.

One typical cell $3\{3\}3\{3\}3\{3\}3$ of the honeycomb is derived from (12·52) by adding 1 to each of u_1, u_2, u_3; another, by adding $i\sqrt{3}$ to u_4. The centres of these cells are, of course, $(1, 1, 1, 0)$ and $(0, 0, 0, i\sqrt{3})$. But these points are vertices. Accordingly, this honeycomb, like $6\{3\}6$ (see page 112), *coincides with its reciprocal.*

EXERCISES

1. How many 4-cells of $3\{3\}3\{3\}3\{3\}3\{3\}3$ come together at each 3-cell, face, edge, or vertex?

2. Which edge of the honeycomb (12·62) has the point $(1, 0, 0, 0)$ for its centre? Obtain coordinates for the vertices of the actual vertex figure at $(0, 0, 0, 0)$.

12·7 CARTESIAN PRODUCTS OF APEIROGONS

The ordinary tessellation of squares, $\{4, 4\}$ or $2\{4\}2\{4\}2$, may be described as a Cartesian product $\{\infty\} \times \{\infty\}$: each vertex, edge, or face is the Cartesian product $\Pi_\mu \times \Pi_\nu$ ($\mu = 0$ or 1, $\nu = 0$ or 1) of a vertex or edge of one $\{\infty\}$ with a vertex or edge of the other. The vertex figure of the tessellation is a square whose diagonals are vertex figures of the two apeirogons. Similarly, any two honeycombs, of m and $n - m$ dimensions, situated in completely orthogonal subspaces and having a common vertex, have as Cartesian product an n-dimensional honeycomb whose typical element $\Pi_\mu \times \Pi_\nu$ $(0 \leqslant \mu \leqslant m,\ 0 \leqslant \nu \leqslant n - m)$ is the Cartesian product of a Π_μ of the first honeycomb and a Π_ν of the second. The vertex figure is a kind of dipyramid based on the concentric vertex figures of the two component honeycombs.

This definition may be extended to the Cartesian product of any finite number of honeycombs and, in particular, to the Cartesian product of n copies of an apeirogon $p\{q\}r$, where

$$\frac{1}{p}+\frac{2}{q}+\frac{1}{r}=1 \tag{12·71}$$

(see (11·61)). Each cell of this Cartesian product is

$$\gamma_n^p = p\{4\}2\{3\}2\ldots\{3\}2,$$

the Cartesian product of n γ_1^p's. The vertex figure, having n diameters β_1^r, is

$$\beta_n^r = 2\{3\}2\ldots\{3\}2\{4\}r.$$

Hence the Cartesian product of n apeirogons $\delta_2^{p,r} = p\{q\}r$ is

$$\delta_{n+1}^{p,r} = p\{4\}2\{3\}2\ldots\{3\}2\{4\}r \tag{12·72}$$

(with $n-1$ twos. The q seems to have disappeared, but it is still given by (12·71). The ordinary n-dimensional cubic honeycomb δ_{n+1} is the special case $p = r = 2$).

We see from §11·6 that the apeirogons $\delta_2^{2,4}$, $\delta_2^{4,4}$, $\delta_2^{4,2}$ all have the same vertices, given by the Gaussian integers; therefore the honeycombs

$$\delta_{n+1}^{2,4},\quad \delta_{n+1}^{4,4},\quad \delta_{n+1}^{4,2}$$

all have the same vertices, given by sets of n Gaussian integers. Again, $\delta_2^{6,6}$ coincides with its reciprocal, and has the same vertices as $\delta_2^{2,6}$, $\delta_2^{3,6}$, $\delta_2^{3,3}$, given by the Eisenstein integers; therefore $\delta_{n+1}^{6,6}$ coincides with its reciprocal, and has the same vertices as

$$\delta_{n+1}^{2,6},\quad \delta_{n+1}^{3,6},\quad \delta_{n+1}^{3,3},$$

given by sets of n Eisenstein integers. Similarly, $\delta_{n+1}^{2,3}$ and $\delta_{n+1}^{6,3}$ have the same vertices, given by sets of n Eisenstein integers not divisible by $i\sqrt{3}$, while $\delta_{n+1}^{3,2}$ and $\delta_{n+1}^{6,2}$ have the same vertices, given by sets of n Eisenstein integers not divisible by 2.

EXERCISES

1. What is the Schläfli symbol for $\delta_3^{p,r}$?
2. Write down the Hermitian form for $\delta_{n+1}^{p,r}$ and deduce the equations

$$u_1 = l,\quad u_1 = u_2,\quad u_2 = u_3,\ \ldots,\ u_{n-1} = u_n,\quad u_n = 0$$

for the $n+1$ mirrors. Obtain the reflections in the cases $p = r = 4$ and $p = r = 3$, choosing $l = (1+i)/2$ for the former and $l = (1-\omega^2)/3$ for the latter.

12·8 CYCLES OF HONEYCOMBS

Among the infinitely many hyperplanes of symmetry of

$$3\{3\}3\{3\}3\{3\}3\{3\}3,$$

five of which appear in (12·61), consider the following eight:

$$u_1 = 0,\qquad u_2 = 0,\qquad u_3 = 0,\qquad u_4 = 1.$$

$$u_2-u_3+u_4 = 0,\quad u_3-u_1+u_4 = 0,\quad u_1-u_2+u_4 = 0,\quad u_1+u_2+u_3 = i\sqrt{3}.$$

The four in each row are mutually orthogonal; also the two in each column are orthogonal. In fact, these eight hyperplanes are related like the vertices of the 'cubic graph'

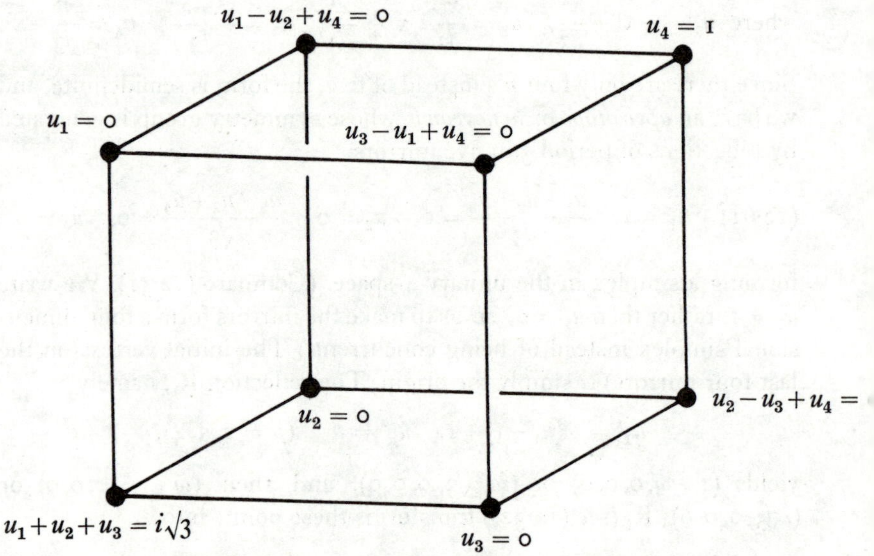

When a 3 has been placed at each of its eight nodes, this graph may be interpreted as a symbol for the group

$$3[3]3[3]3[3]3[3]3$$

with three extra (redundant) generators. Omitting

$$u_2 = 0 \quad\text{and}\quad u_3-u_1+u_4 = 0,$$

we are left with a *cycle* of six hyperplanes

$$u_1 = 0,\quad u_1+u_2+u_3 = i\sqrt{3},\quad u_3 = 0,$$
$$u_2-u_3+u_4 = 0,\quad u_4 = 1,\quad u_1-u_2+u_4 = 0$$

(appearing as a Petrie polygon of the cube) such that reflections of period 3 in *any five* of them will generate the group in the manner indicated by the symbol

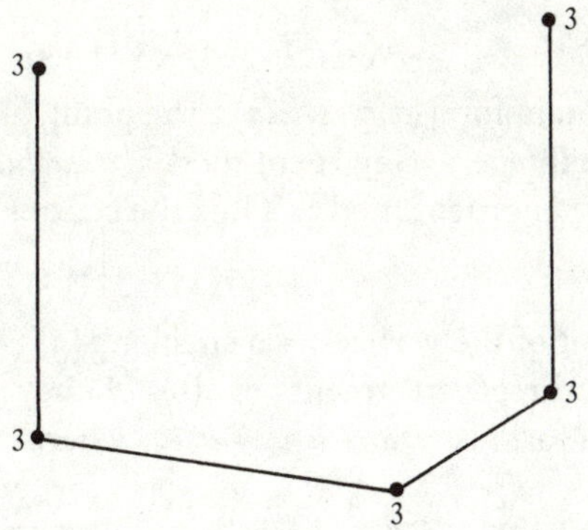

or

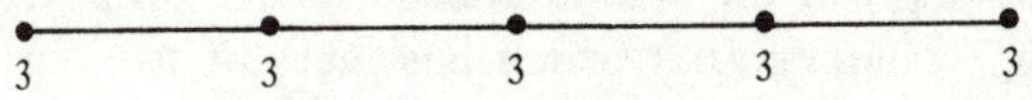

Aside from its perfect symmetry, this cycle of six mirrors resembles the cycles

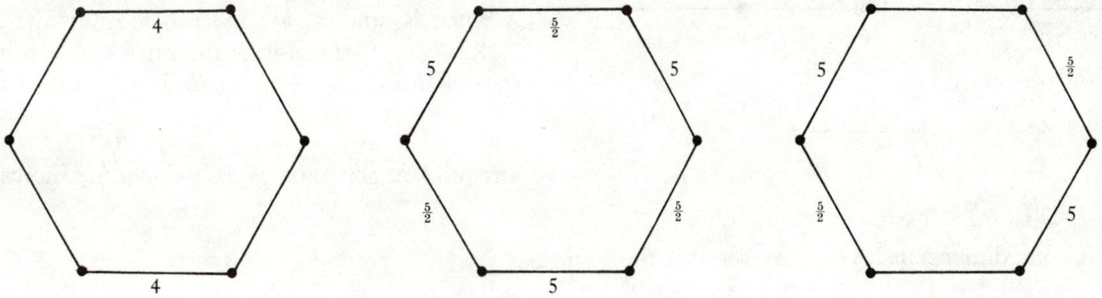

that underlie the frieze patterns (5·11), (5·14), (5·15).

In 3{3}3{3}3{3}3{3}3, the common 3{3}3{3}3 of three adjacent cells 3{3}3{3}3{3}3 lies in a hyperplane of symmetry of the honeycomb and of its vertex figure 3{3}3{3}3{3}3. The section of the vertex figure by this hyperplane is a 3{3}3{4}2 (see (12·54)). Hence the section of the four-dimensional honeycomb itself by such a hyperplane is a three-dimensional honeycomb

3{3}3{3}3{4}2

whose vertices (by (12·62) with $u_4 = 0$) form a lattice consisting of all the points whose three coordinates are Eisenstein integers mutually congruent modulo $i\sqrt{3}$, that is, they satisfy

$$u_1 \equiv u_2 \equiv u_3 \pmod{i\sqrt{3}}.$$

This result may be checked by means of the form $u_1\bar{u}_1 + u_2\bar{u}_2 + u_3\bar{u}_3$, where

$$u_1 = -x^1 + \frac{x^2}{\sqrt{3}}, \quad u_2 = \frac{x^2}{\sqrt{3}} - x^3 + \frac{x^4}{\sqrt{2}}, \quad u_3 = \frac{x^2}{\sqrt{3}} - \frac{x^4}{\sqrt{2}},$$

so that the four mirrors are

(12·81) $$u_1 = l, \quad \frac{u_1+u_2+u_3}{\sqrt{3}} = 0, \quad u_2 = 0, \quad \frac{u_2 - u_3}{\sqrt{2}} = 0,$$

just like (12·61) with 0 for u_4 (so that now R_4, of period 2, simply interchanges u_2 and u_3).

To obtain the reciprocal

2{4}3{3}3{3}3,

we operate on the common point $(1, 0, -1)$ of the first three mirrors (with $l = 1$), obtaining all the points whose three coordinates are Eisenstein integers, mutually *incongruent* mod $i\sqrt{3}$, that is, they are congruent to some permutation of $(1, 0, -1)$ (Hardy and Wright 1960, p. 188, Theorem 222).

In §12·4 we observed two instances ($p = 2$ and 3) of the interesting connection between three polyhedra

(12·82) $$2\{4\}p\{3\}p, \quad p\{3\}p\{3\}p, \quad p\{3\}p\{4\}2.$$

The centres of the edges of the second are the vertices of a 'truncation' (see Coxeter 1963, Chapter VIII) which has faces of two types: vertex figures of $p\{3\}p\{3\}p$, and truncated faces. Since both types are $p\{3\}p$, the polyhedron so obtained is again regular, namely $p\{3\}p\{4\}2$. Reciprocally, $2\{4\}p\{3\}p$ has *vertices* of two types: vertices of the original $p\{3\}p\{3\}p$, and vertices of the reciprocal, which is another $p\{3\}p\{3\}p$. In other words, the three polyhedra (12·82) are so related that the vertices of the first belong to two reciprocal specimens of the second, while the vertices of the third are the centres of the edges of the second. In the notation using graphs (see (12·12)),

(12·83)

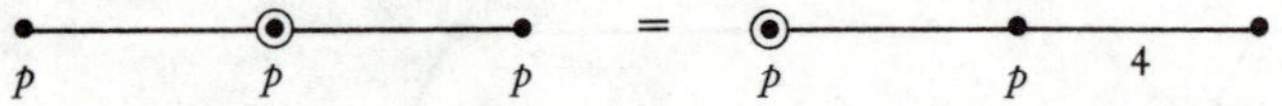

The vertex figure is the Cartesian product

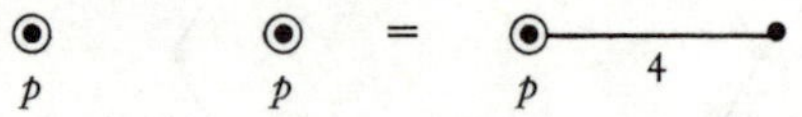

that is, $$\gamma_1^p \times \gamma_1^p = \gamma_2^p.$$

Proceeding similarly in four dimensions (§4·2), we see that the four polytopes

$$2\{3\}2\{4\}p\{3\}p, \quad 2\{4\}p\{3\}p\{3\}p, \quad p\{3\}p\{3\}p\{4\}2, \quad p\{3\}p\{4\}2\{3\}2$$

(the second and third being reciprocals, of such size as to have the same circumradius) are so related that the vertices of the first belong to the second and third together, while the vertices of the fourth are the centres of the edges of the third. Extending this principle from polytopes to honeycombs and applying it with $p = 3$, we deduce that the vertices of

$$2\{3\}2\{4\}3\{3\}3$$

are obtained by combining $2\{4\}3\{3\}3\{3\}3$ and $3\{3\}3\{3\}3\{4\}2$, that is, they are all the points whose coordinates are Eisenstein integers satisfying

$$u_1+u_2+u_3 \equiv 0 \pmod{i\sqrt{3}};$$

and the vertices of $3\{3\}3\{4\}2\{3\}2$, being the edge-centres of

$$3\{3\}3\{3\}3\{4\}2,$$

are the images of the second vertex of the kaleidoscope (12·81), namely the common point $(l, 0, 0)$ of the mirrors for R_1, R_3 and R_4. To avoid fractions we choose $l = i\sqrt{3}$, so that now R_1 replaces u_1 by

$$\omega(u_1 - i\sqrt{3}) + i\sqrt{3} = \omega u_1 - 3\omega^2,$$

and we obtain for $3\{3\}3\{4\}2\{3\}2$ the points whose three coordinates are Eisenstein integers, congruent mod $i\sqrt{3}$, *excluding* those multiples of $i\sqrt{3}$ which are congruent mod 3. Thus the centres of the edges of

$$3\{3\}3\{3\}3\{4\}2$$

are a subset of the vertices of a smaller $3\{3\}3\{3\}3\{4\}2$.

For a direct treatment of this honeycomb $3\{3\}3\{4\}2\{3\}2$, the Hermitian form is $u_1\bar{u}_1 + u_2\bar{u}_2 + u_3\bar{u}_3$, where

$$u_1 = \frac{x^1}{\sqrt{3}} - x^2 + \frac{x^3}{\sqrt{2}}, \quad u_2 = \frac{x^1}{\sqrt{3}} - \frac{x^3 - x^4}{\sqrt{2}}, \quad u_3 = \frac{x^1}{\sqrt{3}} - \frac{x^4}{\sqrt{2}},$$

so that the four mirrors are

$$\frac{u_1+u_2+u_3}{\sqrt{3}} = l, \quad u_1 = 0, \quad \frac{u_1 - u_2}{\sqrt{2}} = 0, \quad \frac{u_2 - u_3}{\sqrt{2}} = 0.$$

Since R_3 and R_4 are the transpositions (1 2) and (2 3), the rotation $R_3R_4 = (1\,3\,2)$ transforms the mirror $u_1 = 0$ into $u_3 = 0$. This suggests that we introduce a fifth mirror $u_3 = l'$, so as to give the group

$$3[3]3[4]2[3]2$$

a redundant generator R_5 in the manner indicated by the graph

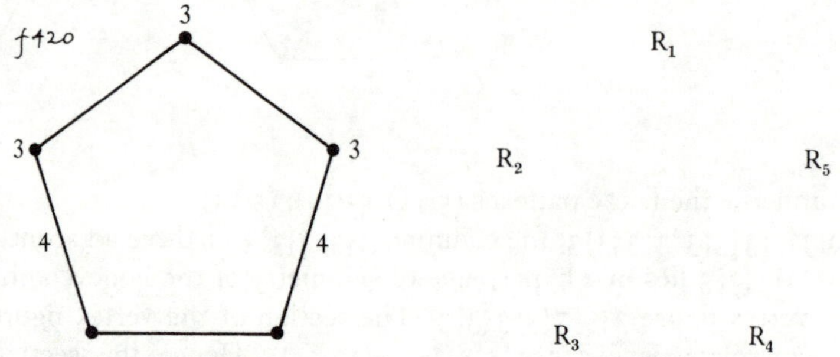

Thus the 'cubic' honeycomb $\delta_4^{3,3} = 3\{4\}2\{3\}2\{4\}3$, whose vertices are given by *all* sets of three Eisenstein integers, serves to complete a cycle of five honeycombs

(12·84) $$3\{3\}3\{3\}3\{4\}2, \quad 3\{3\}3\{4\}2\{3\}2, \quad 3\{4\}2\{3\}2\{4\}3, \quad 2\{3\}2\{4\}3\{3\}3, \quad 2\{4\}3\{3\}3\{3\}3,$$

such that the vertex figure of each is similar to the cell of the next.

Another instance of (12·82) is apparently obtained by setting $p = 4$ (instead of 2 or 3). However, a somewhat unexpected complication arises, as we shall soon see.

For the 'tessellation' $4\{3\}4\{3\}4$, the Hermitian form is

$$(12{\cdot}85) \qquad x^1\bar{x}^1 - \sqrt{\tfrac{1}{2}}(x^1\bar{x}^2 + x^2\bar{x}^1) + x^2\bar{x}^2 - \sqrt{\tfrac{1}{2}}(x^2\bar{x}^3 + x^3\bar{x}^2) + x^3\bar{x}^3 = u_1\bar{u}_1 + u_2\bar{u}_2$$

where
$$u_1 = x^1 - \frac{x^2}{\sqrt{2}}, \quad u_2 = \frac{x^2}{\sqrt{2}} - x^3$$

(the same as for $\delta_3^{4,4} = 4\{4\}2\{4\}4$); so the three mirrors are

$$(12{\cdot}86) \qquad u_1 = l, \quad \frac{u_1 - u_2}{\sqrt{2}} = 0, \quad u_2 = 0.$$

Setting $l = 1$, we find that R_1 replaces u_1 by $iu_1 + 1 - i$ while leaving u_2 unchanged. Also R_2 is

$$u_1' = \tfrac{1}{2}(1+i)\,u_1 + \tfrac{1}{2}(1-i)\,u_2,$$
$$u_2' = \tfrac{1}{2}(1-i)\,u_1 + \tfrac{1}{2}(1+i)\,u_2,$$

and R_3 multiplies u_2 by i. Applying these reflections to the origin we obtain, as the vertices of

$$4\{3\}4\{3\}4,$$

all the points whose two coordinates are Gaussian integers satisfying

$$(12{\cdot}87) \qquad u_1 \equiv u_2 \pmod{1+i}.$$

Since these points include (1, 1) (derived from the origin by applying $R_1R_2R_3$), which lies on the first two mirrors, $4\{3\}4\{3\}4$ *coincides with its reciprocal* (like $3\{3\}3\{3\}3\{3\}3\{3\}3$ and $\delta_{n+1}^{6,6}$). This is the 'somewhat unexpected complication' mentioned above. It means that $2\{4\}4\{3\}4$ has *the same vertices*, namely the points (u_1, u_2) satisfying (12·87).

As for $4\{3\}4\{4\}2$, its vertices are the images of the third vertex of the kaleidoscope (12·86): the common point $(l, 0)$ of the mirrors for R_1 and R_3. To avoid fractions we now choose $l = 1+i$ (so that R_1 replaces u_1 by $iu_1 + 2$), and obtain the pairs of Gaussian integers, congruent mod $1+i$, excluding those multiples of $1+i$ which are congruent mod 2. Thus the centres of the edges of a $4\{3\}4\{3\}4$ are a subset of the vertices of a smaller $4\{3\}4\{3\}4$. Both the $4\{3\}4\{4\}2$ and this smaller $4\{3\}4\{3\}4$ have, as one cell, the $4\{3\}4$ whose 24 vertices are

$$(i^\mu(1+i), 0), \quad (0, i^\nu(1+i)), \quad (i^\mu, i^\nu).$$

The Cartesian product

$$\delta_2^{4,4} \times \delta_2^{4,4} = \delta_3^{4,4} = 4\{4\}2\{4\}4$$

serves to complete a cycle of four tessellations

$$(12{\cdot}88) \qquad 2\{4\}4\{3\}4, \quad 4\{3\}4\{3\}4, \quad 4\{3\}4\{4\}2, \quad 4\{4\}2\{4\}4$$

such that the vertex figure of each is similar to the face of the next, as indicated in the graph

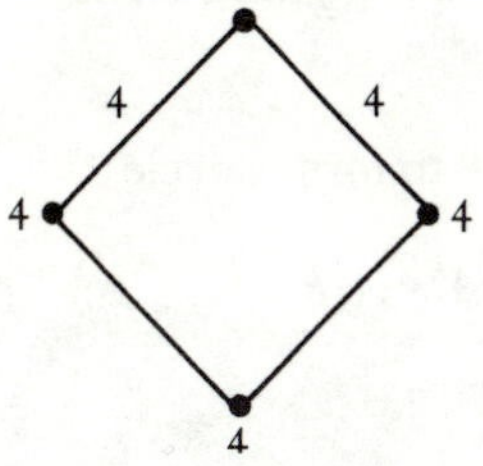

which thus resembles the real cycle

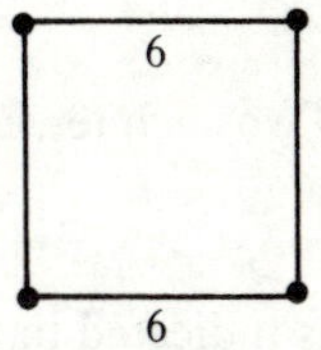

In this real case the cycle is comparatively trivial, consisting of the two reciprocal tessellations

$$2\{6\}2\{3\}2 \quad \text{and} \quad 2\{3\}2\{6\}2$$

repeated. A complex case of the same kind arises from

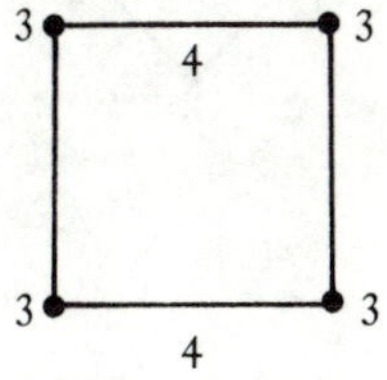

which can be derived from the three-dimensional honeycomb

$$3\{3\}3\{3\}3\{4\}2$$

(see § 12·7) as follows. The common $3\{3\}3$ of two adjacent cells $3\{3\}3\{3\}3$ lies in a plane of symmetry of the honeycomb and of its vertex figure $3\{3\}3\{4\}2$. The section of the vertex figure by this plane is a $3\{4\}3$ (see § 12·4). Hence the section of the honeycomb itself by such a plane is a tessellation

$$3\{3\}3\{4\}3.$$

Choosing the plane $u_1 = u_2$, we see that the vertices of this tessellation are given by sets of three Eisenstein integers satisfying

$$u_1 = u_2, \quad u_2 \equiv u_3 \pmod{i\sqrt{3}}$$

or, in terms of *two* coordinates, all points $(u_2\sqrt{2}, u_3)$ where the Eisenstein integers u_2 and u_3 are congruent mod $i\sqrt{3}$.

Another triad of apeirohedra behaving like (12·82) is

$$2\{4\}p\{4\}2, \quad p\{4\}2\{4\}p, \quad 2\{4\}p\{4\}2.$$

In other words, if $p = 2, 3, 4,$ or 6, there is a complex tessellation

$$2\{4\}p\{4\}2,$$

whose vertices may be described either as belonging to two reciprocal $\delta_3^{p,p}$'s or as the centres of the edges of another $\delta_3^{p,p}$:

(12·89) 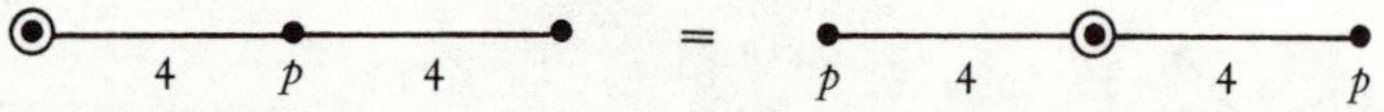

Two such tessellations can be taken with $\delta_3^{p,r}$ and $\delta_3^{r,p}$ to form a cycle

$$2\{4\}p\{4\}2, \quad p\{4\}2\{4\}r, \quad 2\{4\}r\{4\}2, \quad r\{4\}2\{4\}p$$

as indicated in the graph

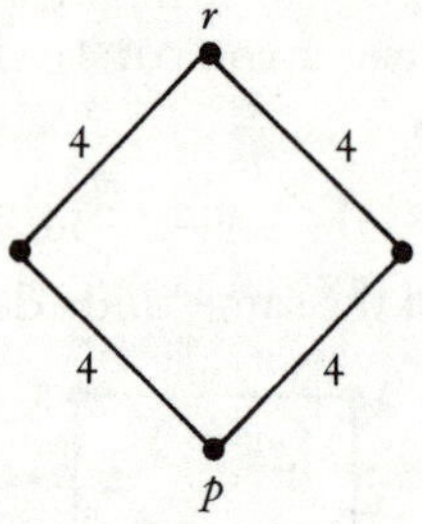

EXERCISES

1. Write down the Hermitian form for $2\{4\}p\{4\}2$, and deduce simple equations for the three mirrors.
2. The same for $3\{3\}3\{4\}3$.
3. How is $2\{4\}6\{4\}2$ related to $6\{4\}2\{4\}6$?

12·9 REMARKS

The notion of a 'medial figure' (§12·1) and the specifications (i) and (ii) for a complex polytope were supplied by McMullen, and he was the first to describe the regular complex honeycombs (except the one in §12·6). He also drew the figures for this chapter.

The history of the 27 lines on the cubic surface (Henderson 1911) provides a classic example of mathematical collaboration: Cayley observed that some finite number of lines must lie on the surface, Salmon showed that this number is 27, and Schläfli invented a notation which indicates quite clearly which pairs of the lines intersect. Although Schläfli's notation

$$a_1, \ldots, a_6; \quad b_1, \ldots, b_6; \quad c_{12}, \ldots, c_{56}$$

is ideal for explaining the connection with the real polytope 2_{21} in Euclidean 6-space†, the connection with the complex polytopes

$$3\{3\}3\{3\}3 \quad \text{and} \quad 3\{3\}3\{3\}3\{3\}3$$

cries out for a quite different notation which happens to have been supplied by Segre (1942).

The polyhedra $2\{4\}3\{3\}3$ and $3\{3\}3\{4\}2$ (§12·4) were discovered by Shephard (see footnote to page 114), who called them $2(18)3(24)3$ and $3(24)3(18)2$. The coordinates (12·55) for the Witting configuration were obtained by Maschke‡. The connection with $PG(3, 2^2)$ was noticed by Frame and rediscovered by Bose, who made the explicit observation that, over $GF[2^2]$, the Hermitian form $\Sigma u\bar{u}$ coincides with the cubic form Σu^3. Although the group of automorphisms of the 27 lines is the same over $GF[2^2]$ as over the field of complex numbers, it was discovered by Hirschfeld that, over any other finite field, the group is smaller (namely, a proper subgroup), because *some* of the 45 tritangent planes are replaced by planes that contain 3 concurrent lines through an Eckardt point, as they *all* are when the field is $GF[2^2]$.

The coordinates (12·62) for $3\{3\}3\{3\}3\{3\}3\{3\}3$ were first obtained by McMullen. It was he also who noticed that a regular honeycomb arises when we take the Cartesian product (§12·7) of n copies of any one of the twelve apeirogons.

We saw, in §12·8, that the vertices of the honeycombs

$$3\{3\}3\{3\}3\{4\}2 \quad \text{and} \quad 2\{3\}2\{4\}3\{3\}3$$

form three-dimensional lattices. Both these lattices first appeared in a discussion of senary quadratic forms.§

† P. H. Schoute, On the relation between the vertices of a definite six-dimensional polytope and the lines of a cubic surface, *Koninklijke Akademie van Wetenschappen te Amsterdam, Proceedings of the Section of Sciences* **13** (1910), 375–83; J. A. Todd, Polytopes associated with the general cubic surface, *Journal of the London Mathematical Society* **7** (1932), 200–5.

‡ H. Maschke, Aufstellung des vollen Formensystems einer quaternären Gruppe von 51840 linearen Substitutionen, *Mathematische Annalen* **33** (1888), 317–44; J. S. Frame, A symmetric representation of the twenty-seven lines on a cubic surface by lines in a finite geometry, *Bulletin of the American Mathematical Society* **43** (1938), 658–61; R. C. Bose, Self-conjugate tetrahedra with respect to the Hermitian variety $x_0^3+x_1^3+x_2^3+x_3^3 = 0$ in $PG(3, 2^2)$ and a representation of $PG(3, 3)$, *Proceedings of Symposia in Pure Mathematics* **19** (1971), 27–37; J. W. P. Hirschfeld, Classical configurations over finite fields: I. The double-six and the cubic surface with 27 lines, *Rendiconti di Matematica* **26** (1967), 1–38; see especially p. 8.

§ H. S. M. Coxeter, Extreme forms, *Canadian Journal of Mathematics* **3** (1951), 391–441; see especially pp. 421, 435–6.

CHAPTER 13

The regular complex polytopes and their symmetry groups

> As long as algebra and geometry travelled separate paths their advance was slow and their applications limited. But when these two sciences joined company, they drew from each other fresh vitality and thenceforward marched on at a rapid pace towards perfection.
>
> J. L. Lagrange (1795, p. 127)

The first two sections of this chapter are devoted to the task of proving that the regular polytopes and honeycombs so far considered are the only ones that can exist in unitary spaces. As by-products of this enumeration, we obtain a 'van Oss polygon' or 'van Oss apeirogon' for each polytope or honeycomb whose vertex figure is r-symmetric. The five Platonic solids are matched by five non-real polyhedra which form a remarkable cycle, any two consecutive members being the face and vertex figure of a honeycomb. We see, in §13·3, that the metrical properties of all these five polyhedra can be read off from one simple frieze pattern having the amusing property that it forms a geometrically symmetrical frieze when Roman numerals are used instead of 1, 2, 3:

I   I   I   I   I   I   ...

...   I   III   I   II   II   I

I   II   II   I   III   I   ...

...   I   I   I   I   I   I

(Try turning the page upside down!) It is pleasant to observe that the frieze pattern for any non-real regular polytope has $(s, s+1) = 1$ and is thus of the simple 'unimodular' type that arose in §3·1, in contrast to the sophisticated kind that was needed in Chapter 5.

Theorem (13·44) is the climax, for which most of the earlier chapters were preparing the way so as to make it seem natural and inevitable. However, it is only 'verified' by consideration of separate cases. Perhaps some reader will be inspired to devise a general proof.

A historical interlude leads to the strange equation (13·62), which provides an implicit formula for the 'exponents' m_λ of the symmetry group in terms of the p's and q's. Numerical properties of the general polytope are found in §13·7, including McMullen's expression (13·72) for the number of mirrors: another challenge, as a general proof is known only in two dimensions.

The next section includes the concise presentation (13·83) for the simple group of order 25920. The final section collects, from various sources, the invariant polynomial forms whose degrees are the numbers $m_\lambda + 1$.

13·1 THE REGULAR POLYTOPES AND THEIR VAN OSS POLYGONS

In Chapter 11 the regular polygons were described and we proved that the list (Table IV) is complete. Chapter 12 dealt with the analogous polytopes in more than two dimensions, but we postponed the more difficult task of making sure that we have not missed any. One possible approach would be to enumerate the groups by means of their Hermitian forms (along the lines of Coxeter 1963, Chapters X and XI) and then appeal to McMullen's extension of Hess's Theorem:†

Every regular star-polytope has the vertices of a unique non-starry regular polytope.

However, the alternative approach used in the present section is to obtain the enumeration as a by-product of another interesting concept: the van Oss polygon (compare Coxeter 1963, pp. 274–5). Any polytope

$$p_1\{q_1\}p_2\{q_2\}\cdots p_n$$

whose vertex figure $p_2\{q_2\}\cdots p_n$ is r-symmetric (see (12·17)) has a *van Oss polygon** which is the section of the polytope by the plane $O_0O_1O_n$ that joins one edge to the centre. This plane contains all the p_1 vertices on the edge O_0O_1, and contains a diameter of the $p_2\{q_2\}\cdots p_n$ whose vertices (such as O_1) are the centres of all the edges through O_0. The plane thus contains many other vertices, a diameter of the vertex figure at each one, and r further edges for each such diameter. Altogether the section is a regular polygon of type $p_1\{q\}r$ (for some q) whose vertices

† Peter McMullen, Regular star-polytopes, and a theorem of Hess, *Proceedings of the London Mathematical Society* (3), **18** (1968), 577–96; see especially pp. 592, 595.

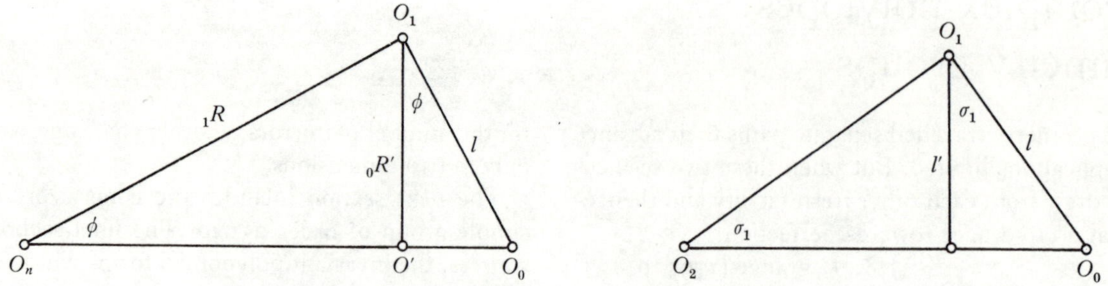

Figure 13·1A: Two faces of the orthoscheme $O_0O_1 \ldots O_n$

and edges are subsets of the vertices and edges of the polytope, while its vertex figure is the β_1^r that constitutes a diameter of the vertex figure of the polytope.

To compute q, we observe that the angle

$$\phi = \angle O_0 O_n O_1$$

of the polytope is equal to the angle σ of the van Oss polygon, given by

$$(13{\cdot}11) \qquad \cos 2\sigma = \left(\cos\frac{\pi}{p_1}\cos\frac{\pi}{r} + \cos\frac{2\pi}{q}\right)\Big/ \sin\frac{\pi}{p_1}\sin\frac{\pi}{r}$$

(see (9·82)).

To compute ϕ, the angle subtended at O_n by the radius l of an edge, let $_0R$ be the circumradius of the polytope, so that

$$l = {}_0R \sin\phi.$$

Let l', ϕ', $_0R'$ denote the properties l, ϕ, $_0R$ of the vertex figure

$$p_2\{q_2\} \cdots p_n,$$

whose centre O' is the common point of the line O_0O_n with the perpendicular hyperplane through O_1, that is, the orthogonal projection of O_1 on O_0O_n. The above expression for l yields

$$l' = {}_0R' \sin\phi'.$$

From the right-angled triangle $O_0O_nO_1$, with O_1O' perpendicular to the hypotenuse O_0O_n (see Figure 13·1A), we find

$$_0R' = l\cos\phi.$$

Similarly, from the right-angled triangle $O_0O_2O_1$,

$$l' = l\cos\sigma_1,$$

where $\sigma_1 = \angle O_0O_2O_1$ is the angle between the two mirrors $O_0O_2 \ldots O_n$ and $O_1O_2 \ldots O_n$. Eliminating l, l' and $_0R'$, we deduce

$$(13{\cdot}12) \qquad \cos\phi = \cos\sigma_1/\sin\phi',$$

that is, $$\cos^2\phi = \cos^2\sigma_1/1-\cos^2\phi'.$$

Similarly, if ϕ'' refers to the vertex figure of the vertex figure,

$$\cos^2\phi' = \cos^2\sigma_2/1-\cos^2\phi''.$$

Continuing thus, we eventually reach the polygon $p_{n-1}\{q_{n-1}\}p_n$, whose ϕ is σ_{n-1}; and $n-2$ such equations produce the continued fraction

$$(13{\cdot}13) \qquad \cos^2\phi = c_1/1-c_2/\ldots/1-c_{n-1}$$

(like (4·38)), where

$$(13{\cdot}14) \quad c_\mu = \cos^2\sigma_\mu = \left\{\cos\left(\frac{\pi}{p_\mu}-\frac{\pi}{p_{\mu+1}}\right)+\cos\frac{2\pi}{q_\mu}\right\}\Big/ 2\sin\frac{\pi}{p_\mu}\sin\frac{\pi}{p_{\mu+1}}$$

(see (9·89) and (12·13)).

Substituting ϕ for σ in (13·11), we conclude that the existence of a van Oss polygon implies the rationality of q where

$$(13{\cdot}15) \qquad \cos\frac{2\pi}{q} = \cos 2\phi \sin\frac{\pi}{p_1}\sin\frac{\pi}{r} - \cos\frac{\pi}{p_1}\cos\frac{\pi}{r},$$

ϕ being given by (13·13) and (13·14).

To obtain the van Oss polygon for a polyhedron, we express (13·13) (with $n = 3$) in the more convenient form

$$(13{\cdot}16) \qquad \cos 2\phi = (1+2\cos 2\sigma_1 + \cos 2\sigma_2)/(1-\cos 2\sigma_2)$$

and replace (13·14) by (12·13) (with $\mu = 1$ or 2). The detailed consideration of cases can be arranged as a table, beginning with the polyhedra that are already known to exist, and continuing with other symbols

$$p_1\{q_1\}p_2\{q_2\}p_3,$$

where $p_1\{q_1\}p_2$ (the face) and $p_2\{q_2\}p_3$ (the r-symmetric vertex figure) both occur in Table IV. Some of the rather formidably many cases can be ruled out quickly. Whenever $2\sigma_1+2\sigma_2 \leqslant \pi$, (13·16) would imply $\cos 2\phi \geqslant 1$. Also we may omit every case (such as $3\{3\}3\{\frac{10}{3}\}2$) in which $p_1 > 2$ and $r > 5$ (because the only polygon $p_1\{q\}r$ with $r > 5$ is $2\{4\}r$); and we may stop after finding $\cos 2\phi \neq 0$, whenever $p_1 = 2$ and $r > 5$ (because then again $q = 4$, and (13·15) shows that this possibility can arise only when $\cos 2\phi = 0$).

$p_1\{q_1\}p_2\{q_2\}p_3$	r	$\cos 2\sigma_1$	$\cos 2\sigma_2$	$\cos 2\phi$	$\cos 2\pi/q$	Van Oss polygon
$3\{3\}3\{3\}3$	2	$-\frac{1}{3}$	$-\frac{1}{3}$	0	0	$3\{4\}2$
$2\{4\}3\{3\}3$	2	0	$-\frac{1}{3}$	$\frac{1}{2}$	$\frac{1}{2}$	$2\{6\}2$
$3\{3\}3\{4\}2$	3	$-\frac{1}{3}$	0	$\frac{1}{3}$	0	$3\{4\}3$
$2\{3\}2\{4\}r$	r	$-\frac{1}{2}$	0	0	0	$2\{4\}r$
$2\{3\}2\{3\}3$	4	$-\frac{1}{2}$	$-\sqrt{\frac{1}{3}}$	$(1-\sqrt{3})/2$	$(1-\sqrt{3})/2\sqrt{2}$	None
$2\{3\}2\{\frac{8}{3}\}3$	6	$-\frac{1}{2}$	$-\sqrt{\frac{2}{3}}$	$2-\sqrt{6}$		None
$2\{3\}2\{5\}3$	12	$-\frac{1}{2}$	$\tau^{-1}/\sqrt{3}$	$\tau^{-3}(\tau^{-1}+\sqrt{3})$		None
$2\{3\}2\{\frac{10}{3}\}3$	12	$-\frac{1}{2}$	$-\tau^{-1}/\sqrt{3}$	$\tau^{-3}(\tau^{-1}-\sqrt{3})$		None
$2\{3\}2\{\frac{5}{2}\}3$	12	$-\frac{1}{2}$	$-\tau/\sqrt{3}$	$\tau^{-3}(\tau-\sqrt{3})$		None

And so on! After the first four cases, q never again takes one of the values for which a polygon $p_1\{q\}r$ occurs in Table IV. To determine which polyhedra $p_1\{q_1\}p_2\{q_2\}p_3$ can exist, it is convenient to separate the *real* cases (when $p_1 = p_2 = p_3 = 2$, see the end of §2·5) from the non-real cases (when at least one $p_\nu > 2$). Every non-real polyhedron must have either a non-real vertex figure $p_2\{q_2\}p_3$ or a non-real face (and then the reciprocal $p_3\{q_2\}p_2\{q_1\}p_1$ has a non-real vertex figure). Also every non-real polygon is r-symmetric for some r. Hence *the only non-real regular polyhedra are*

$$3\{3\}3\{3\}3, \quad 3\{3\}3\{4\}2, \quad 2\{3\}2\{4\}r \quad (r > 2)$$

and their reciprocals. In other words, the only regular star polyhedra are Kepler's

$$2\{\tfrac{5}{2}\}2\{5\}2, \quad 2\{\tfrac{5}{2}\}2\{3\}2$$

and their reciprocals (discovered by Poinsot).

For a four-dimensional polytope

$$p_1\{q_1\}p_2\{q_2\}p_3\{q_3\}p_4,$$

we consider the possible ways of 'telescoping' the symbols

$$p_1\{q_1\}p_2\{q_2\}p_3 \quad \text{and} \quad p_2\{q_2\}p_3\{q_3\}p_4$$

for the cell and vertex figure, and when the latter is r-symmetric, we seek the van Oss polygon $p_1\{q\}r$. (For the real cases, see Coxeter 1963, pp. 274–6.) Since the symbols (12·84) and

$$p\{4\}2\{3\}2\{4\}r$$

(for suitable p and r) represent honeycombs, not polytopes, the only possible non-real polytopes in four dimensions are

$$3\{3\}3\{3\}3\{3\}3, \quad 2\{4\}3\{3\}3\{4\}2, \quad 2\{q_1\}2\{3\}2\{4\}r$$

and the reciprocal of the last. Using $\cos^2\phi = c_1/1-c_2/1-c_3$ and (13·15), we soon find that $2\{4\}3\{3\}3\{4\}2$ has to be discarded because $\cos^2\phi > 1$. Since $2\{q_1\}2\{3\}2\{4\}r$ yields

$$\cos^2\phi = 2\cos^2\frac{\pi}{q_1},$$

we must have $q_1 < 4$, and since $\{q_1, 3\}$ is a real polyhedron this means that $q_1 = 3$ or $\frac{5}{2}$. When $q_1 = \frac{5}{2}$, we have

$$\cos 2\phi = 2\cos\frac{2\pi}{q_1}+1 = 2\cos\frac{4\pi}{5}+1 = -\tau+1 = -\tau^{-1}$$

and, by (13·15), $\quad \cos\dfrac{2\pi}{q} = -\tau^{-1}\sin\dfrac{\pi}{r}.$

Thus $q < 4$ and, since the van Oss polygon is $2\{q\}r$, $r = 2, 3, 4$, or 5; but none of these values for r yields a suitable value for q, so we are left with

$$2\{3\}2\{3\}2\{4\}r = \beta_4^r.$$

The computations are now reduced to a pleasantly short table:

Polytope	r	c_1	c_2	c_3	$\cos^2\phi$	$\cos 2\phi$	$\cos 2\pi/q$	Van Oss polygon
$3\{3\}3\{3\}3\{3\}3$	3	$\frac{1}{3}$	$\frac{1}{3}$	$\frac{1}{3}$	$\frac{2}{3}$	$\frac{1}{3}$	0	$3\{4\}3$
$2\{3\}2\{3\}2\{4\}r$	r	$\frac{1}{4}$	$\frac{1}{4}$	$\frac{1}{2}$	$\frac{1}{2}$	0	0	$2\{4\}r$

Since every non-real regular polyhedron is r-symmetric for some r, we conclude that *the only non-real regular polytopes in four dimensions are* β_4^r, γ_4^r $(r > 2)$, *and the Witting polytope.*

Similarly, since $3\{3\}3\{3\}3\{3\}3\{3\}3$ is not a polytope but a honeycomb, *the only regular polytopes in* $n > 4$ *dimensions are*

$$\alpha_n, \beta_n^r, \gamma_n^r \quad (r \geqslant 2).$$

EXERCISES

1. For the van Oss polygon $p_1\{q\}r$ of a regular polyhedron, express $\cos^2\phi$ directly in terms of $p_1, q_1, p_2, q_2, p_3, r$, and express r in terms of p_2, q_2, p_3.
2. For the regular polytope $p_1\{q_1\}p_2\{q_2\}\ldots p_n$, express the condition

$$\sin^2\phi > 0$$

in terms of the determinant

$$\Delta_{12\ldots n} = \begin{vmatrix} 1 & \sqrt{c_1} & 0 & \ldots & 0 & 0 \\ \sqrt{c_1} & 1 & \sqrt{c_2} & \ldots & 0 & 0 \\ 0 & \sqrt{c_2} & 1 & \ldots & 0 & 0 \\ & \ldots & & \ldots & & \\ 0 & 0 & 0 & \ldots & \sqrt{c_{n-1}} & 1 \end{vmatrix}$$

of the Hermitian form

$$x^1\bar{x}^1 - (x^1\bar{x}^2 + x^2\bar{x}^1)\sqrt{c_1} + x^2\bar{x}^2 - (x^2\bar{x}^3 + x^3\bar{x}^2)\sqrt{c_2} + \ldots$$
$$- (x^{n-1}\bar{x}^n + x^n\bar{x}^{n-1})\sqrt{c_{n-1}} + x^n\bar{x}^n.$$

13·2 THE REGULAR HONEYCOMBS

Just as a regular polytope may have a van Oss polygon, a regular honeycomb may have a 'van Oss apeirogon'. For the complete enumeration of regular honeycombs, this concept will be found helpful, though it does not provide a sufficiently powerful criterion to rule out some such plausible symbols as

$$3\{4\}5\{\tfrac{5}{2}\}3,$$

which require also the consideration of 'special subgroups' and lattices.

If the vertex figure $p_2\{q_2\}\ldots p_n$ of a honeycomb

$$p_1\{q_1\}p_2\{q_2\}\ldots p_n$$

is an r-symmetric polytope, the honeycomb has a *van Oss apeirogon* which is its section by the line that contains one edge. Since the vertex figure is r-symmetric, this line contains infinitely many vertices and edges, forming a regular apeirogon $p_1\{q\}r$ whose vertex figure is the β_1^r that constitutes a diameter of $p_2\{q_2\}\ldots p_n$. To compute q, we simply use (11·61) in the form

$$\frac{2}{q} = 1 - \frac{1}{p_1} - \frac{1}{r}.$$

Since the van Oss apeirogon is $p_1\{q\}r$, we see from (11·62) that p_1 and r can only take the values 2, 3, 4, 6; and if one of them is 4, the other cannot be 3 or 6. Since β_n^r is r-symmetric, we may deduce that, if there is an n-dimensional honeycomb

$$p\{4\}2\{3\}2\ldots\{3\}2\{4\}r,$$

its van Oss apeirogon must be $p\{q\}r$. Hence, for each $n > 1$, *there are just twelve such honeycombs* $\delta_{n+1}^{p,r}$: one for each of the twelve apeirogons $\delta_2^{p,r}$. In other words, the only honeycombs of this type are the Cartesian products (12·72). In particular, these are just twelve tessellations

$$\delta_3^{p,r} = p\{4\}2\{4\}r. \tag{13·21}$$

It follows from (12·89) that *the only tessellations* $2\{4\}p\{4\}2$ *are*

$$2\{4\}2\{4\}2,\quad 2\{4\}3\{4\}2,\quad 2\{4\}4\{4\}2,\quad 2\{4\}6\{4\}2.$$

In each of these four cases, the van Oss apeirogon is simply $2\{\infty\}2$.

As a first step towards a complete list of the remaining non-real tessellations

$$p_1\{q_1\}p_2\{q_2\}p_3,$$

we make a table like the one near the beginning of page 143 (but without the column headed '$\cos 2\pi/q$', which is now superfluous), retaining only the cases where $2\sigma_1 + 2\sigma_2 = \pi$ (so that the figure is a tessellation, not a polyhedron). Since the existence of a tessellation implies the existence of its reciprocal, we save some trouble by listing only one from each pair of reciprocals. Moreover, the van Oss apeirogon enables us to exclude every case (such as $4\{3\}2\{6\}4$ or $4\{\tfrac{8}{3}\}3\{4\}4$), where $p_2\{q_2\}p_3$ or $p_2\{q_1\}p_1$ is r-symmetric with $r > 6$, and every case (such as $3\{3\}2\{6\}3$) where $p_1 = 3$ or $p_3 = 3$ while $r = 4$. We are thus left with the following short list:

$p_1\{q_1\}p_2\{q_2\}p_3$	r	$2\sigma_1$	$2\sigma_2$	Van Oss apeirogon
$3\{3\}3\{4\}3$	6	$\pi - 2\kappa$	2κ	$3\{4\}6$
$4\{3\}4\{4\}2$	4	$\tfrac{1}{2}\pi$	$\tfrac{1}{2}\pi$	$4\{4\}4$
$4\{3\}4\{3\}4$	4	$\tfrac{1}{2}\pi$	$\tfrac{1}{2}\pi$	$4\{4\}4$
$2\{3\}3\{6\}2$	4	$\tfrac{1}{2}\pi + \kappa$	$\tfrac{1}{2}\pi - \kappa$	$2\{8\}4$
$2\{\tfrac{8}{3}\}3\{8\}2$	6	$\pi - \kappa$	κ	$2\{6\}6$
$3\{\tfrac{8}{3}\}2\{8\}3$	6	$\pi - \kappa$	κ	$3\{4\}6$
$3\{\tfrac{5}{2}\}3\{5\}3$	6	$\pi - 2\mu$	2μ	$3\{4\}6$

(In the third and fourth columns we use the notation

$$\kappa = \tfrac{1}{2}\arccos\tfrac{1}{3},\quad \mu = \tfrac{1}{2}\arcsin\tfrac{2}{3},$$

as in Coxeter 1963, p. 290.) We already know from §12·8 that the three symbols $3\{3\}3\{4\}3$, $4\{3\}4\{4\}2$, $4\{3\}4\{3\}4$ represent valid tessellations; but the impossibility of the remaining four symbols cannot be established without the following deeper investigation.

Let $\mathfrak{G}$ be the symmetry group of a regular honeycomb, and $\mathfrak{T}$ the normal subgroup consisting of all the translations in $\mathfrak{G}$. Since $\mathfrak{G}$ is discrete, it contains reflections in only a finite number of families of parallel hyperplanes. By taking, through a fixed point of the unitary n-space, a mirror parallel to a member of each family, we obtain a finite reflection group isomorphic to the quotient group $\mathfrak{G}/\mathfrak{T}$. Instead of an arbitrary fixed point, we may take a vertex of the orthoscheme $O_0 O_1 \ldots O_n$, thus exhibiting the finite reflection group as the subgroup of $\mathfrak{G}$ generated by n of the $n+1$ reflections $R_1, R_2, \ldots, R_{n+1}$. In terms of this subgroup $\mathfrak{S}$, $\mathfrak{G}$ is the semidirect product $\mathfrak{S}\mathfrak{T}$. Being the largest finite subgroup of $\mathfrak{G}$, $\mathfrak{S}$ is generated by either the first n or the last n of the $n+1$ reflections. In other words, the fixed point is either O_n or O_0. Considering the latter of these two possibilities, suppose $\mathfrak{S}$ *is generated by* $R_2, \ldots, R_{n+1}$. Then $\mathfrak{S}$ is the symmetry group of the vertex figure of the honeycomb, the fixed point O_0 is a vertex, $\mathfrak{T}$ is transitive on the vertices, and *the vertices form a lattice.*

Every edge through O_0 belongs to a van Oss apeirogon whose vertices form a one-dimensional lattice: the ordinary integers, Gaussian integers, or Eisenstein integers. Hence, to obtain all the vertices of the honeycomb, or all the vectors of the n-dimensional lattice, we consider the vector sum $\Sigma a_\lambda v_\lambda$, where the coefficients a_λ run over the appropriate ring of integers while the vectors v_λ represent a suitable subset of the vertices of the vertex figure of the honeycomb.

For instance, by applying the technique of §12·4 in two dimensions instead of three, we easily find that the vertices of $4\{3\}4$ are

$$(1+i, 0), \quad (0, i+1), \quad (1, 1)$$

with each coordinate multiplied by any fourth root of unity, that is, by ± 1 or $\pm i$. Since the van Oss apeirogon of $4\{3\}4\{3\}4$ is $4\{4\}4$, it follows that the vertices (u_1, u_2) of this tessellation are given by

$$u_1 = a(1+i)+c, \quad u_2 = b(1+i)+c,$$

where a, b, c run over the Gaussian integers. In other words, u_1 and u_2 are Gaussian integers satisfying (12·87).

Similarly, the vertices of $3\{6\}2$ are

$$(1+i, 0), \quad (0, 1+i), \quad (\omega, \pm\omega), \quad (\omega^2, \pm i\omega^2)$$

with both coordinates multiplied by the same fourth root of unity. The impossibility of $2\{3\}3\{6\}2$ can be seen by observing that, since its van Oss apeirogon would be $2\{8\}4$, its vertices would be

$$a(1+i, 0)+b(0, 1+i)+c\omega(1, 1)+d\omega(1, -1)+f\omega^2(1, i)+g\omega^2(1, -1),$$

where a, b, c, d, f, g are Gaussian integers. Putting $b = 0, c = d, f = g = 0$, we obtain, on the line $u_2 = 0$, points for which

$$u_1 = a(1+i)+2c\omega. \tag{13·22}$$

Since these points are dense on the line, $2\{3\}3\{6\}2$ does not exist.

The vertices of $3\{8\}2$ are

$$(\pm 2, 0), \quad (0, \pm 2), \quad (\pm i\sqrt{2}, \pm i\sqrt{2}), \quad (\pm 1, \pm 1 \pm i\sqrt{2}), \quad (\pm 1 \pm i\sqrt{2}, \pm 1),$$

with both coordinates multiplied by the same power of ω. If $2\{\tfrac{8}{3}\}3\{8\}2$ existed, its van Oss apeirogon would be $2\{6\}6$, and its vertices would include points with $u_2 = 0$ and

$$u_1 = 2(ci\sqrt{2}+d), \tag{13·23}$$

where now c and d are Eisenstein integers. These points again are dense on the line.

The vertices of $2\{8\}3$ are

$$(\pm i\sqrt{3}, 0), \quad (0, \pm i\sqrt{3}), \quad (1, \sqrt{2}), \quad (\sqrt{2}, 1), \quad (-1, -\sqrt{2}), \quad (-\sqrt{2}, -1),$$

with each coordinate multiplied by any power of ω. The impossibility of $3\{\tfrac{8}{3}\}2\{8\}3$ follows by considering the points with $u_2 = 0$ and

$$u_1 = (ai+d\sqrt{6})\sqrt{3}, \tag{13·24}$$

where a and d are Eisenstein integers.

Finally, the vertices of $3\{5\}3$ are

$$(\sqrt{3}, 0), \quad (0, \sqrt{3}), \quad (\tau, -\tau^{-1}), \quad (\tau^{-1}, \tau), \quad (1, \tau+\omega), \quad (1, \tau+\omega^2),$$
$$(\tau+\omega, -1), \quad (\tau+\omega^2, -1),$$

with each coordinate multiplied by any power of ω and both simultaneously multiplied by ± 1. The impossibility of $3\{\tfrac{5}{2}\}3\{5\}3$ follows by considering the points with $u_2 = 0$ and

$$u_1 = (a+b\tau\sqrt{3})\sqrt{3}. \tag{13·25}$$

We conclude that, apart from (13·21) and the real tessellations

$$2\{3\}2\{6\}2, \quad 2\{6\}2\{3\}2,$$

the only regular honeycombs in the plane are (with $p = 3$, 4 or 6):

$$\begin{matrix} [2\{4\}p\{4\}2, & 4\{3\}4\{4\}2, & 2\{4\}4\{3\}4,] \\ 3\{3\}3\{4\}3, & 3\{4\}3\{3\}3, & 4\{3\}4\{3\}4. \end{matrix} \tag{13·26}$$

For three-dimensional honeycombs, we put two polyhedra together and make a table like the one on page 143 (second column), but with

$$c_1/1-c_2/1-c_3 = 1:$$

Honeycomb	r	c_1	c_2	c_3	$\cos^2\phi$	Van Oss apeirogon
$p\{4\}2\{3\}2\{4\}r$	r	$\frac{1}{2}$	$\frac{1}{4}$	$\frac{1}{2}$	1	$\delta_2^{r,r}$
$2\{3\}2\{4\}3\{3\}3$	6	$\frac{1}{4}$	$\frac{1}{2}$	$\frac{1}{3}$	1	$2\{6\}6$
$2\{4\}3\{3\}3\{3\}3$	3	$\frac{1}{2}$	$\frac{1}{3}$	$\frac{1}{3}$	1	$2\{12\}3$

Since $2\{3\}2\{4\}3\{3\}3$ and $2\{4\}3\{3\}3\{3\}3$ are already known to exist (see (12·84)), no 'deeper investigation' is needed to prove that, apart from the twelve honeycombs

$$\delta_4^{p,r} = p\{4\}2\{3\}2\{4\}r,$$

three-dimensional space admits only

$$(13{\cdot}27)\qquad \begin{matrix} 2\{3\}2\{4\}3\{3\}3, & 2\{4\}3\{3\}3\{3\}3, \\ 3\{3\}3\{3\}3\{4\}2, & 3\{3\}3\{4\}2\{3\}2. \end{matrix}$$

Similarly in four dimensions (with $c_1/1-c_2/1-c_3/1-c_4 = 1$), apart from

$$\delta_5^{p,r} = p\{4\}2\{3\}2\{3\}2\{4\}r,$$

we have only

$$(13{\cdot}28)\qquad \begin{matrix} 2\{3\}2\{3\}2\{4\}2\{3\}2, & 2\{3\}2\{4\}2\{3\}2\{3\}2, \\ & 3\{3\}3\{3\}3\{3\}3\{3\}3, \end{matrix}$$

among which the first two are real (Coxeter 1963, p. 153).

In any greater number of dimensions, the twelve honeycombs $\delta_{n+1}^{p,r}$ stand alone.

EXERCISES

1. Justify the above statement that the sets of points (13·22), (13·23), (13·24), (13·25) are *dense* on the complex line.
2. For a three-dimensional honeycomb, why does the condition $c_1/1-c_2/1-c_3 = 1$ have the same effect as $c_3/1-c_2/1-c_1 = 1$?
3. Find the van Oss apeirogons for $3\{3\}3\{4\}2\{3\}2$ and $3\{3\}3\{3\}3\{4\}2$.

13·3 CYCLES AND FRIEZE PATTERNS

In §12·1 we saw how the polytope $p_1\{q_1\}p_2 \ldots \{q_{n-1}\}p_n$ can be derived by Wythoff's construction from the group generated by reflections in a sequence of n hyperplanes forming successive dihedral angles

$$\sigma_1, \sigma_2, \ldots, \sigma_{n-1}$$

(see (13·14)) while all non-consecutive pairs of the hyperplanes are orthogonal. This sequence of n hyperplanes, forming the O_n corner of the orthoscheme $O_0 O_1 \ldots O_n$, can be completed to a cycle of $n+2$ by adding the hyperplane $O_0 O_1 \ldots O_{n-1}$ and one orthogonal to the line $O_0 O_n$. This was done in §3·1 and §5·4 for a real space, but the only change needed for a complex space is to replace

$$\pi/p, \pi/q, \ldots, \quad \text{by} \quad \sigma_1, \sigma_2, \ldots$$

In §5·7, we saw that the radii and angles of a real polytope $2\{q_1\}2 \ldots \{q_{n-1}\}2$ can be written down in terms of symbols (s, t) whenever there is a frieze pattern whose first two rows satisfy (5·72). Changing this to

$$(13{\cdot}31)\qquad \cos^2\sigma_\nu = \frac{(\nu-2,\nu-1)(\nu,\nu+1)}{(\nu-2,\nu)(\nu-1,\nu+1)} \quad (\nu = 1, \ldots, n-1)$$

we can apply the same procedure to the general regular polytope. In particular,

$$(13{\cdot}32)\qquad \psi = \tfrac{1}{2}\pi - \sigma_n, \quad \chi = \sigma_{n+1}, \quad \phi = \tfrac{1}{2}\pi - \sigma_{n+2}.$$

For instance, in the case of the generalized cube γ_n^p we have

$$(-1, t) = 1 \quad (t \geqslant 0), \quad (s, t) = t - s \quad (0 \leqslant s \leqslant t \leqslant n),$$

so that
$$\frac{{}_\nu R}{l} = \sqrt{\frac{(-1,0)(-1,1)(\nu,n)}{(-1,\nu)(-1,n)(0,1)}} = \sqrt{(n-\nu)},$$

$$\sin\psi = \sqrt{\frac{(-1,n)(n-2,n-1)}{(-1,n-1)(n-2,n)}} = \sqrt{\tfrac{1}{2}},$$

$$\cos\chi = \sqrt{\frac{(-1,0)(n-1,n)}{(-1,n-1)(0,n)}} = \sqrt{\frac{1}{n}}, \quad \sin\phi = \sqrt{\frac{(-1,n)(0,1)}{(-1,1)(0,n)}} = \sqrt{\frac{1}{n}},$$

and the dihedral angle is $\pi - 2\psi = \frac{1}{2}\pi$. (As these results are independent of p, they are the same as for the real n-dimensional cube $\gamma_n = \gamma_n^2$.)

On page 60 we saw that the three real polytopes

$$2\{4\}2\{3\}2\{3\}2, \quad 2\{3\}2\{3\}2\{4\}2, \quad 2\{3\}2\{4\}2\{3\}2$$

all yield the same frieze pattern (5·11), beginning at different places. This happens because we are using the same six hyperplanes, differently numbered. In §12·8 we obtained an analogous cycle of six hyperplanes with the peculiarity that the six σ_ν are all equal: it makes no difference where we begin. In fact, the Witting polytope $3\{3\}3\{3\}3\{3\}3$ has $\cos^2\sigma_\nu = \frac{1}{3}$ for each ν, so that

$$\phi = \psi = \kappa, \quad \chi = \tfrac{1}{2}\pi - \kappa.$$

In this case (13·31) yields the Ptolemaic frieze pattern

$$(13{\cdot}33)\qquad \begin{matrix}
1 & & 1 & & 1 & & 1 & & 1 & & 1 & & 1 & & 1 & \\
& \sqrt{3} & & \sqrt{3} & & \sqrt{3} & & \sqrt{3} & & \sqrt{3} & & \sqrt{3} & & \sqrt{3} & & \cdots \\
2 & & 2 & & 2 & & 2 & & 2 & & 2 & & 2 & & 2 & \\
& \sqrt{3} & & \sqrt{3} & & \sqrt{3} & & \sqrt{3} & & \sqrt{3} & & \sqrt{3} & & \sqrt{3} & & \cdots \\
1 & & 1 & & 1 & & 1 & & 1 & & 1 & & 1 & & 1 &
\end{matrix}$$

and $(s, t) = 2\sin(t-s)\pi/6$.

More interestingly, the cycle (12·84), giving for $\cos^2 \sigma_\nu$ the values

$$\tfrac{1}{3}, \tfrac{1}{3}, \tfrac{1}{2}, \tfrac{1}{4}, \tfrac{1}{2}, \tfrac{1}{3}, \tfrac{1}{3}, \tfrac{1}{2}, \tfrac{1}{4}, \tfrac{1}{2}, \tfrac{1}{3}, \ldots$$

yields the pattern

$$(13{\cdot}34)\qquad \begin{array}{cccccccccccccccc} 1 & & 1 & & 1 & & 1 & & 1 & & 1 & & 1 & & 1 & \\ & 1 & & 3 & & 1 & & 2 & & 2 & & 1 & & 3 & & \ldots \\ 1 & & 2 & & 2 & & 1 & & 3 & & 1 & & 2 & & 2 & \\ & 1 & & 1 & & 1 & & 1 & & 1 & & 1 & & 1 & & \ldots \end{array}$$

which we first met in §3·1 in connection with the ordinary cube and octahedron. The present role of this cycle is far more satisfactory, as it supplies metrical properties simultaneously for all the five polyhedra

$$3\{3\}3\{4\}2,\quad 3\{4\}2\{3\}2,\quad 2\{3\}2\{4\}3,\quad 2\{4\}3\{3\}3,\quad 3\{3\}3\{3\}3.$$

In particular, for the Hessian polyhedron $3\{3\}3\{3\}3$, the values of

$$\begin{array}{llllcccccccc} (-1,0) & (0,1) & (1,2) & (2,3) \text{ are} & 1 & & 1 & & 1 & & 1 \\ \quad(-1,1) & (0,2) & (1,3) & & & 1 & & 3 & & 1 & \\ \qquad(-1,2) & (0,3) & & & & & 2 & & 2 & & \\ \qquad\quad(-1,3) & & & & & & & 1 & & & \end{array}$$

yielding

$$\frac{{}_0R}{l} = \sqrt{\frac{(-1,1)(0,3)}{(-1,3)(0,1)}} = \sqrt{\frac{1\times 2}{1\times 1}} = \sqrt{2},$$

$$\frac{{}_1R}{l} = \sqrt{\frac{(-1,0)(1,3)}{(-1,3)(0,1)}} = \sqrt{\frac{1\times 1}{1\times 1}} = 1,$$

$$\frac{{}_2R}{l} = \sqrt{\frac{(-1,0)(-1,1)(2,3)}{(-1,2)(-1,3)(0,1)}} = \sqrt{\frac{1\times 1\times 1}{2\times 1\times 1}} = \sqrt{\tfrac{1}{2}},$$

in agreement with the answer to §12·3, Ex. 1.

EXERCISES

1. Write out a portion of the frieze pattern for γ_5^p.
2. Use (13·33) to check the answer to §12·5, Ex. 1.
3. What diagram, in the style of Figure 5·3A, would be appropriate for (13·33)?
4. What pattern of integers has the same effect as (13·33)?
5. Compute the principal angles for the Hessian polyhedron.
6. Write out a portion of the frieze pattern for the cycle (12·88).

13·4 PRESENTING THE SYMMETRY GROUPS

We saw, in §11·1, that every regular polygon has a symmetry group generated by two reflections. At the end of §9·7 we obtained the complete list of such groups $p_1[q]p_2$, with the presentation (9·81). In §13·1 we saw that every non-real regular polytope of more than two dimensions has a Schläfli symbol (12·12) where the q_μ (as well as the p_ν) are integers and, for each μ, the polygon $p_\mu\{q_\mu\}p_{\mu+1}$ (being non-starry) has $p_\mu[q_\mu]p_{\mu+1}$ for its symmetry group. It follows that the symmetry group for every such polytope is generated by n reflections R_μ, of period p_μ, such that any two consecutive reflections generate a subgroup $p_\mu[q_\mu]p_{\mu+1}$, while all other pairs are commutative; see (12·11) and (12·19). Analogy with (9·81) suggests that these $\frac{1}{2}n(n+1)$ relations may suffice for a presentation. In fact they do, as we shall soon see. However, since the analogous result for apeirogons (§11·6) is false, it would be unreasonable to look for the kind of general treatment that is available for real spaces (Coxeter 1963, pp. 80, 188; Benson and Grove 1971, p. 88); instead, we consider separately each group (or family of groups), beginning with

$$\mathfrak{C}_r \wr \mathfrak{S}_n \quad\text{or}\quad r[4]2[3]2\ldots[3]2 \quad\text{or}\quad 2[3]2\ldots[3]2[4]r$$

($n-1$ twos and $n-2$ threes), of order $r^n n!$

Although this 'generalized symmetric group' was described in §12·2 as the symmetry group of the generalized cube γ_n^r (see (12·23)), we shall find some advantage in regarding it now as the symmetry group of the reciprocal 'generalized octahedron' β_n^r (see (12·24)), in which we may conveniently use the symbol λ_ν to denote the typical vertex

$$(0, \ldots, 0, \epsilon^\lambda, 0, \ldots, 0),$$

where $\epsilon = \exp(2\pi i/r)$ and ϵ^λ is the νth coordinate; in other words, $u_\nu = \epsilon^\lambda$ while every other u_μ is zero. Taking into consideration this change of notation from $R_1, \ldots, R_n$ to $R_n, \ldots, R_1$, we see that the presentation to be established is

$$R_1^2 = R_n^r = 1,$$

$$(13{\cdot}41)\quad R_1R_2R_1 = R_2R_1R_2, \ldots, R_{n-2}R_{n-1}R_{n-2} = R_{n-1}R_{n-2}R_{n-1},^\dagger$$

$$(R_{n-1}R_n)^2 = (R_nR_{n-1})^2,\quad R_\mu \rightleftarrows R_\nu \quad (\mu < \nu - 1).$$

When $n = 2$, this reduces to the known presentation

$$R_1^2 = R_2^r = 1,\quad (R_1R_2)^2 = (R_2R_1)^2$$

† Since the relations $R_1R_2R_1 = R_2R_1R_2, \ldots$ force the first $n-1$ generators to be mutually conjugate, the relation $R_1^2 = 1$ makes it unnecessary to specify the periods of $R_2, \ldots, R_{n-1}$. However, this economy would not be admissible if we used instead the relations

$$(R_1R_2)^3 = \ldots = (R_{n-2}R_{n-1})^3 = 1$$

as G. C. Shephard did on p. 375 of his Unitary groups generated by reflections, *Canadian Journal of Mathematics* 5 (1953), 364–83. His graph 3·7 needs only one small change (from m^2 to 4) to make it identical with our graph for $m\,[4]3[2]3 \ldots$.

for $2[4]r$ (see (9·71)), and we thus have a basis for induction over n. Assuming that the elements $R_2, \ldots, R_n$ of the abstract group (13·41) (with $R_2^2 = 1$ if $n > 2$) generate a subgroup of order $r^{n-1}(n-1)!$, we will deduce that the index of this subgroup is precisely rn.

Letting 1_1 denote the subgroup, so that $1_1 R_\nu = 1_1 (\nu > 1)$, we consider cosets 1_2, 2_1, etc., defined as follows:

$$1_2 = 1_1 R_1, \quad 1_3 = 1_2 R_2, \ldots, \quad 1_n = 1_{n-1} R_{n-1},$$
$$\lambda_n = (\lambda - 1)_n R_n \quad (\lambda = 2, \ldots, r),$$
$$\lambda_{n-1} = \lambda_n R_{n-1}, \quad \lambda_{n-2} = \lambda_{n-1} R_{n-2}, \ldots, \lambda_1 = \lambda_2 R_1.$$

After some rather tricky appeals to induction, the relations (13·41) can be seen to yield

$$\lambda_\mu R_\mu = \lambda_{\mu+1}, \quad \lambda_{\mu+1} R_\mu = \lambda_\mu \quad (\mu < n),$$
$$r_n R_n = 1_n, \quad \lambda_n R_n = (\lambda+1)_n \quad (\lambda < r),$$
$$\lambda_\nu R_\mu = \lambda_\nu \quad \text{for all } \nu < \mu \text{ and for all } \nu > \mu + 1.$$

Thus the rn symbols λ_ν include a complete set of cosets of 1_1, and the index of this subgroup is at most rn. On the other hand, since the symbols λ_ν can be identified with the rn vertices of β_n^r, the index is at least rn. Thus the cosets are all distinct, and the order of the group is indeed

$$r^{n-1}(n-1)!\, rn = r^n n!.$$

As permutations of the cosets or vertices, we have

$$\text{(13·411)} \qquad R_\mu = (1_\mu 1_{\mu+1})(2_\mu 2_{\mu+1}) \ldots (r_\mu r_{\mu+1}) \quad (\mu < n),$$
$$R_n = (1_n 2_n \ldots r_n).$$

In other words, $2[3]2 \ldots [3]2[4]r$ is the group of all permutations (of these rn symbols) commutative with

$$(1_1 2_1 \ldots r_1)(1_2 2_2 \ldots r_2) \ldots (1_n 2_n \ldots r_n).$$

Turning now to one of the few remaining non-real groups, we proceed to establish the presentation

$$\text{(13·42)} \quad R_1^3 = 1, \quad R_1 R_2 R_1 = R_2 R_1 R_2, \quad R_2 R_3 R_2 = R_3 R_2 R_3, \quad R_1 \rightleftarrows R_3$$

for the group $3[3]3[3]3$ of the Hessian polyhedron (see (12·34)). Clearly, R_2 and R_3 generate a binary tetrahedral subgroup $3[3]3$, of order 24 (see (10·62), and §9·7, Ex. 2). By systematic enumeration of cosets (Coxeter and Moser 1972, Chapter 2) we soon find that the index of this subgroup is 27; thus the order of the abstract group (13·42) is 648, in agreement with the geometric aspect. (The 27 cosets can be associated with the 27 vertices of the polyhedron.)

Since the 54 vertices of $2\{4\}3\{3\}3$ (Figure 12·4A) belong to two reciprocal $3\{3\}3\{3\}3$'s, we can derive its symmetry group

$$2[4]3[3]3$$

(or $3[3]3[4]2$) by adjoining to (13·42) an involutory generator R_0 that transforms R_1 into R_3 (and *vice versa*) while leaving R_2 invariant. Writing $R_0 R_1 R_0$ for R_3, we obtain, for this group of order 1296, the presentation

$$R_0^2 = R_1^3 = 1, \quad (R_0 R_1)^2 = (R_1 R_0)^2, \quad R_1 R_2 R_1 = R_2 R_1 R_2, \quad R_0 \rightleftarrows R_2.$$

Finally, the symmetry group

$$3[3]3[3]3[3]3$$

of the Witting polytope (12·52) should have the presentation

$$\text{(13·43)} \quad R_1^3 = 1, \quad R_1 R_2 R_1 = R_2 R_1 R_2, \quad R_2 R_3 R_2 = R_3 R_2 R_3,$$
$$R_3 R_4 R_3 = R_4 R_3 R_4,$$
$$R_1 \rightleftarrows R_3, \quad R_1 \rightleftarrows R_4, \quad R_2 \rightleftarrows R_4.$$

This can be established (easily with the aid of a computer) by enumerating the 240 cosets of the subgroup $3[3]3[3]3$ generated by R_2, R_3, R_4. Thus the order of the whole group is

$$240 \times 648 = 155\,520,$$

in agreement with page 134.

In terms of the convenient abbreviation

$$A \underset{q}{\leftrightarrow} B$$

for the equation $ABA \ldots = BAB \ldots$ with q factors on each side (so that $\underset{2}{\leftrightarrow}$ has the same meaning as $\rightleftarrows$), we may summarize the above results as follows:

(13·44) *Whenever the group $p_1[q_1]p_2[q_2] \ldots p_{n-1}[q_{n-1}]p_n$ is finite, it has the presentation*

$$R_\mu^{p_\mu} = 1, \quad R_\mu \underset{q_\mu}{\leftrightarrow} R_{\mu+1}, \quad R_\mu \underset{2}{\leftrightarrow} R_\nu \quad (\nu > \mu + 1).$$

For instance, when every $q_\mu = 3$ (as in (10·62), (13·42) and (13·43)), we have a group

$$p[3]p \cdots [3]p$$

(with n p's), and when this is finite it has the presentation

$$\text{(13·45)} \qquad R_1^p = 1, \quad R_\mu \underset{3}{\leftrightarrow} R_{\mu+1}, \quad R_\mu \underset{2}{\leftrightarrow} R_\nu \quad (\nu > \mu + 1).$$

By omitting the first relation, we obtain an infinite group

$$R_\mu \underset{3}{\leftrightarrow} R_{\mu+1}, \quad R_\mu \underset{2}{\leftrightarrow} R_\nu \quad (\nu > \mu + 1)$$

or

$$\text{(13·46)} \qquad R_\mu R_{\mu+1} R_\mu = R_{\mu+1} R_\mu R_{\mu+1} \quad (1 \leqslant \mu \leqslant n-1),$$
$$R_\mu \rightleftarrows R_\nu \quad (\mu \leqslant \nu - 2),$$

which is the $(n+1)$-strand *braid group* (see, for instance, Coxeter and Moser 1972, pp. 62–3). Thus the reflection group $p[3]p \dots [3]p$ is a factor group of the braid group (13·46), which Artin† presented in terms of two generators

$$R_1 \quad \text{and} \quad R_1R_2\dots R_n:$$

$$(13{\cdot}461) \qquad R^{n+1} = (RR_1)^n, \quad R_1 \rightleftarrows R^{-\nu}R_1R^{\nu} \quad (2 \leqslant \nu \leqslant \tfrac{1}{2}(n+1)).$$

Restoring the relation $R_1{}^p = 1$, we obtain for $p[3]p \dots [3]p$ the presentation

$$(13{\cdot}47) \qquad R_1{}^p = 1, \quad R^{n+1} = (RR_1)^n, \quad R_1 \rightleftarrows R^{-\nu}R_1R^{\nu} \quad (2 \leqslant \nu \leqslant \tfrac{1}{2}(n+1)).$$

In particular, $p[3]p$ is

$$R_1{}^p = 1, \quad R^3 = (RR_1)^2 \qquad (p < 6)$$

(compare §9·7, Ex. 1), $p[3]p[3]p$ is

$$(13{\cdot}48) \qquad R_1{}^p = 1, \quad R^4 = (RR_1)^3, \quad R_1 \rightleftarrows R^{-2}R_1R^2 \quad (p = 2 \text{ or } 3),$$

and $p[3]p[3]p[3]p$ is

$$(13{\cdot}49) \qquad R_1{}^p = 1, \quad R^5 = (RR_1)^4, \quad R_1 \rightleftarrows R^{-2}R_1R^2 \quad (p = 2 \text{ or } 3).$$

The analogous groups

$$R_1{}^6 = 1, \quad R^3 = (RR_1)^2;$$

$$R_1{}^4 = 1, \quad R^4 = (RR_1)^3, \quad R_1 \rightleftarrows R^{-2}R_1R^2;$$

$$R_1{}^3 = 1, \quad R^5 = (RR_1)^4, \quad R_1 \rightleftarrows R^{-2}R_1R^2$$

are certainly infinite, since they have as factor groups the symmetry groups

$$6[3]6, \quad 4[3]4[3]4, \quad 3[3]3[3]3[3]3[3]3$$

of infinite honeycombs. In other words, $p[3]p \dots [3]p$ (with n p's) is an infinite group when $(p-2)(n-1) = 4$, that is, when $\{p, n+1\}$ is a Euclidean tessellation; but it is a finite group whenever

$$(p-2)(n-1) < 4,$$

that is, whenever $\{p, n+1\}$ is a spherical tessellation. This correspondence will seem less fanciful in §13·6, where we shall obtain a simple formula for the order of the group in terms of the number of vertices of the tessellation.

† E. Artin, Theorie der Zöpfe, *Abhandlungen aus dem Mathematischen Seminar der Universität Hamburg* **4** (1926), 47–72; see especially pp. 52–4. See also Coxeter, Factor groups of the braid group, *Proceedings of the Fourth Canadian Mathematical Congress* (1959), 92–122.

EXERCISES

1. What happens to the group (13·41) when we set $r = 1$?
2. What group is (13·47) when $p = 2$?
3. Verify the equivalence of (13·46) and (13·461).

13·5 A HISTORICAL DIGRESSION

In the spirit of §1·1, an isometry is said to be of *type* ν if it is expressible as the product of ν (but no fewer) reflections. Such an isometry leaves totally invariant an $(n-\nu)$-flat: the intersection of the ν mirrors. Thus the identity is of type 0, a reflection is of type 1, and an isometry of type n leaves only one point invariant.

In a given group generated by n reflections, let b_ν denote the number of isometries of type ν. (Of course, $b_0 = 1$.) According to a remarkable theorem of Shephard and Todd‡, the generating function for these numbers b_ν is the product of n linear factors:

$$(13{\cdot}51) \qquad \sum_{\nu=0}^{n} b_\nu t^\nu = (1+m_1t)(1+m_2t)\dots(1+m_nt).$$

The coefficients m_λ, with $m_1 \leqslant m_2 \leqslant \dots \leqslant m_n$ for convenience, are called the *exponents* of the group because, if the group is irreducible (not a direct product) its n generating reflections can be chosen and named in such a way that their product

$$R = R_1R_2\dots R_n$$

has period $h = m_n + 1$ and characteristic roots

$$(13{\cdot}52) \qquad \epsilon^{m_1}, \ \epsilon^{m_2}, \ \dots, \ \epsilon^{m_n},$$

where $\epsilon = \exp(2\pi i/h)$. The irreducible group has for its centre the cyclic group $\mathfrak{C}_r$ generated by $R^{h/r}$, where

$$(13{\cdot}53) \qquad r = (m_1+1, m_2+1, \dots, m_n+1),$$

meaning the greatest common divisor§. (This r is the $k = g/g'$ of Shephard and Todd.) Setting $t = 1$ in (13·51), we find that the order of the group is equal to the product

$$(13{\cdot}54) \qquad g = (m_1+1)(m_2+1)\dots(m_n+1).$$

‡ G. C. Shephard and J. A. Todd, Finite unitary reflection groups, *Canadian Journal of Mathematics* **6** (1954), 274–304; see especially pp. 279–83, 296–301. Actually they only 'verified' the theorem by separate consideration of each group or family of groups. A general proof, using the powerful tool of exterior differentiation, was supplied nine years later; see L. Solomon, Invariants of finite reflection groups, *Nagoya Mathematical Journal* **22** (1963), 57–64.

§ Coxeter, Groups generated by unitary reflections of period two, *Canadian Journal of Mathematics* **9** (1957), 243–72; see especially p. 255.

Chevalley[†] proved that any reflection group possesses a set of n algebraically independent polynomial invariants forming a polynomial basis for the set of all invariants of the group. Moreover, when this basis is so chosen that the degrees of the n polynomials are as small as possible, the product of the degrees is equal to the order of the group. The story of the identification of these degrees with the numbers $m_\lambda + 1$ has been nicely told by Kostant.[‡]

Steinberg[§] proved that the Jacobian of the n basic invariants is a scalar multiple of a product of linear forms

$$\Pi L^{p-1},$$

such that $L = 0$ is the mirror for a reflection of period p and the product is taken over all the mirrors. It follows that the number of reflections in the group is equal to the degree of the Jacobian, namely

$$m_1 + m_2 + \ldots + m_n,$$

which is b_1 in the notation of (13·51).

These historical remarks may serve to motivate a search for a direct deduction of the exponents from the values of

$$p_1, q_1, p_2, q_2, \ldots, p_{n-1}, q_{n-1}, p_n.$$

Such a deduction is already known[‖] for the cases in which

$$p_1 = p_2 = \ldots = p_n.$$

However, the symmetry group (13·44) of the general regular polytope presents a complication of sufficient interest to deserve careful consideration.

EXERCISE

Given (13·52) and (13·53), prove that $R^{h/r}$ belongs to the centre of the group.

† C. Chevalley, Invariants of finite groups generated by reflections, *American Journal of Mathematics* **77** (1955), 778–82. Chevalley's reflections are stated to be involutory, but this restriction does not seem to be used in his proof. See also Burnside (1911, p. 361).

‡ B. Kostant, The principal three-dimensional subgroup and the Betti numbers of a complex simple Lie group, *American Journal of Mathematics* **81** (1959), 973–1032; see especially p. 1021. Being unaware of this connection, Chevalley called the degrees m_λ instead of $m_\lambda + 1$.

§ R. Steinberg, Finite reflection groups, *Transactions of the American Mathematical Society* **91** (1959), 493–504; see especially p. 500.

‖ See the two papers by Coxeter cited on page 149: pp. 117–20 of the former and 253–4 of the latter.

13·6 PETRIE POLYGONS AND EXPONENTS

To express $R_1 R_2 \ldots R_n$ (the 'Coxeter–Killing transformation' of Kostant) in terms of the oblique coordinates x_λ of §8·8, let the notation be such that R_1 transforms x_λ into x_λ', R_2 transforms x_λ' into x_λ'', and so on. But let us employ Blichfeldt's convention (Miller, Blichfeldt and Dickson 1916, p. 193) whereby the old coordinates are expressed in terms of the new. Then the expression (8·83) for R_ν becomes

$$x_\lambda^{(\nu-1)} = x_\lambda^{(\nu)} + (\epsilon_\nu^2 - 1)\, a_{\nu\lambda} x_\nu^{(\nu)} \quad (\lambda = 1, \ldots, n),$$

where ϵ_ν is given by (8·82), $a_{\nu\nu} = 1$, $a_{\nu\lambda} = a_{\lambda\nu} = 0$ if $|\lambda - \nu| > 1$, and

$$a_{\mu,\mu+1} = a_{\mu+1,\mu} = -\sqrt{c_\mu}$$

in the notation of (13·14). The product $R_1 R_2 \ldots R_n$ transforms the point $(x_1, x_2, \ldots, x_n)$ into the point $(x_1^{(n)}, x_2^{(n)}, \ldots, x_n^{(n)})$ whose coordinates are given indirectly by the n sets of equations

$$\begin{aligned} x_\lambda &= x_\lambda' + (\epsilon_1^2 - 1)\, a_{1\lambda} x_1', \\ x_\lambda' &= x_\lambda'' + (\epsilon_2^2 - 1)\, a_{2\lambda} x_2'', \\ &\ldots \qquad \ldots \\ x_\lambda^{(n-1)} &= x_\lambda^{(n)} + (\epsilon_n^2 - 1)\, a_{n\lambda} x_n^{(n)}. \end{aligned}$$

Since $a_{\nu\lambda} = 0$ when $|\lambda - \nu| > 1$, all these equations are of the form $x_\lambda^{(\mu-1)} = x_\lambda^{(\mu)}$ except the following three:

$$\begin{aligned} x_\lambda^{(\lambda-2)} &= x_\lambda^{(\lambda-1)} + (\epsilon_{\lambda-1}^2 - 1)\, a_{\lambda-1,\lambda} x_{\lambda-1}^{(\lambda-1)} \\ &= x_\lambda^{(\lambda-1)} - \sqrt{c_{\lambda-1}}(\epsilon_{\lambda-1}^2 - 1)\, x_{\lambda-1}^{(\lambda-1)}, \\ x_\lambda^{(\lambda-1)} &= x_\lambda^{(\lambda)} + (\epsilon_\lambda^2 - 1)\, a_{\lambda\lambda} x_\lambda^{(\lambda)} = \epsilon_\lambda^2 x_\lambda^{(\lambda)}, \\ x_\lambda^{(\lambda)} &= x_\lambda^{(\lambda+1)} + (\epsilon_{\lambda+1}^2 - 1)\, a_{\lambda+1,\lambda} x_{\lambda+1}^{(\lambda+1)} \\ &= x_\lambda^{(\lambda+1)} - \sqrt{c_\lambda}(\epsilon_{\lambda+1}^2 - 1)\, x_{\lambda+1}^{(\lambda+1)}. \end{aligned}$$

Thus

$$x_\lambda = x_\lambda' = \ldots = x_\lambda^{(\lambda-2)} = x_\lambda^{(\lambda-1)} - \sqrt{c_{\lambda-1}}(\epsilon_{\lambda-1}^2 - 1)\, x_{\lambda-1}^{(\lambda-1)},$$

$$x_\lambda^{(\lambda-1)} = \epsilon_\lambda^2 x_\lambda^{(\lambda)},$$

$$x_\lambda^{(\lambda)} + \sqrt{c_\lambda}(\epsilon_{\lambda+1}^2 - 1)\, x_{\lambda+1}^{(\lambda+1)} = x_\lambda^{(\lambda+1)} = x_\lambda^{(\lambda+2)} = \ldots = x_\lambda^{(n)}.$$

Now, the characteristic equation for $R_1 R_2 \ldots R_n$ can be obtained by eliminating all the x's from these equations along with $\rho x_\lambda = x_\lambda^{(n)}$, that is, by eliminating $x_1^{(1)}, \ldots, x_n^{(n)}$ from n equations such as

$$\rho\{\epsilon_\lambda^2 x_\lambda^{(\lambda)} - \sqrt{c_{\lambda-1}}(\epsilon_{\lambda-1}^2 - 1)\, x_{\lambda-1}^{(\lambda-1)}\} = x_\lambda^{(\lambda)} + \sqrt{c_\lambda}(\epsilon_{\lambda+1}^2 - 1)\, x_{\lambda+1}^{(\lambda+1)}.$$

Multiplication by $\rho^{-\frac{1}{2}(\lambda+1)}$ yields

$$\begin{aligned} -\sqrt{c_{\lambda-1}}(\epsilon_{\lambda-1} - \epsilon_{\lambda-1}^{-1})\,\epsilon_{\lambda-1}\rho^{-\frac{1}{2}(\lambda-1)} x_{\lambda-1}^{(\lambda-1)} + (\rho^{\frac{1}{2}}\epsilon_\lambda - \rho^{-\frac{1}{2}}\epsilon_\lambda^{-1})\,\epsilon_\lambda \rho^{-\frac{1}{2}\lambda} x_\lambda^{(\lambda)} \\ - \sqrt{c_\lambda}(\epsilon_{\lambda+1} - \epsilon_{\lambda+1}^{-1})\,\epsilon_{\lambda+1}\rho^{-\frac{1}{2}(\lambda+1)} x_{\lambda+1}^{(\lambda+1)} = 0. \end{aligned}$$

Recalling that $\epsilon_\lambda - \epsilon_\lambda^{-1} = 2i \sin \frac{\pi}{p_\lambda}$, we see that this rather formidable equation can be simplified by writing

$$(13{\cdot}61) \qquad C_\lambda^2 = c_\lambda \sin\frac{\pi}{p_\lambda}\sin\frac{\pi}{p_{\lambda+1}} = \tfrac{1}{2}\left\{\cos\left(\frac{\pi}{p_\lambda}-\frac{\pi}{p_{\lambda+1}}\right)+\cos\frac{2\pi}{q_\lambda}\right\},$$

$$(13{\cdot}611) \qquad X_\lambda = (\rho^{\frac{1}{2}}\epsilon_\lambda - \rho^{-\frac{1}{2}}\epsilon_\lambda^{-1})/2i,$$
$$y_\lambda = \sqrt{(\sin\pi/p_\lambda)}\,\epsilon_\lambda \rho^{-\frac{1}{2}\lambda} x_\lambda^{(\lambda)}.$$

In fact, after dividing by $2i$ and multiplying by $\sqrt{(\sin \pi/p_\lambda)}$, we are left with

$$-C_{\lambda-1}y_{\lambda-1} + X_\lambda y_\lambda - C_\lambda y_{\lambda+1} = 0.$$

(Of course, the first term must be omitted if $\lambda = 1$, and the last if $\lambda = n$.) Eliminating the y's from

$$\begin{aligned} X_1y_1 - C_1y_2 \qquad\qquad &= 0,\\ C_1y_1 - X_2y_2 + C_2y_3 \qquad &= 0,\\ -C_2y_2 + X_3y_3 - C_3y_4 &= 0,\\ \ldots \qquad\qquad &\ldots\\ \pm C_{n-1}y_{n-1} \mp X_ny_n &= 0, \end{aligned}$$

we obtain the single equation

$$(13{\cdot}62) \qquad \begin{vmatrix} X_1 & C_1 & 0 & 0 & \ldots & 0 & 0 \\ C_1 & X_2 & C_2 & 0 & \ldots & 0 & 0 \\ 0 & C_2 & X_3 & C_3 & \ldots & & \\ & \ldots & & & \ldots & & \\ 0 & 0 & 0 & 0 & \ldots & C_{n-1} & X_n \end{vmatrix} = 0,$$

where the C's are given by (13·61).

The group being finite, let h denote the period of $R_1R_2\ldots R_n$. Then the characteristic roots are powers of a primitive hth root of unity, as in (13·52). For each such root $\rho = \epsilon^m$, (13·611) yields

$$(13{\cdot}63) \qquad X_\lambda = \sin\left(\frac{m\pi}{h}+\frac{\pi}{p_\lambda}\right).$$

When $n = 2$, the equation (13·62) reduces to $X_1X_2 = C_1^2$, that is,

$$\sin\left(\frac{m\pi}{h}+\frac{\pi}{p_1}\right)\sin\left(\frac{m\pi}{h}+\frac{\pi}{p_2}\right) = \tfrac{1}{2}\left\{\cos\left(\frac{\pi}{p_1}-\frac{\pi}{p_2}\right)+\cos\frac{2\pi}{q}\right\}$$

or

$$-\cos\left(\frac{2m\pi}{h}+\frac{\pi}{p_1}+\frac{\pi}{p_2}\right) = \cos\frac{2\pi}{q}$$

or

$$\frac{2m}{h}+\frac{1}{p_1}+\frac{1}{p_2}-1 \equiv \pm\frac{2}{q} \pmod 2.$$

(In (11·31) we allowed q to be fractional.) Recalling (9·85), we deduce

$$\frac{m+1}{h} \equiv \frac{2}{q} \text{ or } 0 \pmod 1,$$

$$m \equiv \frac{2h}{q} - 1 \text{ or } -1 \pmod h,$$

and we naturally define

$$(13{\cdot}64) \qquad m_1 = \frac{2h}{q} - 1, \quad m_2 = h - 1$$

so that, by (13·54) $\quad g = (m_1+1)(m_2+1) = 2h^2/q,$

in agreement with (9·86).

When $n = 3$, it is convenient to use the temporary abbreviations

$$\eta_\lambda = \frac{m\pi}{h}+\frac{\pi}{p_\lambda}, \quad Q_\lambda = \cos\frac{2\pi}{q_\lambda}$$

so that, by (13·61) and (13·63),

$$2C_\lambda^2 = \cos(\eta_\lambda - \eta_{\lambda+1}) - Q_\lambda, \quad X_\lambda = \sin\eta_\lambda.$$

Now (13·62) equates to zero the expression

$$\begin{aligned} 2(C_1^2X_3 - X_1X_2X_3 + C_2^2X_1) &= \{\cos(\eta_1-\eta_2)+Q_1\}\sin\eta_3 - 2\sin\eta_1\sin\eta_2\sin\eta_3 \\ &\qquad + \{\cos(\eta_2-\eta_3)+Q_2\}\sin\eta_1 \\ &= (\cos\eta_1\cos\eta_2+Q_1)\sin\eta_3 + (\cos\eta_2\cos\eta_3+Q_2)\sin\eta_1 \\ &= \sin(\eta_1+\eta_3)\cos\eta_2 + Q_1\sin\eta_3 + Q_2\sin\eta_1 \\ &= \tfrac{1}{2}\sin(\eta_1+\eta_2+\eta_3) + \tfrac{1}{2}\sin(\eta_1-\eta_2+\eta_3) + Q_1\sin\eta_3 + Q_2\sin\eta_1. \end{aligned}$$

For instance, in the case of the group

$$2[4]3[3]3$$

(§12·4), we have $p_1 = 2$, $p_2 = p_3 = 3$, $Q_1 = 0$, $Q_2 = -\frac{1}{2}$, and $\eta_2 = \eta_3$, so that $\frac{1}{2}\sin(\eta_1 - \eta_2 + \eta_3) + Q_2 \sin\eta_1 = 0$ and the expression to be equated to zero is simply

$$\sin(\eta_1+\eta_2+\eta_3) = \sin\left(\frac{3m\pi}{h}+\frac{\pi}{2}+\frac{2\pi}{3}\right) = -\sin\left(\frac{3m\pi}{h}+\frac{\pi}{6}\right).$$

Hence

$$\frac{3m}{h}+\frac{1}{6} = 1 \text{ or } 2 \text{ or } 3,$$

$$\frac{m}{h} = \frac{5}{18} \text{ or } \frac{11}{18} \text{ or } \frac{17}{18},$$

$$(13{\cdot}65) \qquad \begin{aligned} &h = 18; \quad m_1 = 5, \quad m_2 = 11, \quad m_3 = 17; \\ &r = (m_1+1,\ m_2+1,\ m_3+1) = 6, \\ &g = (m_1+1)(m_2+1)(m_3+1) = 1296. \end{aligned}$$

Since $r = 6$ and $h = 18$, the 6-symmetric polyhedra

$$2\{4\}3\{3\}3 \quad \text{and} \quad 3\{3\}3\{4\}2$$

have skew 18-gons for their Petrie polygons, appearing in Figures 12·4A and 12·4B as the peripheral {18}'s.

For the group

$$2[3]2\ldots[3]2[4]r$$

of the polytopes β_n^r and γ_n^r, we have

$$C_1 = \ldots = C_{n-2} = \tfrac{1}{2}, \quad 2C_{n-1}{}^2 = \sin \pi/r,$$

$$X_1 = \ldots = X_{n-1} = \sin\left(\eta + \frac{\pi}{2}\right) = \cos\eta, \quad X_n = \sin\left(\eta+\frac{\pi}{r}\right),$$

where now

$$\eta = m\pi/h.$$

In terms of the Chebyshev polynomial

(13·66)
$$U_n(x) = \begin{vmatrix} 2x & 1 & 0 & 0 & \ldots & 0 & 0 \\ 1 & 2x & 1 & 0 & \ldots & 0 & 0 \\ 0 & 1 & 2x & 1 & \ldots & 0 & 0 \\ & & \cdots & & \cdots & & \\ 0 & 0 & 0 & 0 & \ldots & 1 & 2x \end{vmatrix} = \frac{\sin(n+1)\eta}{\sin\eta},$$

where $x = \cos\eta$ (Coxeter 1963, p. 222), (13·62) becomes

$$X_n U_{n-1}(\cos\eta)/2^{n-1} - C_{n-1}{}^2 U_{n-2}(\cos\eta)/2^{n-2} = 0,$$

$$\sin\left(\eta+\frac{\pi}{r}\right)\sin n\eta - \sin\frac{\pi}{r}\sin(n-1)\eta = 0,$$

$$-\cos\left((n+1)\eta+\frac{\pi}{r}\right) + \cos\left((n-1)\eta+\frac{\pi}{r}\right) = 0,$$

$$(n+1)\eta+\frac{\pi}{r}+(n-1)\eta+\frac{\pi}{r} = 2\lambda\pi \quad (\lambda = 1, \ldots, n),$$

$$\eta = \left(\lambda-\frac{1}{r}\right)\frac{\pi}{n}, \quad \frac{m_\lambda}{h} = \left(\lambda-\frac{1}{r}\right)\frac{1}{n},$$

(13·67)
$$h = rn, \quad m_\lambda = \lambda r - 1, \quad g = n!\,r^n.$$

For a group $p[q]p[q]\ldots p[q]p$, in which the p's are all equal and the q's are all equal, so that

$$C_1 = \ldots = C_{n-1} = \cos\frac{\pi}{q}, \quad X_1 = \ldots = X_n = \sin\left(\frac{m\pi}{h}+\frac{\pi}{p}\right),$$

the determinant in (13·62) becomes $U_n(x)\cos^n \pi/q$, where

$$x = \tfrac{1}{2}\sin\left(\frac{m\pi}{h}+\frac{\pi}{p}\right)\sec\frac{\pi}{q}.$$

By (13·66), the roots of the equation $U_n(x) = 0$ are

$$x = \cos\eta, \quad \eta = \frac{\lambda\pi}{n+1} \quad (\lambda = 1, \ldots, n).$$

Equating these two expressions for x, we obtain

(13·68)
$$\sin\left(\frac{m_\lambda\pi}{h}+\frac{\pi}{p}\right) = 2\cos\frac{\pi}{q}\cos\frac{\lambda\pi}{n+1}.$$

When $n > 2$, the only finite groups of this type are

$$p[3]p[3]\cdots p[3]p.$$

Setting $q = 3$ in (13·68), we obtain

$$\sin\left(\frac{m_\lambda\pi}{h}+\frac{\pi}{p}\right) = \cos\frac{\lambda\pi}{n+1} = \sin\left(\frac{\pi}{2}+\frac{\lambda\pi}{n+1}\right),$$

(13·681)
$$\frac{m_\lambda}{h} = \tfrac{1}{2}-\frac{1}{p}+\frac{\lambda}{n+1} = \frac{\lambda+1}{n+1} - \left(\frac{1}{p}+\frac{1}{n+1}-\tfrac{1}{2}\right).$$

Since R^{n+1} is central (see (13·47)), $n+1$ is a divisor of h, and h is the denominator of the fraction

$$\frac{1}{p}+\frac{1}{n+1}-\frac{1}{2}$$

when reduced to its simplest form. By (3·14), this fraction is the reciprocal of the number of edges of the spherical tessellation $\{p, n+1\}$; hence h is this number itself. Substituting

$$\frac{1}{p}+\frac{1}{n+1}-\frac{1}{2}=\frac{1}{h}$$

in (13·681), we obtain

$$\frac{m_\lambda+1}{h} = \frac{\lambda+1}{n+1},$$

(13·69)
$$r = \frac{h}{n+1}, \quad m_\lambda = (\lambda+1)r-1, \quad g = (n+1)!\,r^n.$$

But $\{p, n+1\}$, having h edges, has $2h/(n+1) = 2r$ vertices. Thus the r in (13·69) may be described as *half the number of vertices of* $\{p, n+1\}$. (This justifies the remark at the end of §13·4.)

In particular, when $p = 2$, so that the group is $\mathfrak{S}_{n+1}$, we use the hosohedron $\{2, n+1\}$ which has only 2 vertices, and $r = 1$. More interestingly, when $p = 3$, so that the group is

$$3[3]3[3]\cdots 3[3]3 \quad (n = 2, 3, \text{or } 4),$$

the tessellation is $\{3, n+1\}$. By (2·36), $\{3, p\}$ has $12/(6-p)$ vertices, so $\{3, n+1\}$ has $12/(5-n)$. Thus $r = 6/(5-n)$, and (13·69) becomes

(13·691)
$$g = (n+1)!\left(\frac{6}{5-n}\right)^n.$$

Since the octahedron {3, 4} has 6 vertices while the icosahedron {3, 5} has 12, we see again that the Hessian polyhedron is 3-symmetric and the

Witting polytope is 6-symmetric. Since the octahedron has 12 edges while the icosahedron has 30, the Petrie polygons of those complex polytopes are a skew dodecagon and a skew triacontagon, respectively. The dodecagonal projection of the Hessian polyhedron was not used in Figure 12·3B, because it makes 3 of the 27 vertices appear to coincide†; but the frontispiece is a triacontagonal projection of the Witting polytope.

EXERCISES

1. Compute h, m_λ and g for the real groups [3, 5], [3, 4, 3], [3, 3, 5], which we are now calling

$$2[3]2[5]2, \quad 2[3]2[4]2[3]2, \quad 2[3]2[3]2[5]2.$$

2. Apply (13·68) when $n = 2$.

3. Derive (13·691) directly from (13·681), without using the spherical tessellation.

13·7 NUMERICAL PROPERTIES OF THE NON-STARRY POLYTOPES

As we saw in §12·1, Wythoff's construction (Coxeter 1963, pp. 196–9) remains valid when the space is complex: a typical vertex Π_0 lies on the line of intersection of the mirrors for $R_2, \ldots, R_n$; the cyclic group generated by R_1 transforms Π_0 into the p_1 vertices on the edge Π_1; the group $p_1[q_1]p_2$ generated by R_1 and R_2 transforms Π_0 and Π_1 into all the vertices and edges of Π_2, and so on. It follows that the number of elements of each kind can be found by the same rule as in the case of the real polytope

$$2\{q_1\}2\{q_2\}\ldots 2\{q_{n-1}\}2 = \{q_1, q_2, \ldots, q_{n-1}\}$$

(Coxeter 1963, *p.* 131). Since the vertex Π_0 lies on all the generating mirrors except the first, N_0 (the number of vertices) is equal to the index in

$$p_1[q_1]p_2[q_2]\cdots p_{n-1}[q_{n-1}]p_n$$

of the subgroup generated by $R_2, \ldots, R_n$. Being derived from the whole group by deleting the first node (and first branch) of the graph on page 117, this subgroup is

$$p_2[q_2]\ldots p_{n-1}[q_{n-1}]p_n.$$

Dually, N_{n-1} (the number of cells) is equal to the index of the subgroup

$$p_1[q_1]p_2\cdots[q_{n-2}]p_{n-1}$$

generated by $R_1, \ldots, R_{n-1}$ (that is, deleting the *last* node and last branch of the graph). In between we have elements (12·14), and their number N_μ is equal to the index of the subgroup

$$p_1[q_1]\cdots p_{\mu-1}[q_{\mu-1}]p_\mu \times p_{\mu+2}[q_{\mu+2}]\cdots p_{n-1}[q_{n-1}]p_n$$

† See p. 463 of the paper cited on page 119.

generated by all the R's except $R_{\mu+1}$ (that is, deleting the $(\mu+1)$-node and its two attached branches). For instance, when $n = 2$ we have the polygon $p_1\{q\}p_2$ for which $g = 2h^2/q$ (see (9·86)) and the subgroups are cyclic, so that

$$(13·71) \qquad N_0 = \frac{g}{p_2} = \frac{2h^2}{p_2 q}, \quad N_1 = \frac{g}{p_1} = \frac{2h^2}{p_1 q},$$

where h is given by (9·85).

The μ-dimensional elements of the generalized cube γ_n^r consist of

$$N_\mu = \frac{n!\,r^n}{\mu!\,r^\mu(n-\mu)!} = \binom{n}{\mu} r^{n-\mu}$$

generalized cubes γ_μ^r (see §12·2, Ex. 1), while the $(\mu-1)$-dimensional elements of the reciprocal β_n^r consist of

$$N_{\mu-1} = \frac{n!\,r^n}{\mu!\,(n-\mu)!\,r^{n-\mu}} = \binom{n}{\mu} r^\mu$$

regular simplexes $\alpha_{\mu-1} = \{3, 3, \ldots, 3\} = 2\{3\}2\ldots\{3\}2$.

Another numerical property of a regular polytope is M: the number of hyperplanes of symmetry. When the polytope is real, M is equal to the number of reflections

$$b_1 = \Sigma m_\lambda = \tfrac{1}{2}nh$$

(see Coxeter 1963, p. 231). But in other cases $M < b_1$ because, whenever $p_\mu > 1$, $p_\mu - 1$ reflections, such as $R_\mu{}^\lambda$ $(0 < \lambda < p_\mu)$ all share one mirror. McMullen made the remarkable observation that, for any non-starry regular polytope,

$$(13·72) \qquad M = \frac{h}{p_1} + \frac{h}{p_2} + \ldots + \frac{h}{p_n}.$$

No general explanation has so far been discovered for polytopes of more than two dimensions; but trivially $h = p_1$ and $M = 1$ when $n = 1$, and the following proof is valid when $n = 2$.

If q is odd, R_1 and R_2 are conjugate in $p_1[q]p_2$ (that is, in $p[q]p$), so each mirror contains r vertices and r edge-centres, where $r = h/q$ (see (12·18)), and $N_0 + N_1 = 2rM$. But if q is even, half the mirrors contain r vertices (each) while the rest contain r edge-centres, where now $r = 2h/q$; so $N_0 + N_1 = rM$. In either case,

$$N_0 + N_1 = (2h/q)M;$$

hence, by (13·71),

$$(13·73) \qquad M = \frac{h}{p_1} + \frac{h}{p_2}.$$

For an n-dimensional honeycomb, g is infinite. Nevertheless it remains true that the number of μ-dimensional elements in any large (but finite) part of the n-space is inversely proportional to the (finite) order of the subgroup leaving such a Π_μ fixed; i.e., the number of Π_μ's is proportional to ν_μ, where ν_μ is some convenient integer divided by the order of the direct product

$$p_1[q_1]\cdots p_{\mu-1}[q_{\mu-1}]p_\mu \times p_{\mu+2}[q_{\mu+2}]\cdots p_n[q_n]p_{n+1}.$$

For instance, when $n = 1$ we have twelve apeirogons $p\{q\}r = \delta_2^{p,r}$ (see (12·71)), for each of which we naturally write

$$\nu_0 = p, \quad \nu_1 = r.$$

In the case of the n-dimensional 'cubic' honeycomb $\delta_{n+1}^{p,r}$ (see (12·72)) we have, analogously,

$$\nu_\mu = \frac{n!\,p^n r^n}{\mu!\,p^\mu(n-\mu)!\,r^{n-\mu}} = \binom{n}{\mu} p^{n-\mu} r^\mu;$$

thus the numbers of cells γ_μ^p $(\mu = 0, 1, \ldots, n)$ are proportional to the terms of the binomial expansion

$$(p+r)^n.$$

EXERCISES

1. For the n-dimensional polytope $3\{3\}3\ldots\{3\}3$ $(n < 5)$,

$$N_\mu = 6\binom{n+1}{\mu+1}\frac{(5-\mu)^\mu(6-n+\mu)^{n-\mu-1}}{(5-n)^n} \qquad (-1 \leqslant \mu \leqslant n).$$

2. For $3\{3\}3\{3\}3\{3\}3\{3\}3$,

$$\nu_\mu = \binom{5}{\mu+1}(5-\mu)^{\mu-1}(\mu+1)^{4-\mu} \quad (0 \leqslant \mu \leqslant 4);$$

that is, $\nu_0 = \nu_4 = 1, \quad \nu_1 = \nu_3 = 80, \quad \nu_2 = 270.$

13·8 PRESENTING THE COLLINEATION GROUPS

As we saw on page 117, the centre of the symmetry group for an r-symmetric regular polytope is a cyclic group $\mathfrak{C}_r$ generated by the dilatation (12·17) which, by (13·53), is

$$(R_1R_2\ldots R_n)^{h/r}.$$

Hence the central quotient group

$$p_1[q_1]p_2[q_2]\cdots p_{n-1}[q_{n-1}]p_n/\mathfrak{C}_r,$$

which is a collineation group in complex projective $(n-1)$-space, is given by (13·44) with the extra relation

$$(R_1R_2\ldots R_n)^{h/r} = 1.$$

In particular, by (13·69), the group

$$p[3]p[3]\cdots p[3]p$$

has $h = (n+1)r$, so that we can derive its central quotient group from (13·47) by adding the relation $R^{n+1} = 1$ to obtain

(13·81) $R_1^p = R^{n+1} = (R_1R)^n = 1, \quad R_1 \rightleftarrows R^{-\mu}R_1R^\mu \quad (2 \leqslant \mu \leqslant \tfrac{1}{2}(n+1)).$

In particular, the Hessian group $3[3]3[3]3/\mathfrak{C}_3$, of order 216, has the presentation

(13·82) $R_1^3 = R^4 = (R_1R)^3 = 1, \quad R_1 \rightleftarrows R^{-2}R_1R^2.$

Similarly, $3[3]3[3]3[3]3/\mathfrak{C}_6$, the simple group of order $6^2 6! = 25920$, has the presentation†

(13·83) $R_1^3 = R^5 = (R_1R)^4 = 1, \quad R_1 \rightleftarrows R^{-2}R_1R^2.$

EXERCISES

1. Express the generators of (13·82) as permutations of the 9 inflexions of the plane cubic curve.

2. Express the generators of (13·83) as permutations of the 27 lines on the cubic surface.

13·9 INVARIANTS

We see from (13·64), (13·65), (13·67), (13·69) that the degrees of the basic invariant forms for the groups

$$p_1[q]p_2, \quad 2[4]3[3]3, \quad 2[3]2\ldots[3]2[4]r, \quad 3[3]3\cdots[3]3$$

are respectively

$$\frac{2h}{q}, \quad h; \quad 6, 12, 18; \quad r, 2r, \ldots, nr; \quad 2r, 3r, \ldots, (n+1)r,$$

where $h = 2\Big/\left(\frac{1}{p_1}+\frac{1}{p_2}+\frac{2}{q}-1\right)$ in the first case and $r = \frac{6}{5-n}$ in the last case.

Accordingly, it is of some interest to make a list of the invariant forms themselves.

In the case of $p_1[q]p_2$, Shephard and Todd‡ expressed the forms of

† See p. 95 of Factor groups of the braid group, cited on page 149 (first column).

‡ See pp. 284–6 of the paper cited on page 149 (second column). Shephard and Todd gave also analogous expressions for the invariant forms of the groups

$\langle 3, 3, 2\rangle_6$, $GL(2, 3)$, $\langle 4, 3, 2\rangle_2$, $\langle 4, 3, 2\rangle_6$, $\langle 4, 3, 2\rangle_{12}$, $\langle 5, 3, 2\rangle_2$, $\langle 5, 3, 2\rangle_{30}$,

which are their Nos. 7, 12, 13, 15, 11, 22, 19. For the group $G(pq, q, 2)$, they found $u_1^{pq}+u_2^{pq}$ and $u_1^p u_2^p$. For all these 3-generator groups, they still denoted the degrees by m_1+1 and m_2+1; but to cover such cases our (13·62) must be replaced by the more complicated equation 6·1 on p. 253 of Coxeter, Groups generated by unitary reflections of period two, *Canadian Journal of Mathematics* **9** (1957), 243–72.

degrees $2h/q$ and h as powers of Klein's invariants for the polyhedral groups:

$$\Phi = (u_1^2+u_2^2)^2+\omega u_1^2u_2^2, \quad t = u_1u_2(u_1^4-u_2^4);$$

$$W = u_1^8+u_2^8+14u_1^4u_2^4, \quad \chi = u_1^{12}+u_2^{12}-33u_1^4u_2^4(u_1^4+u_2^4);$$

$$f = u_1u_2(u_1^{10}-u_2^{10}+11u_1^5u_2^5),$$

$$H = -(u_1^{20}+u_2^{20})+228u_1^5u_2^5(u_1^{10}-u_2^{10})-494u_1^{10}u_2^{10},$$

$$T = u_1^{30}+u_2^{30}+522u_1^5u_2^5(u_1^{20}-u_2^{20})-10005u_1^{10}u_2^{10}(u_1^{10}+u_2^{10})$$

(Klein 1913, pp. 54–61; Miller, Blichfeldt and Dickson 1916, pp. 225–6). For the details, see Table II on page 156.

The 'generalized symmetric group' $2[3]2\ldots[3]2[4]r$, which is the $G(r,1,n)$ of Shephard and Todd, has for basic invariant forms the n elementary symmetric functions of $u_1^r, u_2^r, \ldots, u_n^r$.

For $3[3]3[3]3$, of order $6\times 9\times 12$, the three forms† are

$$C_6 = (u_1^3+u_2^3+u_3^3)^2-12(u_2^3u_3^3+u_3^3u_1^3+u_1^3u_2^3),$$

$$C_9 = (u_2^3-u_3^3)(u_3^3-u_1^3)(u_1^3-u_2^3),$$

$$C_{12} = (u_1^3+u_2^3+u_3^3)\{(u_1^3+u_2^3+u_3^3)^3+216u_1^3u_2^3u_3^3\}.$$

† See p. 287 of the paper by Shephard and Todd cited on page 149. See also Miller, Blichfeldt and Dickson 1916, p. 253.

For the closely related group $2[4]3[3]3$, of order $6\times 12\times 18$, they are

$$C_6, \quad C_{12}, \quad C_9^2.$$

The four basic invariant forms for $3[3]3[3]3[3]3$, of order

$$12\times 18\times 24\times 30,$$

are so complicated that we may be content to refer to Maschke's formulae (32)‡, remarking that his z_0, z_1, z_2, z_3 are our u_4, u_1, u_2, u_3.

EXERCISES

1. Verify that the generator R_1 of $2[4]3[3]3$ (see (12·41)) alters C_9 but leaves C_9^2 invariant.

2. Find invariants of degrees 2, 6, 10 for [3, 5]; of degrees 2, 6, 8, 12 for [3, 4, 3].

‡ See p. 337 of his paper cited on page 140 (second footnote).

Tables

Table I lists the spherical triangles which can be repeated, by reflections in their sides, so as to cover the sphere a finite number of times. $(p\,q\,r)$ means a triangle whose angles are π/p, π/q, π/r. (See §§2·5, 11·1.)

In Table II, $p_1[q]p_2$ is the group defined in (9·81). Its order is given by (9·85) and (9·86). The order of the centre is $(2, q)h/q$, where $(2, q)$ means the greatest common divisor of 2 and q. By (13·64), $2h/q$ and h are the degrees of the two basic invariant forms.

Table III lists the plane reflection groups that do not belong to infinite families, and shows which of them occur as subgroups of others. The small numbers entered in the body of the table are indices of these subgroups. The symbols for the groups themselves are explained in Chapter 9.

In Table IV, the regular polygons (see §11·1) are listed in the third column. The angle 2σ is given by (9·82). The special angles occurring here are

$$\kappa = \tfrac{1}{2}\operatorname{arc\,sec} 3 \approx 35°, \quad \lambda = \tfrac{1}{2}\operatorname{arc\,tan} 2 \approx 32°, \quad \mu = \tfrac{1}{2}\operatorname{arc\,sin}\tfrac{2}{3} \approx 21°.$$

For h, see (9·85); for t and h', see (11·32) and (11·33).

Table V refers to §13·3 and §13·7. In the notation of page 115,

$$\phi = \angle O_0 O_n O_1, \quad \chi = \angle O_0 O_n O_{n-1}, \quad \phi = \angle O_{n-2} O_n O_{n-1}.$$

The special angles occurring here are those of Table IV along with

$$\eta = \tfrac{1}{2}\operatorname{arc\,sec} 4 \approx 38°.$$

The polytope is 'r-symmetric', g is the order of the symmetry group, and the 'van Oss polygon' is defined on page 141.

Table VI gives, for each honeycomb, the proportional numbers of vertices, edges, etc., in the manner of §13·7 (page 154). The symbol $\delta^{p,r}_{n+1}$ is defined in (12·72). The values p, r are restricted to

$$2, 2;\; 3, 2;\; 2, 3;\; 3, 3;\; 4, 2;\; 2, 4;\; 4, 4;\; 6, 2;\; 2, 6;\; 6, 3;\; 3, 6;\; 6, 6$$

(see (11·62) and §13·2), and the corresponding values of q are

$$\infty,\; 12,\; 12,\; 6,\; 8,\; 8,\; 4,\; 6,\; 6,\; 4,\; 4,\; 3.$$

Similarly, in the case of $2\{4\}r\{4\}2$ we can only have $r = 2, 3, 4, 6$, and then $q = \infty, 12, 8, 6$.

TABLE I. *The Schwarz triangles*

	$(p\ 2\ 2)$		(p rational, greater than 1)			
	$(3\ 3\ 2)$		$(3\ 2\ \frac{3}{2})$		$(2\ \frac{3}{2}\ \frac{3}{2})$	
		$(3\ 3\ \frac{3}{2})$		$(\frac{3}{2}\ \frac{3}{2}\ \frac{3}{2})$		
$(4\ 3\ 2)$		$(4\ 2\ \frac{3}{2})$		$(3\ 2\ \frac{4}{3})$		$(2\ \frac{3}{2}\ \frac{4}{3})$
	$(4\ 4\ \frac{3}{2})$		$(4\ 3\ \frac{4}{3})$		$(\frac{3}{2}\ \frac{4}{3}\ \frac{4}{3})$	
$(5\ 3\ 2)$		$(5\ 2\ \frac{3}{2})$		$(3\ 2\ \frac{5}{4})$		$(2\ \frac{3}{2}\ \frac{5}{4})$
$(5\ \frac{5}{2}\ 2)$		$(5\ 2\ \frac{5}{3})$		$(\frac{5}{2}\ 2\ \frac{5}{4})$		$(2\ \frac{5}{3}\ \frac{5}{4})$
$(3\ \frac{5}{2}\ 2)$		$(3\ 2\ \frac{5}{3})$		$(\frac{5}{2}\ 2\ \frac{3}{2})$		$(2\ \frac{5}{3}\ \frac{3}{2})$
$(5\ 3\ \frac{5}{3})$		$(5\ \frac{5}{2}\ \frac{3}{2})$		$(3\ \frac{5}{2}\ \frac{5}{4})$		$(\frac{5}{3}\ \frac{3}{2}\ \frac{5}{4})$
	$(3\ 3\ \frac{5}{2})$		$(3\ \frac{5}{3}\ \frac{3}{2})$		$(\frac{5}{2}\ \frac{3}{2}\ \frac{3}{2})$	
	$(5\ 5\ \frac{3}{2})$		$(5\ 3\ \frac{5}{4})$		$(\frac{3}{2}\ \frac{5}{4}\ \frac{5}{4})$	
	$(5\ 3\ \frac{3}{2})$		$(3\ 3\ \frac{5}{4})$		$(\frac{3}{2}\ \frac{3}{2}\ \frac{5}{4})$	
	$(3\ \frac{5}{2}\ \frac{5}{3})$		$(\frac{5}{2}\ \frac{5}{2}\ \frac{3}{2})$		$(\frac{5}{3}\ \frac{5}{3}\ \frac{3}{2})$	
		$(\frac{5}{2}\ \frac{5}{2}\ \frac{5}{2})$		$(\frac{5}{2}\ \frac{5}{3}\ \frac{5}{3})$		
		$(5\ 5\ \frac{5}{4})$		$(\frac{5}{4}\ \frac{5}{4}\ \frac{5}{4})$		

TABLE II. *The finite groups generated by two reflections*

Group	Order	Order of centre	Central quotient group	$\frac{2h}{q}$	h	Invariant forms†	
$2[q]2$	$2q$	$(2, q)$	Dihedral	2	q	$u_1^2+u_2^2$	$u_1^q+u_2^q$
$p[4]2$	$2p^2$	p	Dihedral	p	$2p$	$u_1^p+u_2^p$	$u_1^p u_2^p$
$3[3]3$	24	2	Tetrahedral	4	6	Φ	t
$3[6]2$	48	4	Tetrahedral	4	12	Φ	t^2
$3[4]3$	72	6	Tetrahedral	6	12	t	Φ^3
$4[3]4$	96	4	Octahedral	8	12	W	χ
$3[8]2$	144	6	Octahedral	6	24	t	χ^2
$4[6]2$	192	8	Octahedral	8	24	W	χ^2
$4[4]3$	288	12	Octahedral	12	24	χ	W^3
$3[5]3$	360	6	Icosahedral	12	30	f	T
$5[3]5$	600	10	Icosahedral	20	30	H	T
$3[10]2$	720	12	Icosahedral	12	60	f	T^2
$5[6]2$	1200	20	Icosahedral	20	60	H	T^2
$5[4]3$	1800	30	Icosahedral	30	60	T	H^3

† For the symbols Φ, t, W, χ, f, H, T, see page 155.

TABLE III. *The two-dimensional reflection groups and their reflection subgroups, excluding the families $p[2]r$ and $G(pq,q,2)$*

Group	Order \ Sub-group	3[3]3	3[6]2	3[4]3	$\langle 3,3,2\rangle_6$	$GL(2,3)$	4[3]4	$\langle 4,3,2\rangle_2$	3[8]2	4[6]2	4[4]3	$\langle 4,3,2\rangle_6$	$\langle 4,3,2\rangle_{12}$	$\langle 5,3,2\rangle_2$	3[5]3	5[3]5	3[10]2	5[6]2	5[4]3	$\langle 5,3,2\rangle_{30}$
3[3]3	24	1																		
3[6]2	48	2	1																	
3[4]3	72	3		1																
$\langle 3,3,2\rangle_6$	144	6	3	2	1															
$GL(2,3)$	48					1														
4[3]4	96						1													
$\langle 4,3,2\rangle_2$	96					2		1												
3[8]2	144	6		2		3			1											
4[6]2	192					4	2	2		1										
4[4]3	288	12	6	4	2		3				1									
$\langle 4,3,2\rangle_6$	288	12	6	4	2	6		3	2			1								
$\langle 4,3,2\rangle_{12}$	576	24	12	8	4	12	6	6	4	3	2	2	1							
$\langle 5,3,2\rangle_2$	240													1						
3[5]3	360	15		5											1					
5[3]5	600															1				
3[10]2	720	30	15	10	5									3	2		1			
5[6]2	1200													5		2		1		
5[4]3	1800	75		25											5	3			1	
$\langle 5,3,2\rangle_{30}$	3600	150	75	50	25									15	10	6	5	3	2	1

11-2

TABLE IV. *The regular polygons*

Group	Order	Polygon	Vertices	Edges	2σ	h	t	h'
$2[q]2$	$2q$	$2\left\{\frac{q}{d}\right\}2$	q	q	$\frac{2d\pi}{q}$	q	$\frac{q}{2d}$	q
$p[4]2$	$2p^2$	$p\{4\}2$	p^2	$2p$	$\frac{1}{2}\pi$	$2p$	2	$2p$
		$2\{4\}p$	$2p$	p^2				
$3[3]3$	24	$3\{3\}3$	8	8	$\pi-2\kappa$	6	2	4
$3[6]2$	48	$3\{6\}2$	24	16	$\frac{1}{2}\pi-\kappa$	12	3	12
		$2\{6\}3$	16	24				
		$3\{3\}2$	24	16	$\frac{1}{2}\pi+\kappa$	12	$\frac{3}{2}$	12
		$2\{3\}3$	16	24				
$3[4]3$	72	$3\{4\}3$	24	24	2κ	12	3	6
$4[3]4$	96	$4\{3\}4$	24	24	$\frac{1}{2}\pi$	12	3	6
$3[8]2$	144	$3\{8\}2$	72	48	κ	24	4	24
		$2\{8\}3$	48	72				
		$3\{\frac{8}{3}\}2$	72	48	$\pi-\kappa$	24	$\frac{4}{3}$	24
		$2\{\frac{8}{3}\}3$	48	72				
$4[6]2$	192	$4\{6\}2$	96	48	$\frac{1}{4}\pi$	24	3	24
		$2\{6\}4$	48	96				
		$4\{3\}2$	96	48	$\frac{3}{4}\pi$	24	$\frac{3}{2}$	24
		$2\{3\}4$	48	96				
$4[4]3$	288	$4\{4\}3$	96	72	$\frac{1}{2}\pi-\kappa$	24	4	12
		$3\{4\}4$	72	96				
		$4\{\frac{8}{3}\}3$	96	72	$\frac{1}{2}\pi+\kappa$	12	2	24
		$3\{\frac{8}{3}\}4$	72	96				

TABLE IV (*cont.*)

Group	Order	Polygon	Vertices	Edges	2σ	h	t	h'
$3[5]3$	360	$3\{5\}3$	120	120	2μ	30	5	10
		$3\{\frac{5}{2}\}3$	120	120	$\pi-2\mu$	30	$\frac{5}{3}$	10
$5[3]5$	600	$5\{3\}5$	120	120	2λ	30	5	10
		$5\{\frac{5}{2}\}5$	120	120	$\pi-2\lambda$	10	3	6
$3[10]2$	720	$3\{10\}2$	360	240	μ	60	5	60
		$2\{10\}3$	240	360				
		$3\{5\}2$	360	240	$\frac{1}{2}\pi-\mu$	60	$\frac{5}{2}$	60
		$2\{5\}3$	240	360				
		$3\{\frac{10}{3}\}2$	360	240	$\frac{1}{2}\pi+\mu$	60	$\frac{5}{3}$	60
		$2\{\frac{10}{3}\}3$	240	360				
		$3\{\frac{5}{2}\}2$	360	240	$\pi-\mu$	60	$\frac{5}{4}$	60
		$2\{\frac{5}{2}\}3$	240	360				
$5[6]2$	1200	$5\{6\}2$	600	240	λ	60	3	60
		$2\{6\}5$	240	600				
		$5\{5\}2$	600	240	$\frac{1}{2}\pi-\lambda$	20	$\frac{5}{2}$	20
		$2\{5\}5$	240	600				
		$5\{\frac{10}{3}\}2$	600	240	$\frac{1}{2}\pi+\lambda$	20	$\frac{5}{3}$	20
		$2\{\frac{10}{3}\}5$	240	600				
		$5\{3\}2$	600	240	$\pi-\lambda$	60	$\frac{3}{2}$	60
		$2\{3\}5$	240	600				
$5[4]3$	1800	$5\{4\}3$	600	360	$\frac{1}{2}\pi-\lambda-\mu$	60	5	30
		$3\{4\}5$	360	600				
		$5\{\frac{10}{3}\}3$	600	360	$\frac{1}{2}\pi-\lambda+\mu$	15	3	30
		$3\{\frac{10}{3}\}5$	360	600				
		$5\{3\}3$	600	360	$\frac{1}{2}\pi+\lambda-\mu$	30	$\frac{5}{2}$	15
		$3\{3\}5$	360	600				
		$5\{\frac{5}{2}\}3$	600	360	$\frac{1}{2}\pi+\lambda+\mu$	30	2	60
		$3\{\frac{5}{2}\}5$	360	600				

TABLE V. *The non-starry polyhedra and four-dimensional polytopes*

Polytope	N_0	N_1	N_2	N_3	ϕ	χ	ψ	r	g	Van Oss polygon
$\alpha_3 = 2\{3\}2\{3\}2$	4	6	4		$\frac{1}{2}\pi-\kappa$	2κ	$\frac{1}{2}\pi-\kappa$	1	24	None
$\beta_3^p = 2\{3\}2\{4\}p$	$3p$	$3p^2$	p^3		$\frac{1}{4}\pi$	$\frac{1}{2}\pi-\kappa$	κ	p	$6p^3$	$2\{4\}p$
$\gamma_3^p = p\{4\}2\{3\}2$	p^3		$3p$		κ		$\frac{1}{4}\pi$			None
$2\{3\}2\{5\}2$	12	30	20		λ	$\frac{1}{2}\pi-\lambda-\mu$	μ	2	120	None
$2\{5\}2\{3\}2$	20		12		μ		λ			None
$3\{3\}3\{3\}3$	27	72	27		$\frac{1}{4}\pi$	$\frac{1}{3}\pi$	$\frac{1}{4}\pi$	3	648	$3\{4\}2$
$2\{4\}3\{3\}3$	54	216	72		$\frac{1}{6}\pi$	$\frac{1}{4}\pi$	κ	6	1296	$2\{6\}2$
$3\{3\}3\{4\}2$	72		54		κ		$\frac{1}{6}\pi$			$3\{4\}3$
α_4	5	10	10	5	$\frac{1}{2}\pi-\eta$	2η	$\frac{1}{2}\pi-\eta$	1	120	None
β_4^p	$4p$	$6p^2$	$4p^3$	p^4	$\frac{1}{4}\pi$	$\frac{1}{3}\pi$	$\frac{1}{6}\pi$	p	$24p^4$	$2\{4\}p$
γ_4^p	p^4	$4p^3$	$6p^2$	$4p$	$\frac{1}{6}\pi$		$\frac{1}{4}\pi$			None
$2\{3\}2\{4\}2\{3\}2$	24	96	96	24	$\frac{1}{6}\pi$	$\frac{1}{4}\pi$	$\frac{1}{6}\pi$	2	1152	$2\{6\}2$
$2\{3\}2\{3\}2\{5\}2$	120	720	1200	600	$\frac{1}{10}\pi$	$\frac{1}{3}\pi-\eta$	$\eta-\frac{1}{6}\pi$	2	14400	$2\{10\}2$
$2\{5\}2\{3\}2\{3\}2$	600	1200	720	120	$\eta-\frac{1}{6}\pi$		$\frac{1}{10}\pi$			None
$3\{3\}3\{3\}3\{3\}3$	240	2160	2160	240	κ	$\frac{1}{2}\pi-\kappa$	κ	6	155520	$3\{4\}3$

($\kappa = \frac{1}{2}$ arc sec 3, $\lambda = \frac{1}{2}$ arc tan 2, $\mu = \frac{1}{2}$ arc sin $\frac{2}{3}$, $\eta = \frac{1}{2}$ arc sec 4)

TABLE VI. *The regular honeycombs*

Honeycomb	ν_0	ν_1	ν_2	ν_3	ν_4	ν_μ	Van Oss apeirogon
$\delta^{p,r}_{n+1}$						$\binom{n}{\mu} p^{n-\mu} r^{\mu}$	$p\{q\}r$
2{4}r{4}2	2	r^2	2				2{q}r
2{3}2{6}2	1	3	2				2{∞}2
2{6}2{3}2	2		1				None
3{3}3{4}3	1	8	3				3{4}6
3{4}3{3}3	3		1				3{12}2
4{3}4{3}4	1	6	1				4{4}4
2{4}4{3}4	1	12	3				2{8}4
4{3}4{4}2	3		1				4{4}4
3{3}3{3}3{4}2	1	24	27	2			3{4}6
2{4}3{3}3{3}3	2	27	24	1			2{12}3
2{3}2{4}3{3}3	1	27	72	8			2{6}6
3{3}3{4}2{3}2	8	72	27	1			3{6}3
2{3}2{3}2{4}2{3}2	1	12	32	24	3		2{∞}2
2{3}2{4}2{3}2{3}2	3	24		12	1		2{∞}2
3{3}3{3}3{3}3{3}3	1	80	270	80	1	$\binom{5}{\mu}(5-\mu)^{\mu}(\mu+1)^{3-\mu}$	3{4}6

Answers to exercises

§1·1

1. Let m_ν denote the mirror for the reflection R_ν. Suppose R_2 transforms m_1 into m_1'. The three hyperplanes m_1, m_2, m_1' are evenly spaced; R_1R_2 transforms the first into the third.

2. An orthogonal transformation, leaving a point O invariant, is determined by its effect on an arbitrarily small neighbourhood of O. In such a neighbourhood, the distinction between Euclidean and non-Euclidean geometry disappears.

In hyperbolic space, two non-intersecting hyperplanes may or may not have a common perpendicular line. If they do, they are said to be *ultraparallel*, and the product of reflections in them is still a *translation*. If they do not, they are said to be *parallel* and the product of reflections in them is a *parallel displacement* (Coxeter 1965, pp. 185, 202). In this latter case, the images of the mirrors in one another belong to a 'pencil of parallels' and are orthogonal to a family of horocycles. They intersect any one of the horocycles in points that are evenly spaced.

§1·2

1. A rhombus.

2. In both representations, $R_1^2 = \begin{bmatrix} 1 & 0 \\ 0 & 1 \end{bmatrix}$. In the former,

$$R_1R_2R_1 = \begin{bmatrix} 0 & -1 \\ -1 & 0 \end{bmatrix} = R_2R_1R_2;$$

in the latter, $R_1R_2R_1 = \begin{bmatrix} 0 & 1 \\ 1 & 0 \end{bmatrix} = R_2R_1R_2$.

3. $R_1 = \begin{bmatrix} -1 & 0 \\ 0 & 1 \end{bmatrix}$, $R_2 = \begin{bmatrix} -1 & 0 \\ 1 & 1 \end{bmatrix}$. In this case $(R_1R_2)^n = \begin{bmatrix} 1 & 0 \\ n & 1 \end{bmatrix}$.

§1·3

The Great Pyramid, the Pentagon, the Bahai Temple.

§1·4

1. Let U be the matrix for an involutory orthogonal transformation. Then $U^2 = I$ as well as $UU^T = I$; therefore U is symmetric. By Theorem (1·41), any orthogonal matrix S can be expressed as UV, where U and V are symmetric.

2. Yes. The only difference is that one or more of the rotations $Q_\nu Q_\nu'$ may have to be replaced by a parallel displacement. This means that the mirrors for Q_ν and Q_ν', instead of intersecting, are parallel.

§1·5

1. Since an isometry is determined by its effect on a simplex, the desired condition is that $n+1$ consecutive points of the orbit span the whole n-space.

2. These conditions make $A_0A_1A_2 \ldots$ congruent to $A_1A_2A_3 \ldots$.

3. Since the edges and angles of a pentagon (in any number of dimensions) determine all its diagonals, it cannot be both equilateral and equiangular without being regular.

4. In the triangle $A_{\lambda-1}A_\lambda A_{\lambda+1}$, the sides $A_{\lambda-1}A_\lambda$, $A_\lambda A_{\lambda+1}$ and the angle $A_{\lambda-1}(A_\lambda)A_{\lambda+1}$ determine $A_{\lambda-1}A_{\lambda+1}$; in the tetrahedron $A_{\lambda-1}A_\lambda A_{\lambda+1}A_{\lambda+2}$, the faces $A_{\lambda-1}A_\lambda A_{\lambda+1}$ and $A_\lambda A_{\lambda+1}A_{\lambda+2}$ and the dihedral angle $A_{\lambda-1}(A_\lambda A_{\lambda+1})A_{\lambda+2}$ determine the edge $A_{\lambda-1}A_{\lambda+2}$; and so on. For further details see Coxeter, On Schläfli's generalization of Napier's Pentagramma Mirificum, *Bulletin of the Calcutta Mathematical Society* **28** (1936), 123–44, especially p. 129.

5. Consider five mutually equidistant points (the vertices of a regular simplex). These, in any cyclic order, are the vertices of a regular skew pentagon. More generally, consider the orbit of an arbitrary point for the transformation in Exercise 6 with $\alpha = 2\pi/5$, $\beta = 4\pi/5$.

6. $$\begin{bmatrix} 1 & 0 & 0 & 0 \\ 0 & -1 & 0 & 0 \\ 0 & 0 & 1 & 0 \\ 0 & 0 & 0 & -1 \end{bmatrix} \begin{bmatrix} \cos\alpha & \sin\alpha & 0 & 0 \\ \sin\alpha & -\cos\alpha & 0 & 0 \\ 0 & 0 & \cos\beta & \sin\beta \\ 0 & 0 & \sin\beta & -\cos\beta \end{bmatrix}.$$

These matrices represent half-turns. When $\beta = \pm\alpha$, their product is a Clifford displacement, and the orbit of any point lies in a plane; for instance, when $\beta = \alpha$, the orbit of (a_1, a_2, a_3, a_4) lies in the plane

$$\frac{x_1}{a_1} = \frac{x_3}{a_3}, \quad \frac{x_2}{a_2} = \frac{x_4}{a_4}.$$

§1·6

1. The mirror for U is the perpendicular bisector of A_0A_1.

2. $l \operatorname{cosec} d\pi/p$, $l \cot d\pi/p$. The sum of the areas of the p triangles that join the centre to the edges, namely $pl^2 \cot d\pi/p$.

3. Everything, except that, when U and V are reflections in non-intersecting mirrors, u and v may be either parallel or ultraparallel. In the former case S is a parallel displacement and the apeirogon $A_0A_1A_2\ldots$ is inscribed in a horocycle. In the latter, S is a translation and the apeirogon is either straight or inscribed in a horocycle. In the latter, S is a translation and the apeirogon is either straight or inscribed in one branch of an equidistant curve (Coxeter 1969, p. 300).

§1·7

1. A parallelepiped can be derived by drawing, through each edge of the given tetrahedron, a plane parallel to the opposite edge. If the parallelepiped

is rectangular (so that its faces are rectangles) each edge of the tetrahedron has the same length as the opposite edge. Such a tetrahedron is, in general, a *rhombic disphenoid* (belonging to the 'rhombic' system of crystallography). Its unfolded net is a triangle with creases joining the mid-points of pairs of sides. The *tetragonal* disphenoid is the special case when two opposite faces of the rectangular parallelepiped are squares, so that the brick reduces to a square prism and the triangle is isosceles.

2. Since every finite group of isometries leaves invariant at least one point, every isometry of finite period is an orthogonal transformation. In three dimensions, this means that such an isometry is either a rotation or a rotatory-reflection (including a reflection as a degenerate case). Thus every properly three-dimensional regular polygon is antiprismatic: its vertices lie alternately on two circles in parallel planes.

3. Use Ex. 3 of §1·5. (For another treatment of this theorem, see B. L. van der Waerden, Ein Satz über räumliche Fünfecke, *Elemente der Mathematik* **25** (1970), 73–8.)

4. Yes. Take two squares with a common edge in perpendicular planes (like two adjacent faces of a cube) and remove the common edge. Alternate vertices form equilateral triangles (just as they would for a *regular* skew hexagon), but these triangles are not in parallel planes.

5. $$\frac{p}{d} > \frac{3}{2}.$$

This may be proved as follows. In terms of the circumradius of the 'base' $A_0A_2A_4\ldots$ as unit, the triangle $A_0A_1A_2$ has its horizontal side $A_0A_2 = 2\sin d\pi/p$ while its oblique side A_0A_1 projects into a segment of length $2\sin d\pi/2p$. Thus the condition $A_0A_1 = A_0A_2$ requires

$$\sin\frac{d\pi}{2p} < \sin\frac{d\pi}{p}.$$

§1·8

1. (i) A glide.
   (ii) A glide (or, if the axis and centre are incident, a reflection).
   (iii) A half-twist (or, if the centre is on the mirror, a half-turn).

2. When considering a displacement UV as the product of half-turns about two lines u and v, we assumed these two axes to have a common perpendicular line l. In elliptic space, as in the Euclidean plane, two lines may have an infinity of common perpendiculars (Coxeter 1965, p. 134). In hyperbolic space they may have none. This happens when they are parallel (*ibid.* p. 202); so that the product of the two half-turns is a parallel displacement and $A_0A_1A_2\ldots$ is a *horocyclic apeirogon* whose infinitely many vertices are evenly spaced along a horocycle. By expressing each of the half-turns as the product of reflections in two perpendicular planes, and choosing one of the planes to be the same for both, we can express the parallel displacement as the product of reflections in two parallel planes. The set of all planes perpendicular to both these mirrors is a *pencil* of parallel planes (*ibid* p. 185). All these planes are invariant for the parallel displacement, and any point A_0 in any one of them yields a horocyclic apeirogon.

This case when u and v are parallel is quite exceptional. In any other case, if u and v do not intersect they have a shortest distance, and any segment PQ that measures this shortest distance is perpendicular to both lines (for, if the angle at P were not a right angle, a shorter segment could be found along the perpendicular from Q to u). Clearly the half-turn about the common perpendicular reverses both lines. If u and v are skew lines in hyperbolic space, the plane uQ bisects the dihedral angle formed by the two half-planes through u parallel to v (that is, the two half-planes through u that just fail to meet v). Similarly, vP bisects the dihedral angle formed by the half-planes through v parallel to u. The common perpendicular PQ can be constructed as the line of intersection of these two angle-bisectors. (In the Euclidean case the 'dihedral angles' are π, and the 'bisectors' are planes perpendicular to the two parallel planes through u and v.)

In §1·7 we considered a rotatory-reflection. This is the product of a half-turn and a reflection. The axis of the half-turn intersects the mirror in an invariant point, which is the centre of the antiprismatic polygon. If, instead, the axis and mirror had a common perpendicular line, we could take the section by any plane through this line and obtain, in that plane, a glide. In hyperbolic space there is a third possibility: the axis and mirror may be parallel. We now have a *parallel reflection*, which is the product of a parallel displacement and the reflection in one of its invariant planes.

§2·1

1. Since great circles intersect one another in pairs of antipodal points, and antipodal points on the equator project into diametrically opposite points on the corresponding circle ϵ, each of the circles in Figure 2·1A intersects a certain circle ϵ (not drawn) in diametrically opposite points (of ϵ).

2. This can be seen by dissecting both p-gons (in corresponding ways) into $p-2$ (or more) triangles.

3. The single edge of $\{1, 2\}$ is a whole great circle; its two 'ends' coincide at the single vertex. The single edge of $\{2, 1\}$ is a great semicircle.

4. The face of $\{p, q\}$ is a spherical $\{p\}$ whose angle $2\pi/q$ must be greater than the angle $\pi-2\pi/p$ of a Euclidean $\{p\}$.

§2·2

1. There are p edges for each face, q for each vertex. By Euler's formula (Coxeter 1969, p. 152),

$$(q^{-1}-\tfrac{1}{2}+p^{-1})\,N_1 = \tfrac{1}{2}(N_0-N_1+N_2) = 1.$$

2. The values $p = q = 1$ would yield the absurd number of edges $N_1 = \frac{2}{3}$. Moreover, if $p = 1$, q can only be 2, since

$$N_1^{-1} = \tfrac{1}{2}+q^{-1}.$$

The symbol $\{0, 0\}$ is appropriate for the edgeless monohedron that has $N_1 = 0$, $N_2 = N_0 = 1$, like the chart in Lewis Carroll's *The Hunting of the Snark*.

3. Congruent regular faces of an irregular polyhedron can only be triangles. The number of triangles may be 6, 10, 12, 14 or 16. The first two of these *deltahedra* are the triangular and pentagonal dipyramids; the remaining three were discovered by Hans Freudenthal and B. L. van der Waerden (Cundy and Rollett 1961, p. 136).

4. $\{h\}$.

5. One less than the number of vertices. Consider the reversed process of cutting the complete solid before unfolding it to produce the net. The edges

that have to be cut form a spanning tree or 'scaffolding' whose vertices are all the vertices of the polyhedron. If a tree has N_0 vertices, it has $N_0 - 1$ edges.

§2·3

1. For the first (or classical) rhombic dodecahedron, whose faces have diagonals in the ratio $\sqrt{2}:1$, the symmetry group is [4, 3]. For Bilinski's second rhombic dodecahedron, whose faces have diagonals in the ratio $\tau:1$, it is [2, 2].

2.

p	A	B	C
2	(ab)	(abc)	(ac)
3	(abc)	(abd)	$(ad)\,(bc)$
4	$(abcd)$	(adb)	(bc)
5	$(abcde)$	(adb)	$(bc)\,(de)$

(Coxeter 1970, p. 5).

3. The axes of the given rotations, say OP and OQ, meet the unit sphere round O in pairs of antipodal points. Choose P from the first pair and Q from the second so that the rotations about P and Q are counterclockwise, through angles α and β ($\alpha, \beta \leqslant \pi$). Draw great circles PR and QR so that PQR is a clockwise spherical triangle with angles $\frac{1}{2}\alpha$ at P, $\frac{1}{2}\beta$ at Q. Suppose the angle at R is $\frac{1}{2}\gamma$. Since the first rotation is the product of reflections in RP and PQ while the second is the product of reflections in PQ and QR, the product of the two rotations is the clockwise rotation through γ about R.

4. More generally, for any spherical triangle XYZ, a 'sufficiently small' rotation about Z can be expressed as a product ABC, where A and C are rotations about X while B is a rotation about Y. To see this, let R and U denote the reflections in XZ and YZ, and suppose the given rotation about Z is RV, its angle being so small that the mirror for V intersects the circle with centre X and radius XY. Let P be either of the points of intersection, and let S be the reflection that interchanges Y and P; let the mirrors for V and S intersect at Q, and let T be the reflection in YQ. Then SV = TS = TU US and

$$\text{RV} = \text{RS}\,\text{SV} = \text{RS TU US} = \text{ABC},$$

where A = RS, B = TU, C = US. In particular, if $XY = \frac{1}{2}\pi$, the circle with centre X and radius XY is a great circle and so cannot fail to intersect the mirror for V.

For an interesting extension of this result, see Franklin Lowenthal, Uniform finite generation of the isometry groups of Euclidean and non-Euclidean geometry, *Canadian Journal of Mathematics* **23** (1971), 364–73.

§2·4

1. (i) A line-segment, whose two ends are an arbitrary point and its image in a single mirror.
(ii) A square, formed by the images, in two perpendicular mirrors, of a point equidistant from them.
(iii) A cube. (Three mutually perpendicular mirrors.)
(iv) A triangle, formed by the images, in two mirrors at 60°, of a point on the 'second' mirror.
(v) A regular hexagon.
(vi) A regular octagon.
(vii) A triangular prism.
(viii) Another cube.
(ix) A hexagonal prism.
(x) A regular tetrahedron.
(xi) A regular octahedron.
(xii) A square antiprism.

2.

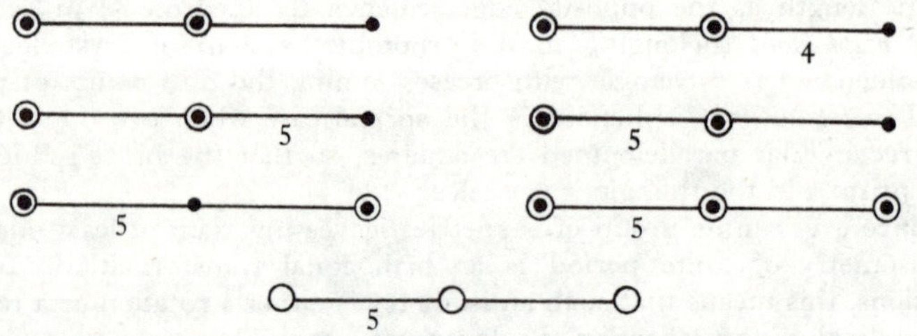

3. (i) Octahedron. (ii) Cuboctahedron.
(iii) Truncated octahedron. (iv) Icosahedron.

4. By the classical formula $\sin a = \sin c \sin A$ for a spherical triangle ABC, right-angled at C, the sine of the inradius may be expressed equally well by each of the three given expressions.

5. The chosen points inside the three white triangles adjacent to one black triangle are the vertices of an equilateral triangle whose sides η_ν satisfy

$$\sin\frac{\eta_1}{2} = \sin\xi_1 \sin\frac{\pi}{p}, \ldots .$$

§2·5

1. For $\left(\frac{n}{d}\frac{n}{d}\,1\right)$ the group is $[n]$; for $\left(\frac{n}{d}\,2\,2\right)$ it is $[n, 2]$. Otherwise it is [3, 3] or [4, 3] or [5, 3] according as the greatest numerator that occurs is 3 or 4 or 5.

2. The areas of the triangles are respectively

$$(p^{-1}+q^{-1}+r^{-1}-1)\,\pi \quad \text{and} \quad s^{-1}\pi.$$

3. The resulting 'pattern' will cover the sphere infinitely many times.

4. Extend the arc RP to T so that $\angle PQT = \pi/q$ (and QP bisects $\angle RQT$), and define $\theta = \angle PTQ$. Comparing the spherical triangles PQR and PQT, we see that

$$\cos\frac{\pi}{p}\cos\frac{\pi}{q}+\cos\frac{\pi}{r} = \cos PQ \sin\frac{\pi}{p}\sin\frac{\pi}{q}$$

$$= \cos PQ \sin\left(\pi-\frac{\pi}{p}\right)\sin\frac{\pi}{q} = \cos\left(\pi-\frac{\pi}{p}\right)\cos\frac{\pi}{q}+\cos\theta,$$

whence

$$\cos\theta = 2\cos\frac{\pi}{p}\cos\frac{\pi}{q}+\cos\frac{\pi}{r}.$$

Since QT is one of the great circles in the pattern generated by the Schwarz triangle PQR, PQT is another Schwarz triangle, and $\theta = \pi/t$, where t is one of the numbers (2·51).

5. 2, 3, 6, 15. The group [4, 3] contains [3, 3], [4, 2] and [2, 2] as subgroups of indices 2, 3, 6, and [5, 3] contains [2, 2] as a subgroup of index 15.

§3·1

1.

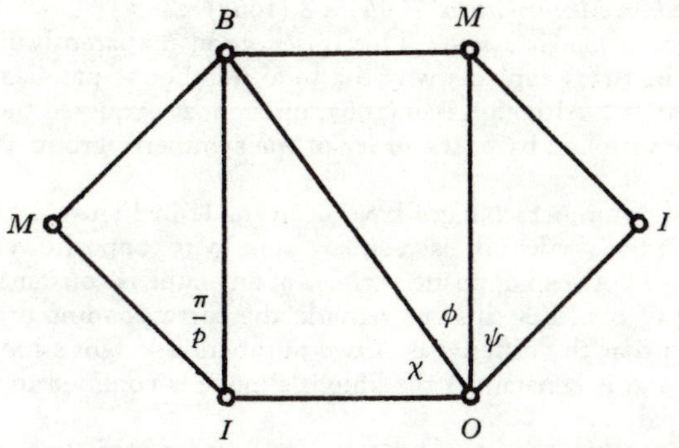

2. Two of the classical formulae for a right-angled spherical triangle are

$$\sin c \sin A = \sin a, \quad \cos a \sin B = \cos A$$

(Coxeter 1965, p. 234). Applying the former to the triangle CBD, we obtain

$$\sin a \sin B = \sin CD.$$

Hence, by squaring and adding,

$$\sin^2 B = \cos^2 A + \sin^2 CD.$$

3. cosec $\pi/h = \sqrt{(1+\gamma+\epsilon)}$. In fact, by (2·22),

$$\sin^2\frac{\pi}{h} = \sin^2\frac{\pi}{p} - \cos^2\frac{\pi}{q} = \frac{\alpha}{\gamma\delta} - \frac{1}{\delta\epsilon} = \frac{\epsilon\alpha-\gamma}{\gamma\delta\epsilon} = \frac{1}{\gamma\delta\epsilon}$$

and $\gamma\delta\epsilon = \gamma(1+\beta) = \gamma+1+\epsilon$.

4. When the two reciprocal solids are placed in corresponding positions, the roles of P and Q are interchanged. Thus the arc $\chi = PQ$ (which is the angle subtended at O by the circumradius of a face) is the same for both. (Notice that (3·12) involves p and q symmetrically.) The surface area can be expressed as

$$S = 2sCM \times MI = 2s\left(CI \sin\frac{\pi}{p}\right)\left(CI \cos\frac{\pi}{p}\right) = s\, CI^2 \sin\frac{2\pi}{p}.$$

Since s (the number of edges) and CI (the circumradius of a face) are the same for $\{q, p\}$ as for $\{p, q\}$, it is $\{p, q\}$ that has the greater area†. (This looks wrong at first, till we recall that $2\pi/q$ is obtuse!)

5. Let ρ_2 be $Xx+Yy+Zz = 0$, where $X^2+Y^2+Z^2 = 1$. Then

$$X = \cos\theta_{12} = \cos\frac{\pi}{p} = \sqrt{\frac{1}{\gamma\delta}}, \quad Z = \cos\theta_{23} = \cos\frac{\pi}{q} = \sqrt{\frac{1}{\delta\epsilon}}$$

and, by (2·22),

$$Y^2 = 1 - X^2 - Z^2 = 1 - \cos^2\frac{\pi}{p} - \cos^2\frac{\pi}{q} = \sin^2\frac{\pi}{h} = \frac{1}{\gamma\delta\epsilon}.$$

Thus ρ_2 has the equation $\quad (\sqrt{\epsilon})\, x + y + (\sqrt{\gamma})\, z = 0.$

For instance, in the case of (5 3 2) we have the plane $\tau^2 x + y + \tau z = 0$ or

$$\tau x + \tau^{-1} y + z = 0,$$

which contains the four points

$$(1, 0, -\tau), \quad (0, \tau, -1), \quad (-1, 0, \tau), \quad (0, -\tau, 1),$$

and is thus a plane of symmetry of the icosahedron whose vertices are the cyclic permutations of $(\pm\tau, \pm 1, 0)$ (Coxeter 1969, p. 163).

6. If a pattern of integers includes

$$\begin{matrix} & b & \\ a & & d \\ & c & & f \\ & & e & \end{matrix}$$

the 'unimodular' rule yields $ad-bc = 1 = cf-de$, whence $d(a+e) = (b+f)\,c$. Since d and c are relatively prime, c divides $a+e$. Thus, in any diagonal (in either direction), each number divides the sum of its two neighbours. If the numbers in a diagonal are 1, a, c, 1, then a and c must be 1 and 1, or 1 and 2, or 2 and 3, or 3 and 2, or 2 and 1.

For patterns having three non-trivial rows, we find

1		1		1		1		1
	1		3		1		3	
2		2		2		2		2
	3		1		3		1	
1		1		1		1		1

1		1		1		1		1		1
	1		2		3		1		2	
2		1		5		2		1		5
	1		2		3		1		2	
1		1		1		1		1		1

and the latter reversed (right to left). One diagonal determines the whole pattern. Wider patterns admit far more variety and provide an entertaining pastime for children, who can use crayons to adorn the patterns with a different colour for each number. When diagonal dividing lines have been drawn, each number stands in the middle of its coloured 'diamond' and the surprising periodicity gives the pattern an artistic appeal.

7. Assuming $\alpha, \beta, \ldots$ to be positive, we see that the equations

$$\begin{vmatrix} \alpha & 1 & 0 \\ 1 & \beta & 1 \\ 0 & 1 & \gamma \end{vmatrix} = 1 \quad \text{and} \quad \begin{vmatrix} \beta & 1 & 0 \\ 1 & \gamma & 1 \\ 0 & 1 & \delta \end{vmatrix} = 1$$

imply $\quad \dfrac{\beta+1}{\delta} = \beta\gamma - 1 = \beta\dfrac{\alpha+1}{\alpha\beta-1} - 1 = \dfrac{\beta+1}{\alpha\beta-1},$

whence $1+\delta = \alpha\beta$, as in (3.13). Thus we obtain Gauss's $\alpha, \beta, \gamma, \delta, \epsilon$ over again, and the pattern is of period 5.

This is the case $m = 3$ of the following theorem (Coxeter, Frieze patterns, *Acta Arithmetica* **18** (1971), 307): *Let $m-1$ positive numbers $a_1, \ldots, a_{m-1}$ be given, and let the sequence be continued so that*

$$\begin{vmatrix} a_{r+1} & 1 & 0 & \ldots & 0 & 0 \\ 1 & a_{r+2} & 1 & \ldots & 0 & 0 \\ & \ldots & & \ldots & & \\ 0 & 0 & 0 & \ldots & 1 & a_{r+m} \end{vmatrix} = 1 \quad (r = 0, 1, \ldots);$$

then the sequence is periodic: $a_r = a_{r+m+2}$.

See also D. P. Roselle, Generalized frieze patterns, *Duke Mathematical Journal* **39** (1972), 637–48 and my review thereof in *Mathematical Reviews*.

† See an Editorial Note in the *American Mathematical Monthly* **76** (1969), 192.

§3·2

1. (i) A rod perpendicular to the mirror *COI*, with its other end at *A* (or any other point on *OM*).
(ii) A line-segment perpendicular to *OC* on the mirror *COI*, with its other end at *B* (or any other point on *OI*).

2. In the demonstration of $\{5, \frac{5}{2}\}$, the images of *AB* are true edges, belonging also to $\{3, 5\}$, but the images of *C′B* are 'false edges': intersections of non-adjacent face-planes. The great icosahedron has the same edges as $\{\frac{5}{2}, 5\}$, but it also has a bewildering variety of false edges, many of which do not lie in any plane of symmetry.

3. τ^{-2}, τ^{-3}.

§3·3

1. (i) $A^4 = C^2 = (AC)^3 = 1$, the truncated octahedron.
(ii) $B^3 = C^2 = (BC)^4 = 1$, the truncated cube.

2. The diagram for [4, 3] consists of the 48 vertices and 72 edges (without arrows) of the truncated cuboctahedron. More generally, for each reflection group the Cayley diagram is the polyhedron or tessellation obtained by putting rings round all the nodes in the graph.

§3·4

1. Comparing (3·32) with (3·34), we see that $CBA = (R_3R_1R_2)^2 = S^2$. (See (3·43).) Of course, the other commutators, $B^{-1}C^{-1}BC$ and $C^{-1}A^{-1}CA$, have the same period.

2. When $\frac{1}{2}p$ is odd, so that

$$(R_1R_2R_3)^{p/2} = (R_1R_2)^{p/2}\, R_3{}^{p/2} = (R_1R_2)^{p/2}\, R_3.$$

3. $U_\nu{}^2 = 1$, $(U_1U_2)^{p/2} = (U_2U_1)^{p/2} = U_3$. (These relations suffice to make U_3 commute with U_1 and U_2.)

§4·1

1.

	{5, 3, 2}	{2, 3, 5}	{3, 5, 2}	{2, 5, 3}	{5, 2, 3}	{3, 2, 5}
N_0	20	2	12	2	5	3
N_1	30	12	30	20	5	3
N_2	12	30	20	30	3	5
N_3	2	20	2	12	3	5

2. By (4·13),

$$N_0:N_1:N_2:N_3 = \frac{1}{t}:\frac{1}{r}:\frac{1}{p}:\frac{1}{s} = \frac{1}{q}+\frac{1}{r}-\frac{1}{2}:\frac{1}{r}:\frac{1}{p}:\frac{1}{p}+\frac{1}{q}-\frac{1}{2}.$$

§4·2

1. Equilateral. To prove this let the angles at the point *2* be 150°, α, β, α, Suppose, if possible, that $\beta \gtrless 60°$. Then $13 = 35 \gtrless 23$, and the triangle *123* yields $\alpha \gtrless 75°$, whence

$$150° + 2\alpha + \beta \gtrless 360°,$$

which is absurd. (See Coxeter 1969, p. 9, Ex. 3.)

2. $2+\sqrt{2}+\sqrt{6}$. The centres of the eight discs are arranged like the points *1*, ..., *8* in Figure 4·2A. (See J. Schaer and A. Meir, On a geometric extremum problem, *Canadian Mathematical Bulletin* **8** (1965), 21–7.)

3. A *pandiagonal magic square*. This observation (apparently first made by J. H. Conway in 1957) explains why the total number of pandiagonal squares of order 4 is just 384. Although Ball (1967, pp. 202, 204) quoted the number 48, this has to be multiplied by 8, the order of the symmetry group $\mathfrak{D}_4$ of the geometric square.

After the same number-changes have been made in Figure 4·2B, we have a 'magic 4-cube': the vertices of each square sum to 30, opposite vertices sum to 15, the difference between opposite vertices of any cube is constant for the cube and is a power of two. For this last remark, the corresponding property of the magic square is that the difference of two numbers a knight's move apart is a power of two, and is constant if the knight's move is confined to two adjacent rows or columns.

4. The configuration formed by three tetrahedra, any two of which are quadruply perspective, in fact, perspective from each vertex of the third. This configuration was discovered independently by C. Stephanos (in 1879) and T. Reye (see Hilbert and Cohn-Vossen 1952, pp. 134–43, 154–7).

5. Because the vertices of each occur among the points of a *lattice* (Coxeter 1963, pp. 122, 123, 158, 205). Among the points in Euclidean 5-space that have 5 integral Cartesian coordinates with sum zero,

$$(1, 0, 0, 0, -1),\quad (0, 1, 0, 0, -1),\quad (0, 0, 1, 0, -1),\quad (0, 0, 0, 1, -1),$$
$$(0, 0, 0, 0, 0),$$

form a regular simplex {3, 3, 3}. The case of {4, 3, 3} is obvious. {3, 3, 4} belongs to a lattice {3, 3, 4, 3} consisting of points that have 4 integral Cartesian coordinates with an even sum. {3, 4, 3} belongs to the honeycomb {3, 4, 3, 3} whose vertices are a subset of this lattice, as their coordinates are two even and two odd (Coxeter 1963, p. 158).

6. Yes. The $n+1$ points

$$(n, 0, 0, \ldots, 0),\quad (0, n, 0, \ldots, 0),\quad \ldots,\quad (0, 0, 0, \ldots, n),\quad (x, x, x, \ldots, x)$$

are equidistant if $x = 1 \pm \sqrt{(n+1)}$. This is an integer whenever $n+1$ is a perfect square. A more complicated choice of coordinates will solve the problem whenever $\frac{1}{4}(n+1)$ is an integer less than 47 (see Ball 1967, p. 110).

§4.3

1. After (1, 1, 1), the vertices are (2, 1, 1), (2, 2, 1), (2, 2, 2), (3, 2, 2), ..., derived from one another by means of the twist

$$x' = z+1,\quad y' = x,\quad z' = y.$$

Calling this helical polygon *ABCDEF* ..., we observe that *ABCD, BCDE, CDEF* are three directly congruent orthoschemes which fit together to form an oblique triangular prism *ABCDEF* (whose edges *AD*, *BE*, *CF* are congruent and parallel). But any triangular prism, right or oblique, is a space-filler. Hence this simplest orthoscheme is a space-filler.

2. We can derive ψ from the triangle $P_1P_2P_3$, or by reciprocation from (4·34). From the triangle $P_0P_1P_3$,

$$\cos\chi = \cot\phi_{p,q}\cot\chi_{q,r}$$
$$= \left(\cos\frac{\pi}{p}\Big/\sin\frac{\pi}{h_{p,q}}\right)\left(\cos\frac{\pi}{q}\cos\frac{\pi}{r}\Big/\sin\frac{\pi}{h_{q,r}}\right).$$

§4·4

1. As we see from Figure 4·4A, the area is that of a square of side $(\sqrt{2})\,\pi$, namely $2\pi^2$.

2. (i) $\dfrac{x_4}{x_1} = \tan\alpha = \dfrac{x_3}{x_2}$ or $\dfrac{x_2}{x_1} = \tan\beta = \dfrac{x_3}{x_4}$.

(ii) $x_1 = \sqrt{\tfrac{1}{2}}\cos\xi$, $x_2 = \sqrt{\tfrac{1}{2}}\sin\xi$, or $x_3 = \sqrt{\tfrac{1}{2}}\cos\eta$, $x_4 = \sqrt{\tfrac{1}{2}}\sin\eta$.

3. Elliptic space (like real projective space) is derived from spherical space by identifying antipodes, that is, by regarding the four x_ν as homogeneous coordinates. This kind of identification reduces the spherical torus (4·41) or (4·44) to the rectangular Clifford surface

$$x_1x_3 = x_2x_4 \quad \text{or} \quad {x_1}^2+{x_2}^2 = {x_3}^2+{x_4}^2$$

(which is, of course, a ruled quadric in the projective space). The two systems of great circles on the spherical torus yield the two systems of generators on the Clifford surface. Any two generators of the same system are Clifford parallel to each other. The same identification reduces the outer square in Figure 4·4A to the inner square $(\pm\tfrac{1}{2}\pi, \pm\tfrac{1}{2}\pi)$, whose pairs of opposite sides coincide to form a smaller torus. Thus the area of the Clifford surface is π^2.

Clifford parallel lines (the elliptic counterpart of isocline planes) were first described in 1873 (see Clifford 1882, p. 193).

§4·5

Since the four vertices have the real coordinates

$$\left(\sqrt{\tfrac{1}{2}}\cos\frac{\pi}{p},\quad \pm\sqrt{\tfrac{1}{2}}\sin\frac{\pi}{p},\quad \sqrt{\tfrac{1}{2}}\cos\frac{\pi}{p},\quad \pm\sqrt{\tfrac{1}{2}}\sin\frac{\pi}{p}\right),$$

their plane has the equations

$$x_1 = x_3 = \sqrt{\tfrac{1}{2}}\cos\frac{\pi}{p}.$$

§4·6

1. When $\mu+\nu$ is odd, the vertices adjacent to $2\mu|2\nu$ are

$$|2\nu\pm 1,\quad 2\mu\pm 4|2\nu,\quad 2\mu\pm 2|2\nu,\quad 2\mu\pm 2|2\nu\pm 2,\quad 2\mu|2\nu\pm 2.$$

Hence, in the case of $4|2$ they are

$$|1,\ \ |3,\ \ 0|2,\ \ 2|2,\ \ 6|2,\ \ 8|2,\ \ 2|0,\ \ 4|0,\ \ 6|0,\ \ 2|4,\ \ 4|4,\ \ 6|4.$$

2. The *grand antiprism* discovered by J. H. Conway: the only four-dimensional convex uniform polytope that cannot be obtained by Wythoff's construction. Since each vertex belongs to 10 edges, 21 triangles, 1 pentagon, 12 tetrahedra and 2 pentagonal antiprisms (such as the antiprism formed by the points

$$0|0,\ \ 2|2,\ \ 0|4,\ \ 2|6,\ \ 0|8,\ \ 2|10,\ \ 0|12,\ \ 2|14,\ \ 0|16,\ \ 2|18$$

in Figure 4·6A), and there are 100 vertices, we easily deduce that there are 500 edges, 700 triangles, 20 pentagons, 300 tetrahedra, and 20 pentagonal antiprisms. Ex. 2 of §4·1 provides a check:

$$N_0-N_1+N_2-N_3 = 100-500+720-320 = 0.$$

3. A pattern consisting of 200 triangles and 100 squares: triangles which are sections of the tetrahedra (4·62), (4·63), (4·65), (4·66), and squares which are sections of (4·64). Superposing Figures 4·6B and 4·6C, we see that these triangles and squares are arranged (on the torus) like the faces of the infinite Euclidean tessellation 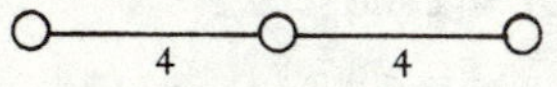 which is the Cayley diagram for the group (4, 4, 2) (see §3·3).

4. Equidistant from the two opposite vertices $(\pm i, 0)$ or $(0, \pm 1, 0, 0)$, we find the thirty vertices

$$|2\nu+1,\quad 0|2\nu,\quad 10|2\nu \quad (\nu = 0, 1, \ldots, 9),$$

all lying in the hyperplane $x_2 = 0$ or $u = \bar{u}$. These thirty points are the vertices of an icosidodecahedron $\left\{\begin{matrix}5\\3\end{matrix}\right\}$ or $\left\{\begin{matrix}3\\5\end{matrix}\right\}$ (Coxeter 1963, pp. 53, 298). They appear in Figure 4·6D as

$$1, 3, 5, \ldots;\quad 0, 2', 4, \ldots;\quad 0', 2, 4', \ldots;$$

and we recognize a familiar projection of the icosidodecahedron (Coxeter 1963, p. 19, Fig. 2·3A).

§4·7

$-\tau^6$, as we see by comparing the values of ${}_3R/l$ for $\{3, 3, 5\}$ and $\{3, 3, \tfrac{5}{2}\}$ in Coxeter 1963, pp. 293, 295, Table I (ii).

§4·8

1. Among the 24 vertices of $\{3, 4, 3\}$, those equidistant from two opposite vertices belong to an 'equatorial' octahedron, whose Petrie polygon is a regular skew hexagon. The 24 omitted edges form four such skew hexagons, appearing in the figure as four concentric $\{6\}$'s, two large and two small.

2. Twelve regular decagons, appearing in the figure as three of each of four sizes, all concentric.

§5·1

1.

Order 2:	a		a		a		a		a	
Order 3:	a		b		c		a		b	
		c		a		b		c		a
Order 4:	a		b		c		d		a	
		f		g		f		g		f
	c		d		a		b		c	

where a, b, c, d are arbitrary and $g = (ac+bd)/f$.

The line with Plücker coordinates

$$(p_{01}, p_{02}, p_{03}, p_{23}, -p_{13}, p_{12})$$

(Coxeter 1965, p. 88) can be represented by the frieze pattern

$$\begin{array}{ccccccccc}
p_{01} & & p_{12} & & p_{23} & & p_{03} & & p_{01} \\
 & p_{02} & & p_{13} & & p_{02} & & p_{13} & & \cdots \\
p_{23} & & p_{03} & & p_{01} & & p_{12} & & p_{23}
\end{array}$$

2. When the order is 4, the two rows of 1's are separated by a single row of $\sqrt{2}$'s. When the order is 5, the number repeated in the second and third rows satisfies the equation $x^2 - x = 1$, so it is τ. When the order is 6, the number in the second row is $\sqrt{3}$; in the third, 2; and in the fourth, $\sqrt{3}$ again.

3. Yes. If $(0, 1) = (0, 6) = 1$ and $(0, 2) = (0, 5) = x$ and $(0, 3) = (0, 4) = y$, we have

$$x^2 = y+1, \quad y^2 = xy+1,$$
$$x^3 - x^2 - 2x + 1 = 0.$$

The expression on the left takes the value -1 when $x = 1$, and 1 when $x = 2$. Thus there is a root x that yields a positive value for $y = x^2 - 1$.

4. It consists of $\sqrt{5}+2$, $\sqrt{5}+1$, $\sqrt{5}$, $\sqrt{5}-1$, $\sqrt{5}-2$.

§5·2

1. Setting $s = t = u = v$ in (5·24), we obtain $(v, v) = 0$. Setting $s = u$, $t = v$, then instead $s = v$, $t = u$, and adding the results,

$$\{(u, v) + (v, u)\}^2 = 0.$$

2. In the row above (s, s) we have $(s, s-1)$, $(s+1, s)$, which are equal to $-(s-1, s)$, $-(s, s+1)$, respectively. Thus the row of zeros acts like a magic mirror, repeating everything that occurs below but reversing all the signs. For instance, the general frieze of order 3 (§5·1, Ex. 1) yields the wallpaper

$$\begin{array}{ccccccccccccc}
0 & & 0 & & 0 & & 0 & & 0 & & 0 & & 0 \\
a & & b & & c & & a & & b & & c & & a \\
& c & & a & & b & & c & & a & & b & & c \\
0 & & 0 & & 0 & & 0 & & 0 & & 0 & & 0 \\
& -c & & -a & & -b & & -c & & -a & & -b & & -c \\
-a & & -b & & -c & & -a & & -b & & -c & & -a \\
& 0 & & 0 & & 0 & & 0 & & 0 & & 0 & & 0 \\
a & & b & & c & & a & & b & & c & & a \\
& c & & a & & b & & c & & a & & b & & c \\
0 & & 0 & & 0 & & 0 & & 0 & & 0 & & 0
\end{array}$$

3. By (5·23) and (5·25), $(s, t) + (s, t+n) = 0$. Thus

$$f_{s+n} = -f_s, \quad g_{s+n} = -g_s.$$

4. Both are zero: the former because it is skew-symmetric, the latter because it is equal to the square of

$$(0, 1)(2, 3) + (0, 2)(3, 1) + (0, 3)(1, 2).$$

§5·3

1. The 'triangle inequality'

$$(-a+b+c)(a-b+c)(a+b-c) > 0.$$

2. A regular n-gon $X_0X_1 \ldots X_{n-1}$ of edge $X_0X_1 = 1$ yields a frieze pattern whose νth row consists of repetitions of the number

$$X_0X_\nu = \sin\frac{\nu\pi}{n} \Big/ \sin\frac{\pi}{n}.$$

§5·4

1. $$O_1O_4 = l\cot\phi = l\sqrt{\frac{(-1, 0)(1, 4)}{(-1, 4)(0, 1)}},$$

$$O_2O_4 = O_3O_4\sec\psi = l\sqrt{\frac{(-1, 0)(-1, 1)(2, 4)}{(-1, 2)(-1, 4)(0, 1)}}.$$

These results can be combined as

$$O_sO_t = l\sqrt{\frac{(-1, 0)(-1, 1)(s, t)}{(-1, s)(-1, t)(0, 1)}}.$$

2. Since $\{p, q, r\}$ can be dissected into g orthoschemes, each having content

$O_0O_1 \cdot O_1O_2 \cdot O_2O_3 \cdot O_3O_4/4!$

$$= \frac{l^4}{24}\sqrt{\frac{(-1, 0)(1, 2)}{(-1, 2)(0, 1)}}\sqrt{\frac{(-1, 0)(-1, 1)(2, 3)}{(-1, 2)(-1, 3)(0, 1)}}\sqrt{\frac{(-1, 0)(-1, 1)(3, 4)}{(-1, 3)(-1, 4)(0, 1)}},$$

the whole content of the polytope is

$$\frac{gl^4}{24}\,\frac{(-1, 0)^{\frac{3}{2}}(-1, 1)(1, 2)^{\frac{1}{2}}(2, 3)^{\frac{1}{2}}(3, 4)^{\frac{1}{2}}}{(-1, 2)(-1, 3)(-1, 4)^{\frac{1}{2}}(0, 1)^{\frac{3}{2}}}.$$

§5·5

1. The 'natural' pattern for the regular tetrahedron $\{3, 3\}$ is easily transformed into the one given in §3·1:

$$\begin{array}{cccccccccc}
& 1 & & 1 & & 1 & & 1 & & 4 \\
3 & & 2 & & 2 & & 2 & & 3 & \\
& 2 & & 3 & & 3 & & 2 & & 2 \\
1 & & 1 & & 4 & & 1 & & 1 &
\end{array}
\quad \text{into} \quad
\begin{array}{cccccccccc}
& 1 & & 1 & & 1 & & 1 & & 1 \\
3 & & \frac{1}{2} & & 8 & & \frac{1}{2} & & 3 & \\
& \frac{1}{2} & & 3 & & 3 & & \frac{1}{2} & & 8 \\
1 & & 1 & & 1 & & 1 & & 1 &
\end{array}$$

In fact, the equations

$$a_{-1}a_0 = a_0a_1 = a_1a_2 = a_2a_3 = 1, \quad a_3a_4 = \tfrac{1}{4}, \quad a_4 = a_{-1}$$

have the solution $a_{-1} = a_1 = a_3 = \frac{1}{2}$, $a_0 = a_2$. But nothing like this can be done for the regular simplex $\{3, 3, 3\}$ (see (5·12)), as the equations

$$a_{-1}a_0 = a_0a_1 = a_1a_2 = a_2a_3 = a_3a_4 = 1, \quad a_4a_5 = \tfrac{1}{5}, \quad a_5 = a_{-1}$$

have no solution.

2. No. According to (5·43), the diagram for such a Ptolemaic pattern would be formed by five collinear points X_{-1}, X_0, X_1, X_2, X_3. The answer to §5·4, Ex. 1 shows that the equation $(O_sO_t)^2 = (s, t)$ implies

$$(-1, 0) = (-1, 1) = (-1, 2) = (-1, 3).$$

Thus X_{-1} must be the point at infinity, as in Figure 5·4A, and the 'pattern' would have $(-1, t) = \infty$. (The pattern (5·11) is not Ptolemaic.)

§5·6

For a pattern of order 6 and period 3 to be Ptolemaic, we must have

$$(0, 3) = (1, 4) = (2, 5),$$

each being equal to the circumdiameter of the hexagon in Figure 5·6A. To transform (5·15) into a Ptolemaic pattern (namely, the one in §5·6 with

$$A = \tfrac{1}{2}\pi-\lambda+\mu, \quad B = 2\lambda, \quad C = \tfrac{1}{2}\pi-\lambda-\mu),$$

we must replace each (s, t) by $a_s a_t(s, t)$ where

$$a_0 = 5^{-\frac{1}{4}}\tau^{\frac{1}{2}}, \quad a_1 = 3^{-\frac{1}{2}}, \quad a_2 = 5^{-\frac{1}{4}}\tau^{-\frac{1}{2}}, \quad a_{s+3} = a_s.$$

For instance, the first τ has to be multiplied by $a_0 a_1$ so as to become

$$3^{-\frac{1}{2}}5^{-\frac{1}{4}}\tau^{\frac{3}{2}} = \sin(\lambda+\mu).$$

§5·7

Since the n edges $O_0O_1, O_1O_2, O_2O_3 \ldots, O_{n-1}O_n$ are mutually orthogonal, the orthoscheme has n-dimensional content

$$O_0O_1\cdot O_1O_2\cdot O_2O_3 \ldots O_{n-1}O_n = \frac{l^n}{n!}\sqrt{\frac{(-1,0)(1,2)}{(-1,2)(0,1)}}$$
$$\times\sqrt{\frac{(-1,0)(-1,1)(2,3)}{(-1,2)(-1,3)(0,1)}}\cdots\sqrt{\frac{(-1,0)(-1,1)(n-1,n)}{(-1,n-1)(-1,n)(0,1)}}.$$

Hence the content of the whole polytope is

$$\frac{gl^n}{n!(-1,0)(-1,1)(-1,2)\ldots(-1,n-1)}$$
$$\times\left(\frac{(-1,0)(0,1)(1,2)\ldots(n-1,n)}{(-1,n)}\right)^{\frac{1}{2}}\left(\frac{(-1,0)(-1,1)}{(0,1)}\right)^{\frac{1}{2}n}.$$

§6·1

1. $(s, t)(u, v).(y, z) = (su-t\bar{v}, sv+t\bar{u})(y, z)$

$$= (suy-t\bar{v}y-sv\bar{z}-t\bar{u}\bar{z},\ suz-t\bar{v}z+sv\bar{y}+t\bar{u}\bar{y}),$$

$(s, t).(u, v)(y, z) = (s, t)(uy-v\bar{z}, uz+v\bar{y})$

$$= (suy-sv\bar{z}-t\bar{u}\bar{z}-t\bar{v}y,\ suz+sv\bar{y}+t\bar{u}\bar{y}-t\bar{v}z).$$

2. The order is 8. In fact, we have $i^2 = iji = j^2$ and

$$i = jij = j^2ij^2 = i^2ii^2 = i^5, \quad i^4 = 1,$$

so the eight elements are $\quad 1, i, j, i^2, ij, ji, i^3, j^3.$

§6·2

1. (i) -4. (ii) -8. (iii) -2. (iv) -4.

2. Since the conjugate of ab is $\tilde{b}\tilde{a}$, we see from (6·15) and (6·17) that the equation $x^2 = -1$ implies

$$\tilde{x}^2 = -1, \quad (x\tilde{x})^2 = 1, \quad \mathsf{N}x = 1,$$
$$(x+\tilde{x})^2 = x^2+\tilde{x}^2+2\mathsf{N}x$$
$$= -1-1+2 = 0, \quad \mathsf{S}x = 0.$$

3. The quaternion conjugate of $\tilde{a}xa$ is $\tilde{a}\tilde{x}a$. But $2\mathsf{S}x = x+\tilde{x}$. Hence

$$2\mathsf{S}(\tilde{a}xa) = \tilde{a}xa+\tilde{a}\tilde{x}a = \tilde{a}(x+\tilde{x})a$$
$$= 2\tilde{a}(\mathsf{S}x)a = 2\mathsf{S}x.$$

4. $a = \exp(y\alpha) = \{\exp(\tfrac{1}{2}\pi y)\}^{2\alpha/\pi} = y^{2\alpha/\pi}$.

5. The quaternion

$$e^{\alpha i}e^{\beta j}e^{\gamma i} = (\cos\alpha+i\sin\alpha)(\cos\beta+j\sin\beta)(\cos\gamma+i\sin\gamma)$$
$$= \{\cos(\alpha+\gamma)+i\sin(\alpha+\gamma)\}\cos\beta$$
$$+\{j\cos(\alpha-\gamma)+k\sin(\alpha-\gamma)\}\sin\beta$$

can be identified with a given unit quaternion x by choosing α, β, γ so that

$$\tan(\alpha+\gamma) = x_1/x_0, \quad \tan(\alpha-\gamma) = x_3/x_2,$$
$$\cos^2\beta = x_0{}^2+x_1{}^2, \quad \sin^2\beta = x_2{}^2+x_3{}^2$$

(compare (4·42)). Finally, we can take

$$a = 2\alpha/\pi, \quad b = 2\beta/\pi, \quad c = 2\gamma/\pi.$$

Theorem (6·42) shows that unit quaternions can be interpreted as rotations. Thus we have here an alternative treatment for Ex. 4 of §2·3.

§6·3

First suppose that $a \neq b$. The relation $\mathsf{S}a = \mathsf{S}b$ implies, in turn,

$$a+\tilde{a} = b+\tilde{b}, \quad \tilde{a}-\tilde{b} = b-a,$$
$$\tilde{a}(\tilde{a}-\tilde{b})b = \tilde{a}(b-a)b,$$
$$\tilde{a}(\tilde{a}b-\mathsf{N}b) = (\tilde{a}b-\mathsf{N}a)b.$$

In this case the problem can be solved by taking

$$y = \mathsf{U}(\tilde{a}b-\mathsf{N}a) = \mathsf{U}(\tilde{a}b-\mathsf{N}b), \quad z = yb.$$

For, these definitions imply $\tilde{a}y = yb = z$; and since $\mathsf{N}y = 1$ it follows that $\tilde{a} = z\tilde{y}$ and $b = \tilde{y}z$.

If, on the other hand, $a = b = \mathsf{S}a+x$, where $x = \mathsf{V}a \neq 0$ (since, if a were real we could use $y = a$ and $z = 1$), let w be any pure quaternion independent of x (that is, not a real multiple of x) and define

$$y = \mathsf{UV}(wx), \quad z = ya.$$

Since $\mathsf{S}a$ is real, $\quad (\mathsf{S}a)y = y\,\mathsf{S}a.$

Since the vector $\mathbf{y}$, being an outer product, is perpendicular to $\mathbf{x}$, (6·33) tells us that $-xy = yx$, that is,

$$-(\mathsf{V}a)y = y\mathsf{V}a.$$

By addition, $\quad \tilde{a}y = ya = z,$

whence $\tilde{a} = z\tilde{y}$ and $a = \tilde{y}z$, as desired.

§6·4

According to Theorem (6·42) as it stands, the positive quarter-turn about the point k transforms i into

$$\left(\cos\frac{\pi}{4}-k\sin\frac{\pi}{4}\right)i\left(\cos\frac{\pi}{4}+k\sin\frac{\pi}{4}\right) = \tfrac{1}{2}(1-k)\,i(1+k)$$
$$= \tfrac{1}{2}(i-j)(1+k) = -j;$$

therefore we are actually using a *left-handed* frame of reference. This may be regarded as a consequence of our convention whereby group elements are multiplied from left to right. Du Val (1964, p. 38) is able to use a right-handed frame because he prefers the opposite convention, which yields $e^{\alpha y}xe^{-\alpha y}$ instead of $e^{-\alpha y}xe^{\alpha y}$.

§6·5

1. $$A^p = B^q = Z, \quad (AB)^r = Z^{r-1}, \quad Z^2 = 1.$$

2. Yes. Since
$$B = ABA = A^2BA^2 = \ldots = A^pBA^p = B^2BB^2 = B^5,$$
the relation $Z^2 = 1$ in (6·62) (with $q = 2$) is a consequence of the other relations.

§6·6

The reflection in the great circle QR, being ixi, transforms
$$P = k\cos\psi + i\sin\psi \quad \text{into} \quad P' = k\cos\psi - i\sin\psi.$$
The reflection in PR, being jxj, transforms
$$Q = k\cos\phi + j\sin\phi \quad \text{into} \quad Q' = k\cos\phi - j\sin\phi.$$
The product of these two reflections, being the half-turn $-kxk$ about R, transforms P into P′, and Q into Q′.

§6·7

1. The diagonals of a rhombus are perpendicular.

2. If x and y are perpendicular, so are Px and Py, xQ and yQ. By (6·72), both P and Q are perpendicular to 1 and therefore also to PQ and to both $1 \pm PQ$. These properties of P and Q are shared by any linear combination of them, such as $P \pm Q$. Since
$$(P+Q)(-P+Q)+(P-Q)(-P-Q) = 0$$
(or by Ex. 1), $P+Q$ and $P-Q$ are perpendicular. Left multiplication by $-P$ shows that this is the case also for $1-PQ$ and $1+PQ$.

3. Since $a+\tilde{a} = b+\tilde{b}$, we have
$$\tilde{a}(1-ab)\,b = (\tilde{a}-b)\,b = (\tilde{b}-a)\,b = 1-ab,$$
$$\tilde{a}(\tilde{a}-b)\,b = \tilde{a}(\tilde{b}-a)\,b = \tilde{a}-b.$$
Thus $\tilde{a}xb$ leaves invariant $1-ab$ and $\tilde{a}-b$, as well as 0. Alternatively, if
$$a = \cos\alpha + P\sin\alpha \quad \text{and} \quad b = \cos\alpha + Q\sin\alpha,$$
we find
$$\tilde{a}-b = -(P+Q)\sin\alpha \quad \text{and} \quad 1-ab = (1-PQ)\sin^2\alpha - (P+Q)\sin\alpha\cos\alpha;$$
therefore the plane of 0, $1-PQ$, $P+Q$ is the same as the plane of 0, $1-ab$, $\tilde{a}-b$.

§6·8

The product of $-y\tilde{x}y$ and $-z\tilde{x}z$ is $z\tilde{y}x\tilde{y}z$.

§7·1

1. Use (7·13) and (7·131) with $p = 2$.

2. When the unit pure quaternion $x = x_1 i + x_2 j + x_3 k$ (with $Nx = 1$) is represented by the point (x_1, x_2, x_3) on the sphere $\Sigma x^2 = 1$, the points i and j are interchanged by the half-turn about the mid-point $h = 2^{-\frac{1}{2}}(i+j)$ of the great-circle arc that joins them, and we can verify directly that $h^{-1}ih = -hih = j$. The relations
$$i^2 = j^2 = (ij)^2 = Z = h^2 \quad \text{and} \quad h^{-1}ih = j$$
imply $Z^2 = 1$ and $i^2 = h^2 = (ih)^4 = Z = (ih)^{-4}$ whence, in terms of $f = Z(ih)^{-1}$,
$$f^4 = i^2 = h^2 = fih.$$
Thus the enlarged group is $\langle 4, 2, 2\rangle$.

3. The quaternions in $\langle p, 2, 2\rangle$ are 1, A^q, A^{2q}, ... and j, A^qj, $A^{2q}j = \ldots$, where $A = \exp(\pi i/pq)$. The element A of $\langle pq, 2, 2\rangle$ transforms j into
$$j\cos\frac{2\pi}{pq} - k\sin\frac{2\pi}{pq} = A^{-2}j,$$
which belongs to $\langle p, 2, 2\rangle$ only if $q = 1$ or 2.

4. $\exp\{\frac{1}{2}\pi i(\frac{1}{2}-1/p)\}$.

§7·2

1. $a^3 = b^3 = c^2 = abc = (11')(22')(33')(44')$.

(i) $a^3 = abab = b^3$.
(ii) $l = a^{-1}, \quad m = b$.
(iii) $l^{-1}ml = m^{-1}l^{-1} = mlm^{-1} = n$.
(iv) $nm = ln = ml = k^{-1}$.
(v) $K = k, \quad L = l^{-1}, \quad M = m, \quad N = n^{-1}$.

The relations (v) evidently imply $MLM = L^2$, $LML = M^2$, in agreement with (i). (See also Seifert & Threlfall 1947, p. 218.)

2. $\mathfrak{C}_5$.

§7·3

1. $A^4 = B^3 = C^2 = ABC = (11')(22')(33')(44')(55')(66')(77')(88')$.

(i) $A' = A^{-1}B, \quad B = AA'$.
(ii) $A'' = BA^{-1}, \quad B = AA' = A'A'' = A''A$,
$$A = 2^{-\frac{1}{2}}(1+i), \quad A' = 2^{-\frac{1}{2}}(1+j), \quad A'' = 2^{-\frac{1}{2}}(1+k).$$

See Symmetrical definitions for the binary polyhedral groups, *Proceedings of Symposia in Pure Mathematics*, American Mathematical Society, 1 (1959), 64–87; especially pp. 79–80.

2. No. The factor group $A^2 = A'^2 = 1$ is the infinite dihedral group $(\infty, 2, 2)$.

§7·4

1. In the case of $\langle p, 3, 2\rangle$ we have $A = B^{-1}A^{-1}B^2$ and $B = A^{p-1}B^{-1}A^{-1}$, whence
$$Z = A^p = (A^{-1}B)^p = (A^{p-2}B^{-1}A^{-1})^p = (A^{p-3}B^{-1})^p.$$

Thus, when $p = 3$, $Z = B^{-3} = Z^{-1}$, and when $p = 4$,

$$Z = (AB^{-1})^4 = (B^{-1}A)^4 = (A^{-1}B)^{-4} = Z^{-1}.$$

2. No. If we allowed $p \neq q$ with $r = 1$, we could present any cyclic group $\mathfrak{C}_{p+q}$ as

$$A^p = B^q = C = ABC$$

(implying $B = A^{-1}$), and then Z (which is C) could easily have a period greater than 2.

3. No. If we admitted the 'equals' sign, we could have the relations

$$A^3 = B^3 = C^3 = ABC = Z.$$

These allow Z to be aperiodic; for, the factor group derived by setting

$$A = B = C$$

is the infinite cyclic group (that is, the free group with one generator).

Similarly, the two groups

$$A^6 = B^3 = C^2 = ABC \quad \text{and} \quad A^4 = B^4 = C^2 = ABC$$

can be seen to have infinite centre by setting $B = A^2$, $C = A^3$ in the former, and $B = A$, $C = A^2$ in the latter.

4. Given $A^5 = B^3 = (AB)^2 = Z$, we define

$$a = A, \quad b = BAB^{-1}, \quad c = B^{-1}AB$$

and deduce

$$bc = BABABZ = B,$$

$$\begin{aligned} cb^{-1}c^{-1}b &= B^{-1}ABBA^{-1}B^{-1}B^{-1}A^{-1}BBAB^{-1} \\ &= B^{-1}AB^{-1}A^{-1}BA^{-1}B^{-1}AB^2 \\ &= B^{-1}AABBBAAB^2 = B^{-1}A^{-1}B^2 = A = a. \end{aligned}$$

Since $b = BaB^{-1}$ and $c = B^{-1}aB$, we can transform by B to obtain the other relations

$$ac^{-1}a^{-1}c = b, \quad ba^{-1}b^{-1}a = c.$$

Conversely, given these three relations involving a, b, c in a cyclically symmetric manner, we define $A = a$, $B = bc$, and deduce

$$a^{-1}cb^{-1}\cdot c^{-1}b = 1 = ac^{-1}\cdot a^{-1}cb^{-1}, \quad c^{-1}b = ac^{-1},$$

$$bc = ca = ab, \quad aba = bca = bab,$$

$$bac = baba^{-1}b^{-1}a = abaa^{-1}b^{-1}a = a^2,$$

$$(abc)^2 = ab\,ca\,bc = (bc)^3 = (ca)^3 = (ab)^3 = a\,abaca = a^5,$$

$$A^5 = B^3 = (AB)^2.$$

§7·5

1. $U_1^{-2} = U_2^{2} = (U_1U_2)^p$. ($A = U_2U_1$, $B = U_1^{-1}$, $C = U_2$.)
2. $2(p-1)$. In fact,

$$\begin{aligned} Z = (U_2U_1)^p &= (U_2^{-1}ZZU_1^{-1})^p = Z^{2p}(U_1U_2)^{-p} \\ &= Z^{2p-1}. \end{aligned}$$

3. When $p = 4$, we have $U_1BACB = (U_1U_2U_3)^3 = -1$, so

$$U_1 = BACB = BA^2B^2.$$

Similarly, when $p = 5$, $U_1 = B(ACB)^2 = B(A^2B^2)^2$.

§7·6

1. When $p = 3$,

$$(A^2BA^{-3}B)^2 = (A^{-1}B^2)^2 = (A^{-1}B^{-1}Z)^2 = (BA)^{-2}Z^2 = Z^{-1}Z^2 = Z.$$

When $p = 5$,

$$(A^3BA^{-3})^2 = (A^3BAZ^{-1}AB)^2 = (A^3A^{-1}B^{-1}ZB^{-1}A^{-1})^2 = (A^2BA^{-1})^2 = Z.$$

2. In the notation of Coxeter and Moser (1972, pp. 73–5), it is $\langle -2, 3 \mid p \rangle$. When $p = 3$ this is $SL(2, 3)$; when $p = 5$ it is $SL(2, 5) \times \mathfrak{C}_5$. Thus the last relation in (7·67) cannot be omitted even in these first two cases.

3. The given relations, with $A \rightleftarrows B$, imply $B = A^2$, $A^5 = A^6$, $A = 1$, $B = 1$.

4. By direct computation, if $ad - bc = 1$,

$$\begin{bmatrix} a & c \\ b & d \end{bmatrix} = \begin{cases} S^{(b-1)/a}TS^{-a}TS^{-(c+1)/a}T & (a \neq 0), \\ TS^{-c}\,TS^{b}\,TS^{c(d-1)}T & (a = 0). \end{cases}$$

The number of solutions of $ad - bc \equiv 1 \pmod{p}$ is $p(p^2-1)$, the order of $SL(2, p)$. Since any three of a, b, c, d determine the remaining one, the number of solutions of the same congruence modulo p^c is

$$p^{3(c-1)}\,p(p^2-1) = p^{3c-2}(p^2-1) = p^{3c}(1-1/p^2).$$

Multiplying together such expressions for all the factors p^c of m, we obtain, for the number of solutions modulo $m = \prod p^c$,

$$m^3 \prod_{p \mid m} \left(1 - \frac{1}{p^2}\right).$$

For instance, the order of $SL(2, 15)$ is $15^3(1-\frac{1}{9})(1-\frac{1}{25}) = 2880$. In fact, $SL(2, 15) \cong SL(2, 3) \times SL(2, 5)$. Similarly, the general $SL(2, m)$ is the direct product of such groups corresponding to the relatively prime divisors p^c of m.

According to Sunday (see the footnote on page 80), for any odd $m = 2q - 1$, $SL(2, m)$ has the presentation

$$S^m = 1, \quad T^2 = (ST)^3 = (S^qTS^4T)^2$$

or, equally well,

$$A^m = B^3 = (AB)^2 = (A^qBA^{-3}B)^2.$$

§7·7

By writing $a = R_1R_2$, $c = R_1R_2R_3$;

$$R_1 = c^{-3}ac^2, \quad R_2 = c^{-2}ac, \quad R_3 = c^{-1}a.$$

§7·8

1. The powers of the commutator

$$\begin{aligned} K = B^{-1}C^{-1}BC &= \begin{bmatrix} 2 & 3 \\ -1 & -1 \end{bmatrix}\begin{bmatrix} 0 & 1 \\ -1 & 0 \end{bmatrix}\begin{bmatrix} -1 & -3 \\ 1 & 2 \end{bmatrix}\begin{bmatrix} 0 & -1 \\ 1 & 0 \end{bmatrix} \\ &= \begin{bmatrix} 13 & -5 \\ -5 & 2 \end{bmatrix} \end{aligned}$$

include

$$K^2 = \begin{bmatrix} 194 & -75 \\ -75 & 29 \end{bmatrix}, \quad K^3 = \begin{bmatrix} 2897 & -1120 \\ -1120 & 433 \end{bmatrix},$$

$$K^5 = \begin{bmatrix} 646018 & -249755 \\ -249755 & 96557 \end{bmatrix}.$$

These have the desired residues modulo 5, 7, 11, respectively.

2. (i) Modulo 2, the matrices

$$A = \begin{bmatrix} 1 & 1 \\ 0 & 1 \end{bmatrix}, \quad B = \begin{bmatrix} 1 & 1 \\ 1 & 0 \end{bmatrix}, \quad C = \begin{bmatrix} 0 & 1 \\ 1 & 0 \end{bmatrix}$$

satisfy $A^2 = B^3 = C^2 = ABC = 1$, and thus generate (3, 2, 2), the dihedral (or symmetric) group of order 6.

(ii) Since

$$A^6 = \begin{bmatrix} 571 & 780 \\ 1560 & 2131 \end{bmatrix} \equiv \begin{bmatrix} -1 & 0 \\ 0 & -1 \end{bmatrix} \quad (\text{mod } 13),$$

we are now considering a factor group of the infinite group $\langle 6, 3, 2\rangle$. Since, modulo 13,

$$(A^2B^{-1})^3 A^{-2}B = \begin{bmatrix} 2 & 5 \\ 5 & 0 \end{bmatrix}^3 \begin{bmatrix} -2 & -2 \\ -2 & 4 \end{bmatrix} = \begin{bmatrix} 4 & 2 \\ 2 & -2 \end{bmatrix}\begin{bmatrix} -2 & -2 \\ -2 & 4 \end{bmatrix} = \begin{bmatrix} 1 & 0 \\ 0 & 1 \end{bmatrix},$$

this finite factor group has the presentation

$$A^6 = B^3 = (AB)^2 = Z, \quad Z^2 = (A^2B^{-1})^3 A^{-2}B = 1.$$

It was proved by J. M. Kingston (*Transactions of the Royal Society of Canada*, Third Series, Section III, **35** (1941), 37) that the order of the group

$$A^6 = B^3 = (AB)^2 = Z, \quad Z^r = (A^2B^{-1})^b (A^{-2}B)^c = 1$$

is $6r(b^2+bc+c^2)$. In the present case

$$b = 3, \quad c = 1, \quad r = 2,$$

so the order is 156, and the group is a subgroup of index 14 in $SL(2, 13)$.

§8·1

$$\left(\frac{x_1+y_1+z_1}{3}, \ldots, \frac{x_n+y_n+z_n}{3}\right).$$

§8·2

1. Indefinite. It is $(x_1-\sqrt{2}x_2)(\bar{x}_1-\sqrt{2}\bar{x}_2)-x_2\bar{x}_2$.
2. Definite. It is

$$(x_1-\sqrt{\tfrac{1}{3}}x_2)(\bar{x}_1-\sqrt{\tfrac{1}{3}}\bar{x}_2)+\tfrac{1}{3}x_2\bar{x}_2+(\sqrt{\tfrac{1}{3}}x_2-x_3)(\sqrt{\tfrac{1}{3}}\bar{x}_2-\bar{x}_3).$$

3. Definite. It is

$$(x_1-\sqrt{\tfrac{1}{3}}x_2)(\bar{x}_1-\sqrt{\tfrac{1}{3}}\bar{x}_2)+(\sqrt{\tfrac{1}{3}}x_2-x_3+\sqrt{\tfrac{1}{3}}x_4)(\sqrt{\tfrac{1}{3}}\bar{x}_2-\bar{x}_3+\sqrt{\tfrac{1}{3}}\bar{x}_4)$$
$$+\tfrac{1}{3}(x_2-x_4)(\bar{x}_2-\bar{x}_4)+\tfrac{1}{3}x_4\bar{x}_4.$$

4. Semidefinite. It is

$$(x_1-\sqrt{\tfrac{1}{3}}x_2)(\bar{x}_1-\sqrt{\tfrac{1}{3}}\bar{x}_2)+(\sqrt{\tfrac{1}{3}}x_2-x_3+\sqrt{\tfrac{1}{3}}x_4)(\sqrt{\tfrac{1}{3}}\bar{x}_2-\bar{x}_3+\sqrt{\tfrac{1}{3}}\bar{x}_4)$$
$$+\tfrac{1}{3}(x_2-x_4)(\bar{x}_2-\bar{x}_4)+(\sqrt{\tfrac{1}{3}}x_4-x_5)(\sqrt{\tfrac{1}{3}}\bar{x}_4-\bar{x}_5).$$

§8·4

The tree can be built up, one branch at a time, so that each new branch joins an old node to a new node. At each stage, the new node represents a line whose vector can be suitably multiplied.

§8·5

1. From (8·561) we deduce $B^{-1} = \bar{B}^T$; therefore the transformation

$$u' = uB$$

can be expressed as $u = u'B^{-1} = u'\bar{B}^T$. Also, since $\bar{B}^{-1} = B^T$, $B^T\bar{B} = I_n$.

2. In terms of ϵ, one of the square roots of the determinant $b_{11}b_{22}-b_{12}b_{21}$, let us define a and c by

$$b_{11} = \epsilon a, \quad b_{12} = \epsilon c.$$

The first of the three given relations implies $a\bar{a}+c\bar{c} = 1$ and the other two may be written as the matrix equation

$$(b_{21}, b_{22})\begin{bmatrix} \bar{a} & -c \\ \bar{c} & a \end{bmatrix} = (0, \epsilon),$$

which yields

$$(b_{21}, b_{22}) = (0, \epsilon)\begin{bmatrix} a & c \\ -\bar{c} & \bar{a} \end{bmatrix} = (-\epsilon\bar{c}, \epsilon\bar{a}).$$

3. $$(u', v') = \epsilon(u, v)\begin{bmatrix} a & c \\ -\bar{c} & \bar{a} \end{bmatrix}.$$

§8·7

Since $p = 2$, $\mathbf{X} = \mathbf{x} - 2\dfrac{\mathbf{x}\cdot\mathbf{y}}{\mathbf{y}\cdot\mathbf{y}}\mathbf{y}$ (Coxeter 1963, p. 182).

§8·8

Since $\epsilon_\nu^2 = -1$, $X_\lambda = x_\lambda - 2a_{\lambda\nu}x_\nu$ (Coxeter 1963, equation (10·63)).

§9·1

1. The complex numbers ϵ for which there is some transformation $\epsilon x\kappa$ in $\mathfrak{G}$ themselves form a finite group containing -1. Since the only finite groups of complex numbers containing -1 are the cyclic groups of even order, this group is $\mathfrak{C}_{2m}$ for some m. Moreover, the complex numbers ϵ for which the transformation ϵx is in $\mathfrak{G}$ form a subgroup $\mathfrak{C}_f$. Similarly, the quaternions κ for which there is some transformation $\epsilon x\kappa$ in $\mathfrak{G}$ form a finite group $\mathfrak{R}$ which contains -1 and which, therefore (by §6·5), must be one of the groups $\langle p, q, r\rangle$. The quaternions κ for which the transformation $x\kappa$ is in $\mathfrak{G}$ form a normal subgroup $\mathfrak{S}$. Now, if κ belongs to $\mathfrak{R}$, there is some ϵ in $\mathfrak{C}_{2m}$ such that the transformation $\epsilon x\kappa$ is an element of $\mathfrak{G}$. The mapping that takes κ to the coset $\epsilon\mathfrak{C}_f$ is a well defined homomorphism of $\mathfrak{R}$ onto $\mathfrak{C}_{2m}/\mathfrak{C}_f$ with kernel $\mathfrak{S}$, thus establishing an isomorphism between $\mathfrak{R}/\mathfrak{S}$ and $\mathfrak{C}_{2m}/\mathfrak{C}_f$. Finally, every element of $\mathfrak{G}$ may be expressed as $\epsilon x\kappa$ and as $(-\epsilon)x(-\kappa)$, where ϵ is an element of $\mathfrak{C}_{2m}$ and κ is any element of the coset of $\mathfrak{S}$ in $\mathfrak{R}$ which is the image of the coset $\epsilon\mathfrak{C}_f$ under the isomorphism between $\mathfrak{C}_{2m}/\mathfrak{C}_f$ and $\mathfrak{R}/\mathfrak{S}$.

2. No; the coset of $\mathfrak{C}_m$ in $\mathfrak{C}_{2m}$ is not a subset of the coset of $\mathfrak{C}_{2m}$ in $\mathfrak{C}_{4m}$.

§9·2

1. $x,\ \epsilon x\epsilon^3,\ \epsilon^2 x\epsilon^6,\ \epsilon^3 x\epsilon^9,\ \epsilon^4 x\epsilon^2$.
   (Since $\epsilon^5 = -1$, $\ \epsilon^5 x\epsilon^5 = x$.)
2. When $\mu = m$ and $\nu = 0$, $\ e^{\mu\pi i/m} x\, e^{d\nu\pi i/n} = -x$.

§9·3

1. It is the direct product of the $\langle p, q, r\rangle_{2^e}$ generated by A, B, C, Z^n and the $\mathfrak{C}_n$ generated by $Z^{2^{e+1}}$. In particular, n being odd,

$$\langle p, q, r\rangle_n \cong \langle p, q, r\rangle \times \mathfrak{C}_n.$$

2. Defining $GF[3^2]$ by the moduli 3 and i^2+1, we may regard its nine elements as the 'complex numbers' $a+bi$, where a and b belong to $GF[3]$. Then $1+i$ is a primitive root; and if $m = 2$ or 4, $\langle 3, 3, 2\rangle_m$ is generated by the four matrices

$$A = \begin{bmatrix} 1 & 1 \\ -1 & 0 \end{bmatrix},\quad B = \begin{bmatrix} -1 & 0 \\ 1 & -1 \end{bmatrix},\quad C = \begin{bmatrix} 0 & -1 \\ 1 & 0 \end{bmatrix} \pmod 3$$

and $Z = iI_2$ or $(1+i) I_2$, respectively.

§9·4

1. $\epsilon = \exp(\pi i/p),\ \kappa = \exp(\pi P/p)$, where

$$P = \frac{(1-t\bar t)\,i + 2tk}{1+t\bar t}.$$

2. $\epsilon = \exp(\pi i/p),\quad \kappa = \exp(\pi i/p)\ $ or $\ \exp(-\pi i/p)$.

The former is obtained directly from Ex. 1 by setting $t = 0$. The latter comes from the second of the equations (9·41), which can be expressed as

$$cu = (\bar\epsilon - \bar a)\,v = (a-\epsilon)\,v$$

or
$$(\zeta - i\eta)\,u = (\xi - 1)\,v\ \text{ or }\ v = \frac{\zeta - i\eta}{\xi - 1}\,u.$$

Since $\xi^2 + \eta^2 + \zeta^2 = 1$, we can obtain $v = 0$ by setting $\xi = -1$ and $\eta = \zeta = 0$.

3. Setting $a_1 = \sin\sigma$ and $a_2 = \cos\sigma$ in (8·77), we obtain

$$u_1' = u_1 + (\epsilon^2 - 1)\sin\sigma\,(u_1 \sin\sigma + u_2 \cos\sigma)$$
$$= \epsilon\left\{\left(\cos\frac{\pi}{p} - i\sin\frac{\pi}{p}\cos 2\sigma\right) u_1 + \left(i\sin\frac{\pi}{p}\sin 2\sigma\right) u_2\right\},$$

$$u_2' = u_2 + (\epsilon^2 - 1)\cos\sigma\,(u_1 \sin\sigma + u_2 \cos\sigma)$$
$$= \epsilon\left\{\left(i\sin\frac{\pi}{p}\sin 2\sigma\right) u_1 + \left(\cos\frac{\pi}{p} + i\sin\frac{\pi}{p}\cos 2\sigma\right) u_2\right\}.$$

§9·5

1. $A = R_1 R_2,\ \ B = R_2 R_3,\ \ C = R_1 R_3,\ \ Z = (R_1 R_2 R_3)^{-\frac12 h}$.
2. $ixj,\ \ ix(j-k)/\sqrt2,\ \ ix(i-k)/\sqrt2$.
3. Since the $8p$ transformations in $\langle p, 2, 2\rangle_2$ consist of

$$i^\mu x \exp(\nu\pi i/p)\ \text{ and }\ i^\mu xj \exp(\nu\pi i/p)$$

($\mu = 0$ or 1; $\nu = 0,\ 1,\ ...,\ 2p-1$), the only reflections among them when p is odd are $ixj \exp(\nu\pi i/p)$. These do not generate the whole group but only the subgroup

$$(\mathbb{C}_4/\mathbb{C}_2;\ \langle p, 2, 2\rangle/\mathbb{C}_{2p}) \cong [2p],$$

of order $4p$.

4. No; for then the products $A = R_1 R_2$ and $B = R_2 R_3$ would no longer satisfy $A^p = B^q = (AB)^2$. In fact, the effect of this change would be to multiply the order of the abstract group by

$$\left(\frac1p + \frac1q + \frac12\right)\Big/\left(\frac1p + \frac1q - \frac12\right).$$

See Coxeter, The binary polyhedral groups and other generalizations of the quaternion group, *Duke Mathematical Journal* 7 (1940), 374.

5. Since $a = A = R_1 R_2$ and $c = R_1 R_2 R_3$, (9·57) shows that these two generators are
$$x(1+i+j+k)/2\ \text{ and }\ ix(-1+k)/\sqrt2.$$

§9·6

For $\langle 3, 3, 2\rangle_6$, over $GF[5^2]$,

$$R_1 = \rho^{-2}\begin{bmatrix} -1 & -1 \\ -2 & 2 \end{bmatrix},\quad R_2 = \rho^{-2}\begin{bmatrix} -1 & 2 \\ 1 & 2 \end{bmatrix},\quad R_3 = \rho^{-2}\begin{bmatrix} 0 & 1 \\ -1 & 0 \end{bmatrix};$$

for the other two groups, over $GF[7^2]$ and $GF[11^2]$, respectively,

$$R_1 = \rho^{-2}\begin{bmatrix} 1 & 1 \\ 2 & 3 \end{bmatrix},\quad R_2 = \rho^{-2}\begin{bmatrix} -1 & -3 \\ 1 & 2 \end{bmatrix},\quad R_3 = \rho^{-2}\begin{bmatrix} 0 & -1 \\ 1 & 0 \end{bmatrix}.$$

§9·7

1. (i) Given $R_1 R_2 R_1 = R_2 R_1 R_2$, we define $R_3 = R_2^{-1} R_1 R_2$ and deduce $R_2 R_3 = R_1 R_2,\ \ R_2 R_1 R_2 = R_1 R_2 R_1 = R_2 R_3 R_1,\ \ R_1 R_2 = R_3 R_1$. Conversely, $R_1 R_2 = R_2 R_3 = R_3 R_1$ implies $R_1 R_2 R_1 = R_2 R_3 R_1 = R_2 R_1 R_2$.

(ii) Given $R_1 R_2 R_1 = R_2 R_1 R_2$, we define

$$B = (R_1 R_2)^{-1},\quad C = R_1 R_2 R_1 = R_2 R_1 R_2,$$

and deduce $BC = R_1,\ B^3 C^2 = (R_1 R_2)^{-3} R_1 R_2 R_1 \cdot R_2 R_1 R_2 = 1$.
Conversely, given $B^3 C^2 = 1$, we define $R_1 = BC$, $R_2 = CB$ and deduce

$$R_1 R_2 R_1 = BC^2 B^2 C = B^{1-3+2} C = C,\quad R_2 R_1 R_2 = CB^2 C^2 B = CB^{2-3+1} = C.$$

(Coxeter and Moser 1972, p. 78.)

2. Given $A^3 = B^3 = C^2 = ABC = Z$, so that $Z^2 = 1$, we deduce $A^2 = BC$ and $(BC)^3 = A^6 = Z^2 = B^3 C^2$. Conversely, given $B^3 C^2 = (BC)^3 = 1$, we have

$$B^2 C = B^{-1} C^{-1} = (CB)^{-1},$$

$$C^{-1} B^2 C = C^{-1} B^{-1} C^{-1} = BCB = B(B^2 C)^{-1} = BC^{-1} B^{-2} = B(BC)^{-1} B^{-1},$$

$$C^{-1} B^6 C = B(BC)^{-3} B^{-1} = 1,\quad B^6 = 1,\quad C^2 = B^{-3} = B^3 = Z,$$

say. Defining $A = Z(BC)^{-1}$, so that $ABC = Z$, we deduce

$$Z^2 = 1,\quad A^3 = Z = B^3 = C^2 = ABC.$$

3. $3[3]3$ is generated by the elements R_1 and $R_3 R_1 R_3$ of $\langle 3, 3, 2\rangle_6$, that is, by

$$\rho^{-2}\begin{bmatrix} -1 & -1 \\ -2 & 2 \end{bmatrix}\ \text{ and }\ \rho^{-6}\begin{bmatrix} 1 & 2 \\ 1 & -2 \end{bmatrix}\ \text{ over }\ GF[5^2].$$

3[4]3 is generated by the elements R_1 and R_2 of $\langle 3, 3, 2\rangle_6$, that is, by

$$\rho^{-2}\begin{bmatrix} -1 & -1 \\ -2 & 2 \end{bmatrix} \quad\text{and}\quad \rho^{-2}\begin{bmatrix} -1 & 2 \\ 1 & 2 \end{bmatrix} \quad\text{over}\quad GF[5^2].$$

Yes, 3[5]3 can be so generated. As a by-product of his investigation of the finite two-dimensional unitary group $U(2, 5^2)$, Crowe discovered that 3[5]3 is generated by

$$\begin{bmatrix} \rho^{10} & \rho^{-5} \\ \rho^{-5} & \rho^{10} \end{bmatrix} \quad\text{and}\quad \begin{bmatrix} 1 & 0 \\ 0 & \rho^{8} \end{bmatrix} \quad\text{over}\quad GF[5^2].$$

(See Generating reflections for $U(2, p^{2n})$, *Canadian Mathematical Bulletin* **7** (1964), 213–7.) In an earlier paper (Regular polygons over $GF[3^2]$, *American Mathematical Monthly* **68** (1961), 761–5), he identified 4[3]4 with $U(2, 3^2)$ and found the generators

$$\begin{bmatrix} -1+i & 1+i \\ 1+i & -1+i \end{bmatrix} \quad\text{and}\quad \begin{bmatrix} 1 & 0 \\ 0 & -i \end{bmatrix} \quad\text{over}\quad GF[3^2].$$

§9·8

The *cyclic* group generated by a single reflection of period p.

§10·2

1. For an example where $f = g$ is even while $m = n$ is odd, consider $(\mathfrak{C}_6/\mathfrak{C}_2; \mathfrak{C}_6/\mathfrak{C}_2)_1$ of order 6, consisting of $\pm e^{\nu\pi i/3}xe^{\nu\pi i/3}$ $(\nu = 0, 1, 2)$. Among these six transformations, only $e^{\pi i/3}xe^{\pi i/3}$ and $e^{2\pi i/3}xe^{2\pi i/3}$ are reflections. The group generated by them is not the whole group but only the subgroup $(\mathfrak{C}_6/\mathfrak{C}_1; \mathfrak{C}_6/\mathfrak{C}_1)_1$. Thus, although $(\mathfrak{C}_6/\mathfrak{C}_2; \mathfrak{C}_6/\mathfrak{C}_2)_1$ is a group of unitary transformations (namely, the cyclic group $\mathfrak{C}_6$ generated by $-e^{\pi i/3}xe^{\pi i/3}$), and although it *contains* reflections, it is not *generated* by reflections.

2. 60[2]2, 12[2]10, 20[2]6, 30[2]4.

§10·3

No! $\langle p, 2, 2\rangle_{2p}$ contains the reflections

$$e^{\pi i/p}xe^{\pi i/p}, \quad ixj, \quad ixe^{\pi i/p}j,$$

which generate the group $\langle p, 2, 2\rangle_p$ or $(\mathfrak{C}_{4p}/\mathfrak{C}_{2p}; \langle p, 2, 2\rangle/\mathfrak{C}_{2p})$ according as p is even or odd. But no product of them can yield $\exp(\pi i/2p)\,x$, which belongs to $\langle p, 2, 2\rangle_{2p}$.

§10·4

1. (i) $m[4]2$. (ii) $[mn]$.
2. (i) When $m = 1$, we have $G(n, n, 2) \cong [n]$, and Shephard's presentation becomes $R_1 = 1 = R_2{}^2 = R_3{}^2 = (R_2R_3)^n$.
   (ii) When $n = 1$, we have $G(m, 1, 2) \cong m[4]2$. The relations

$$R_2{}^2 = R_3{}^2 = 1, \quad R_2R_1R_2 = R_3R_1R_3 = R_1R_2R_3$$

clearly imply

$$R_3 = R_1R_2R_1{}^{-1}, \quad R_2R_1R_2 = (R_1R_2)^2R_1{}^{-1}, \quad (R_2R_1)^2 = (R_1R_2)^2.$$

   (iii) When $n = 2$ (and $m = p$), we have $G(2p, 2, 2)$ with the presentation (10·33). In the presence of $R_2{}^2 = R_3{}^2 = 1$, the relation

$$R_3R_1R_3 = R_1(R_2R_3)^2 \quad\text{implies}\quad R_3R_1R_2 = R_1R_2R_3,$$
$$\text{while}\quad R_2R_1R_2 = R_3R_1R_3 \quad\text{implies}\quad R_1R_2R_3 = R_2R_3R_1.$$

§10·7

1. No! The left-multipliers in $\langle 4, 3, 2\rangle_6$, being powers of $\exp(\pi i/6)$, cannot include $A = \exp(\pi i/4)$.
2. (i) [4]. (ii) 3[6]2.

§10·8

1. (i) [6]. (ii) [4].
2. 3[6]2, 4[3]4, 3[8]2, 4[4]3 (orders 48, 96, 144, 288).

§10·9

1. No, only when p, q, r are relatively prime, that is, when they are 5, 3, 2 (permuted).
2.

$$\begin{aligned} \langle 3, 3, 2\rangle_6 &\cong \mathfrak{C}_3\times 3[6]2, & p[2]r &\cong \mathfrak{C}_p\times\mathfrak{C}_r,\\ 3[8]2 &\cong \mathfrak{C}_3\times GL(2, 3), & 3[4]3 &\cong \mathfrak{C}_3\times 3[3]3,\\ \langle 4, 3, 2\rangle_6 &\cong \mathfrak{C}_3\times\langle 4, 3, 2\rangle_2, & 4[4]3 &\cong \mathfrak{C}_3\times 4[3]4,\\ 3[5]3 &\cong \mathfrak{C}_3\times\langle 5, 3, 2\rangle, & \langle 4, 3, 2\rangle_{12} &\cong \mathfrak{C}_3\times 4[6]2,\\ 3[10]2 &\cong \mathfrak{C}_3\times\langle 5, 3, 2\rangle_2, & 5[3]5 &\cong \mathfrak{C}_5\times\langle 5, 3, 2\rangle,\\ 5[4]3 &\cong \mathfrak{C}_3\times\mathfrak{C}_5\times\langle 5, 3, 2\rangle, & 5[6]2 &\cong \mathfrak{C}_5\times\langle 5, 3, 2\rangle_2,\\ & & \langle 5, 3, 2\rangle_{30} &\cong \mathfrak{C}_3\times\mathfrak{C}_5\times\langle 5, 3, 2\rangle_2. \end{aligned}$$

§11·1

Since this real reflection reverses the sense of the rotations about P_1 and P_2, it transforms the complex reflections into their inverses. This transformation is an automorphism of the complex polygon, but it is not regarded as an element of the symmetry group because it is not unitary. For instance, when $p_1\{q\}p_2$ is a non-starry polygon, its symmetry group (9·81) evidently possesses an automorphism that replaces R_ν by $R_\nu{}^{-1}$ $(\nu = 1, 2)$. In the notation of §8·9, this is the semilinear transformation

$$(x^1, x^2) \to (\bar{x}^1, \bar{x}^2).$$

§11·2

No. Since $\langle 4, 3, 2\rangle$ has no subgroup $\mathfrak{C}_{12}$, the group

$$3[6]2 \cong (\mathfrak{C}_{12}/\mathfrak{C}_4; \langle 3, 3, 2\rangle/\langle 2, 2, 2\rangle)$$

cannot be a subgroup of $[4, 3, 3]^+$ or $[3, 4, 3]^+$. Similarly, since $\langle 5, 3, 2\rangle$ has no subgroup $\mathfrak{C}_{20}$, the group $5[6]2 \cong \langle 5, 3, 2\rangle_{10}$ cannot be a subgroup of $[5, 3, 3]^+$.

§11·3

1. When the polygon is non-starry, the two fractions (11·31) are, in the notation of (9·85),

$$\left(1+h-\frac{2h}{q}\right)\frac{1}{h} \quad\text{and}\quad \frac{1}{h}.$$

Since the order of the centre of $p_1[q]p_2$ is either $2h/q$ or h/q, $2h/q$ is an integer. Thus h serves as a 'common denominator'.

2. Obvious by (11·32).

3. Obvious by (11·33).

4. When the polygon is non-starry, the greater of the two fractions (11·33) is a multiple of the smaller, and the smaller has numerator 1. Hence

$$\frac{2}{h'} = \frac{1}{t} - \left|\frac{1}{p_1} - \frac{1}{p_2}\right|.$$

5. (*1 4*) (*3 7*) (*5 8*) or (*1 8*) (*2 6*) (*4 5*). (Being *odd* permutations, these are *outer* automorphisms of 3[3]3.)

§11·4

1. In the group $p[3]p$, defined by

$$R_1^p = 1, \quad R_1R_2R_1 = R_2R_1R_2,$$

R_1 and its conjugate $R_2^{-1}R_1R_2$ generate the whole group, since

$$R_2 = R_1^{-1}(R_2^{-1}R_1R_2)\,R_1.$$

In $p[5]p$, defined by $R_1^p = 1$, $R_1R_2R_1R_2R_1 = R_2R_1R_2R_1R_2$, the whole group is generated by R_1, $R_2^{-1}R_1R_2$, $R_2R_1R_2^{-1}$, since

$$R_2 = R_1^{-1}(R_2^{-1}R_1R_2)R_1(R_2R_1R_2^{-1})R_1^{-1}.$$

2. Since

$$\begin{aligned}(R_1\cdot R_2^{-1}R_1R_2)^2 &= R_1R_2\cdot R_2R_1R_2R_1R_2\cdot R_2R_1R_2 = (R_1R_2)^5 = (R_2R_1)^5\\ &= R_2\cdot R_2R_1R_2R_1R_2\cdot R_2R_1R_2R_1 = (R_2^{-1}R_1R_2\cdot R_1)^2,\end{aligned}$$

this subgroup of 3[5]3 is 3[4]3, of index 5. Therefore the 120 vertices of 3{5}3 belong also to five 3{4}3's.

3. Since $(R_1R_2^2)^2 = (R_1R_2)^3 = (R_2R_1)^3 = (R_2^2R_1)^2$, this subgroup of 4[3]4 is 4[4]2, of index 3. Therefore the 24 vertices of 4{3}4 belong also to three 2{4}4's.

4. Since

$$\begin{aligned}(R_1R_2^2)^3 &= R_1R_2\cdot R_2R_1R_2\cdot R_2R_1R_2\cdot R_2\\ &= R_1R_2(R_1R_2)^2\,R_1^{-1}\cdot R_1^{-1}(R_2R_1)^2\,R_2 = (R_1R_2)^6 = (R_2R_1)^6\\ &= R_2\cdot R_2R_1R_2\cdot R_2R_1R_2\cdot R_2R_1 = (R_2^2R_1)^3,\end{aligned}$$

this subgroup of 3[4]4 is 3[6]2, of index 6. Therefore the 96 vertices of 4{4}3 belong also to six 2{6}3's.

§11·5

See Figure 11·51. Notice that each equilateral triangle represents $kji = 1$, while each isosceles triangle represents $jk = i$ or $ki = j$ or $ij = k$.

§11·6

1. (i) 2{8}4; (ii) 2{12}3.

2. No. In every case q is an integer, and when q is odd, $p_1 = p_2$.

3. No. The symmetry groups of the regular apeirogons, being groups of Euclidean rotations and translations, are:

for 2{∞}2, $\mathfrak{D}_\infty$ (generated by two half-turns);
for 3{6}3, (3, 3, 3) (see Figure 3·3A);
for 4{8}2 and 2{8}4 and 4{4}4, (4, 4, 2);
for all the remaining seven, (6, 3, 2).

4. No. The valency is still p, but the girth is only $\frac{1}{2}q$.

§11·7

1. (i) The diagram consists of a $2p$-gon with its edges blown up into digons (representing U_1 and U_2 alternately). Since there are altogether p digons of each kind and 2 negatively oriented p-gons, the period of Z is $2p-2$.
(ii) The diagram consists of $2s/p$ p-gons, $2s/q$ q-gons, and $2s/r$ negatively oriented $2r$-gons, so the period of Z is $2s(p^{-1}+q^{-1})$.
(iii) Writing $A = a^{-1}$, $B = b$, we have $A^{-p} = B^q = Z$, $(AB)^r = 1$ or $A^p = Z^{-1}$, $B^q = Z$, $(AB)^r = 1$. Now the p-gons and q-gons are oppositely oriented, so the period of Z is $2s|p^{-1}-q^{-1}|$. This group is $\langle -p, q\,|\,r\rangle$ in the notation of Coxeter and Moser (1972, pp. 73–5).

2.

$\langle \pm p, 2, 2\rangle$,	1				
$\langle \pm p, -2, 2\rangle$,	$p\mp 1$	$\langle 4, 3, 2\rangle$,	1	$\langle 5, 3, 2\rangle$,	1
$\langle \pm p, -2, -2\rangle$,	$2p\mp 1$	$\langle -4, 3, 2\rangle$,	5	$\langle -5, 3, 2\rangle$,	11
$\langle 3, 3, 2\rangle$,	1	$\langle 4, -3, 2\rangle$,	7	$\langle 5, -3, 2\rangle$,	19
$\langle -3, 3, 2\rangle$,	3	$\langle 4, 3, -2\rangle$,	11	$\langle 5, 3, -2\rangle$,	29
$\langle 3, 3, -2\rangle$,	5	$\langle -4, -3, 2\rangle$	13	$\langle -5, -3, 2\rangle$,	31
$\langle -3, -3, 2\rangle$,	7	$\langle -4, 3, -2\rangle$,	17	$\langle -5, 3, -2\rangle$,	41
$\langle -3, 3, -2\rangle$,	9	$\langle 4, -3, -2\rangle$,	19	$\langle 5, -3, -2\rangle$,	49
$\langle -3, -3, -2\rangle$,	13	$\langle -4, -3, -2\rangle$,	25	$\langle -5, -3, -2\rangle$,	61

3. In every case except $\langle \pm p, -2, 2\rangle$ (p odd), of order $4p(p\mp 1)$, $\langle -3, 3, 2\rangle$, of order 72, and $\langle -3, 3, -2\rangle$, of order 216. See Coxeter, The binary polyhedral groups and other generalizations of the quaternion group, *Duke Mathematical Journal* **7** (1940), 367–79.

4. The 'snub $\{p, q\}$' considered in Ex. 5 of §2·4 has N_2 p-gons, N_0 q-gons, N_1 digons and $2N_1$ negatively oriented triangles.

§11·8

No; $\mathfrak{S}_5$ has no subgroup of order 10.

§12·1

1. There are at least two vertices, two edges, ..., and two cells.

2. Each face is a polygon (see condition (ii) of §11·1). Among the cells incident with a Π_{n-3}, any two are connected by a 'chain' of successively adjacent Π_{n-2}'s and Π_{n-1}'s.

3. No. Consider, for instance, Kepler's *stella octangula*: two reciprocal tetrahedra whose edges are the diagonals of the faces of a cube (Coxeter 1963, pp. 47–8). We would not wish to call this a single regular polytope, because it

is not connected. But it is the medial figure of the (-1)-flat and the 3-flat, not of a $(\nu-3)$-flat and a ν-flat.

4.
$$\begin{bmatrix} 24 & 8 & 12 & 6 \\ 2 & 96 & 3 & 3 \\ 3 & 3 & 96 & 2 \\ 6 & 12 & 8 & 24 \end{bmatrix}.$$

5. Being the property $N_{\mu-2,\mu-1}$ or $N_{\mu-2,\mu}$ of a $\Pi_{\mu+1}$,

$$q_\mu = \frac{N_{\mu+1,\mu-1}N_{\mu-1,\mu-2}}{N_{\mu+1,\mu-2}} = \frac{N_{\mu+1,\mu}N_{\mu,\mu-2}}{N_{\mu+1,\mu-2}}.$$

In particular, $q_1 = N_{20} = N_{21}$, $q_{n-1} = N_{n-3,n-2} = N_{n-3,n-1}$.

6. Being the property $N_{\mu-2,\mu-1}$ of a Π_μ,

$$p_\mu = N_{\mu,\mu-1}N_{\mu-1,\mu-2}/N_{\mu,\mu-2}.$$

In particular, $p_1 = N_{10}$, $p_n = N_{n-2,n-1}$.

§12·2

1. In the case of γ^p_n, Π_μ is γ^p_μ, so $N_0 = p^n$ and $N_{\mu 0} = p^\mu$. Since the vertex figure is a simplex α_{n-1}, which has $\binom{n}{\mu}$ elements $\alpha_{\mu-1}$, $N_{0\mu} = \binom{n}{\mu}$. But $N_\mu N_{\mu 0} = N_0 N_{0\mu}$. Hence $N_\mu = \binom{n}{\mu} p^{n-\mu}$ $(\mu \geqslant 0)$, $N_{\mu\nu} = \binom{\mu}{\nu} p^{\mu-\nu}$ $(\mu > \nu \geqslant 0)$ and $N_{\nu\mu} = \binom{n-\nu}{\mu-\nu}$.

2. No. In fact,

$$\sum_{-1}^{n}(-1)^\mu N_\mu = -1 + \sum_{0}^{n}(-1)^\mu \binom{n}{\mu} p^{n-\mu} = -1+(p-1)^n,$$

but this is zero only when $p = 2$.

3. A regular simplex α_n. For, if $0 < |\mu-\nu| \leqslant n$, $\epsilon^{\mu-\nu}$ is an $(n+1)$th root of 1, and therefore the angle θ subtended at the origin by the points

$$(\epsilon^\mu, \epsilon^{2\mu}, \ldots, \epsilon^{n\mu}) \quad \text{and} \quad (\epsilon^\nu, \epsilon^{2\nu}, \ldots, \epsilon^{n\nu})$$

is given by

$$\cos\theta = \frac{\epsilon^{\mu-\nu}+\epsilon^{2(\mu-\nu)}+\ldots+\epsilon^{n(\mu-\nu)}}{n} = -\frac{1}{n}.$$

§12·3

1. From the vertex $(1, -1, 0)$, edge $(\omega^\mu, -1, 0)$ with centre $(0, -1, 0)$ and face (12·35) with centre $(0, -\frac{1}{2}, \frac{1}{2})$, we see that the polyhedron (12·33) with centre $(0, 0, 0)$ has

$${}_0R = \sqrt{2}, \quad {}_1R = 1, \quad {}_2R = \sqrt{\tfrac{1}{2}}.$$

2. Since, in $GF[2^2]$, $-1 = 1$, the 27 vertices (12·32) are now

$$(0, \omega^\mu, \omega^\nu), \quad (\omega^\nu, 0, \omega^\mu), \quad (\omega^\mu, \omega^\nu, 0),$$

but the concise symbols (12·33) remain appropriate, and so do the permutations (12·34). As for the cubic surface in $PG(3, 2^2)$, we may take its equation to be

$$u_1{}^3+u_2{}^3+u_3{}^3+u_4{}^3 = 0$$

and then there are just 45 points on it: the 18 permutations of $(\omega^\lambda, 1, 0, 0)$, and the 27 points $(\omega^\lambda, \omega^\mu, \omega^\nu, 1)$. (Since the coordinates are homogeneous, $(1, \omega^2, 0, 0)$ is the same as $(\omega, 1, 0, 0)$, and so on.)

3. The six-dimensional polytope 2_{21} has a symmetry group $[3^{2,2,1}]$ generated by six involutory reflections as indicated by the graph

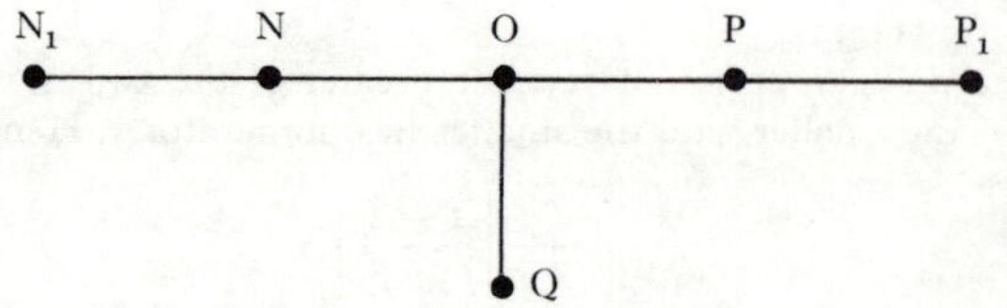

N_1 replaces u_1 by $\bar{u}_1$ (leaving u_2 and u_3 unaltered), P_1 replaces u_2 by $\bar{u}_2$, N replaces u_1 by $\omega\bar{u}_1$, P replaces u_2 by $\omega\bar{u}_2$, Q replaces u_3 by $\omega\bar{u}_3$, and O subtracts

$$(u_1+\bar{u}_1+u_2+\bar{u}_2+u_3+\bar{u}_3)/3$$

from each of u_1, u_2, u_3. The initial vertex of 2_{21}, invariant for all the reflections except N_1, is $(\omega^2, 0, -\omega^2)$. Choosing instead $(0, 0, i\sqrt{3})$, invariant for all the reflections except Q, we exhibit 1_{22} as the real six-dimensional counterpart of the complex polyhedron

$$3\{3\}3\{4\}2$$

(Figure 12·4c). For the analogous treatment of 4_{21} and the Witting polytope, see pp. 479–85 of the paper cited on page 119.

4. The 8 vertices and centre are 9 points; the 8 edges and 4 diameters are 12 lines. Each line contains 3 of the points; each point lies on 4 of the lines. (See also Coxeter, 1969, p. 237, Ex. 3.)

§12·4

1.
$$\begin{bmatrix} 54 & 8 & 8 \\ 2 & 216 & 3 \\ 6 & 9 & 72 \end{bmatrix}, \quad \begin{bmatrix} 27 & 8 & 8 \\ 3 & 72 & 3 \\ 8 & 8 & 27 \end{bmatrix}, \quad \begin{bmatrix} 72 & 9 & 6 \\ 3 & 216 & 2 \\ 8 & 8 & 54 \end{bmatrix}.$$

2. $(R_1R_2R_3)^9$.

3. In the unitary space over $GF[2^2]$, where $-x = x$, $2\{4\}3\{3\}3$ has the same 27 vertices as $3\{3\}3\{3\}3$, and $3\{3\}3\{4\}2$ has 36:

$$(\omega^\lambda, \omega^\mu, \omega^\nu), \quad (\omega^\lambda, 0, 0), \quad (0, \omega^\mu, 0), \quad (0, 0, \omega^\nu).$$

But the unitary space (being affine) has only 64 points altogether; thus the two polyhedra and their common centre account for all of them.† In other words, the points whose coordinates involve just one zero belong to $2\{4\}3\{3\}3$ or $3\{3\}3\{3\}3$, while those whose coordinates involve two zeros or none belong to $3\{3\}3\{4\}2$. Similarly, the 84 planes in the unitary space consist of the 36 face-planes of $2\{4\}3\{3\}3$ the 27 face-planes of $3\{3\}3\{4\}2$, and their $12+9$ planes of symmetry.

§12·5

1. Since the cell (12·53) has centre $(0, 0, 0, 1)$, ${}_3R = 1$, and we can obtain ${}_\nu R^2$ for the Witting polytope by adding 1 to ${}_\nu R^2$ for the Hessian polyhedron (see Ex. 1 of §12·3). Thus

$${}_0R = \sqrt{3}, \quad {}_1R = \sqrt{2}, \quad {}_2R = \sqrt{\tfrac{3}{2}}, \quad {}_3R = 1.$$

2. Whereas the cycles in (12·34) are triads of skew lines, those in (12·58) are triads of intersecting lines (lying in tritangent planes of the cubic surface).

† Compare the remark about $PG(5, 2)$ in Coxeter, Polytopes over $GF[2]$ and their relevance for the cubic surface group, *Canadian Journal of Mathematics* **11** (1959), 648.

The latter arrangement (in which the columns likewise form triads of intersecting lines) exhibits, for each generator, one of the 40 triads of trihedral pairs.

3. Since the complex conjugate of ω is ω^2, this collineation, which transforms each R into its inverse, is

$$(011\ 022)\ (101\ 202)\ (110\ 220)\ (012\ 021)\ (201\ 102)\ (120\ 210)\ (013\ 023)$$
$$\cdot(301\ 302)\ (130\ 230)\ (031\ 032)\ (103\ 203)\ (310\ 320).$$

§12·6

1. At each 3-cell, 3 (the final 3 in the symbol $3\{3\}3\{3\}3\{3\}3\{3\}3$); at each face, 8 (because the polygon $3\{3\}3$ has 8 edges); at each edge, 27 (because the polydron $3\{3\}3\{3\}3$ has 27 faces); at each vertex, 240 (because the polytope $3\{3\}3\{3\}3\{3\}3$ has 240 cells).

2. A typical edge of the polytope (12·52) is $(-\omega^\lambda, -1, -1, 0)$ $(\lambda = 0, 1, 2)$. Adding 1 to each of u_1, u_2, u_3, we obtain $(1-\omega^\lambda, 0, 0, 0)$ as an edge of the honeycomb. The centre of this edge is $(1, 0, 0, 0)$. Thus the vertex figure is given by (12·52) with all the coordinates multiplied by $i\sqrt{3}$.

§12·7

1. $p\{4\}2\{4\}r$.

2. The form is

$$x^1\bar{x}^1 - \sqrt{\tfrac{1}{2}}(x^1\bar{x}^2 + x^2\bar{x}^1) + x^2\bar{x}^2 - \tfrac{1}{2}(x^2\bar{x}^3 + x^3\bar{x}^2) + \ldots - \tfrac{1}{2}(x^{n-1}\bar{x}^n + x^n\bar{x}^{n-1})$$
$$+ x^n\bar{x}^n - \sqrt{\tfrac{1}{2}}(x^n\bar{x}^{n+1} + x^{n+1}\bar{x}^n) + x^{n+1}\bar{x}^{n+1}$$
$$= u_1\bar{u}_1 + u_2\bar{u}_2 + \ldots + u_{n-1}\bar{u}_{n-1} + u_n\bar{u}_n, \quad \text{where}$$

$$u_1 = x^1 - \frac{x^2}{\sqrt{2}}, \quad u_2 = \frac{x^2 - x^3}{\sqrt{2}}, \ldots, u_{n-1} = \frac{x^{n-1} - x^n}{\sqrt{2}}, \quad u_n = \frac{x^n}{\sqrt{2}} - x^{n+1}.$$

Thus the $n+1$ mirrors are

$$u_1 = l, \quad \frac{u_1 - u_2}{\sqrt{2}} = 0, \quad \frac{u_2 - u_3}{\sqrt{2}} = 0, \ldots, \frac{u_{n-1} - u_n}{\sqrt{2}} = 0, \quad u_n = 0$$

(for a suitable complex number l).

When $p = r = 4$, the reflection R_1 appears, in the Argand plane representing u_1, as a quarter-turn about the centre of the square

$$0,\ 1,\ 1+i,\ i,$$

that is, about the point $(1+i)/2$. Thus R_1, being a reflection of period 4 in the mirror $u_1 = (1+i)/2$, changes u_1 into

$$i\left(u_1 - \frac{1+i}{2}\right) + \frac{1+i}{2} = iu_1 + 1$$

while leaving the rest of the coordinates unaltered; $R_2, \ldots, R_n$ are the transpositions $(1\ 2), \ldots, (n-1\ n)$; and R_{n+1} multiplies u_n by i. Clearly, Gaussian integers are transformed into Gaussian integers.

Similarly, when $p = r = 3$, the apeirogon $3\{6\}3$ suggests that we should choose $l = (1-\omega^2)/3$, this being the centre of the triangle

$$0,\ 1,\ -\omega^2$$

in the Argand diagram. Now R_1, being a reflection of period 3 in the mirror $u_1 = (1-\omega^2)/3$, changes u_1 into

$$\omega\left(u_1 - \frac{1-\omega^2}{3}\right) + \frac{1-\omega^2}{3} = \omega u_1 + 1$$

while leaving $u_2, \ldots, u_n$ unaltered; $R_2, \ldots, R_n$ are transpositions, as before; and R_{n+1} multiplies u_n by ω.

§12·8

1. The form is again (12·85), so the mirrors are again (12·86) or, still more simply,

$$u_1 = l, \quad u_1 = u_2, \quad u_2 = 0.$$

2. The form is

$$x^1\bar{x}^1 - \sqrt{\tfrac{1}{3}}(x^1\bar{x}^2 + x^2\bar{x}^1) + x^2\bar{x}^2 - \sqrt{\tfrac{2}{3}}(x^2\bar{x}^3 + x^3\bar{x}^2) + x^3\bar{x}^3 = u_1\bar{u}_1 + u_2\bar{u}_2,$$

where $u_1 = x^1 - \sqrt{\tfrac{1}{3}}x^2$, $u_2 = \sqrt{\tfrac{2}{3}}x^2 - x^3$; thus the mirrors are

$$u_1 = l, \quad \sqrt{\tfrac{1}{3}}u^1 - \sqrt{\tfrac{2}{3}}u^2 = 0, \quad u_2 = 0, \quad \text{or} \quad u_1 = l, \quad u_1 = u_2\sqrt{2}, \quad u_2 = 0.$$

3. We would expect $2\{4\}6\{4\}2$ to have the vertices of two reciprocal $\delta_3^{6,6}$'s. However, as we saw at the end of §12·7, $\delta_3^{6,6}$ coincides with its reciprocal, and has the same vertices as $\delta_3^{2,6}$, $\delta_3^{3,6}$, and $\delta_3^{3,3}$. Hence the vertices of $2\{4\}6\{4\}2$ are these same points, having for their coordinates the pairs of Eisenstein integers.

§13·1

1. By (13·12), (9·89), and (9·891),

$$\cos^2\phi = \cos^2\sigma_1/\sin^2\sigma_2$$

$$= -\frac{\cos\left(\dfrac{\pi}{p_1} - \dfrac{\pi}{p_2}\right) + \cos\dfrac{2\pi}{q_1}}{2\sin\dfrac{\pi}{p_1}\sin\dfrac{\pi}{p_2}} \Bigg/ \frac{\cos\left(\dfrac{\pi}{p_2} + \dfrac{\pi}{p_3}\right) + \cos\dfrac{2\pi}{q_2}}{2\sin\dfrac{\pi}{p_2}\sin\dfrac{\pi}{p_3}}$$

$$= -\frac{\cos\left(\dfrac{\pi}{p_1} - \dfrac{\pi}{p_2}\right) + \cos\dfrac{2\pi}{q_1}}{\cos\left(\dfrac{\pi}{p_2} + \dfrac{\pi}{p_3}\right) + \cos\dfrac{2\pi}{q_2}} \, \frac{\sin\dfrac{\pi}{p_3}}{\sin\dfrac{\pi}{p_1}}.$$

By (9·85) and (12·18),

$$r = 2(2, q_2) \Big/ q_2\left(\frac{1}{p_2} + \frac{1}{p_3} + \frac{2}{q_2} - 1\right).$$

2. By (13·13),
$$\sin^2\phi = 1 - c_1/1 - c_2/\ldots/1 - c_{n-1}$$
$$= \Delta_{12\ldots n}/\Delta_{2\ldots n}.$$

Since the Hermitian form is positive definite, both these determinants are positive. Hence, for a polytope whose vertex figure $p_2\{q_2\}\ldots p_n$ is known, p_1 and q_1 must satisfy the condition

$$\Delta_{12\ldots n} > 0.$$

Similarly, if $p_1\{q_1\}p_2\{q_2\}\ldots p_n$ is a honeycomb,

$$\Delta_{12\ldots n} = 0.$$

§13·2

1. In (13·22), taking $a \equiv 0 \pmod{1+i}$ we get a set of points (in the Argand plane) whose symmetry group includes rotations of period 12 about several points, such as 0 and 2. In (13·23) and (13·24) we have a similar situation, as the coefficients run over the Eisenstein integers. In (13·25), since $\tau\sqrt{3}$ is irrational, a dense set on the real line is already obtained by letting a and b run over the set of ordinary integers.

2. In the notation of §13·1, Ex. 2,

$$1-c_1/1-c_2/1-c_3 = \Delta_{1234}/\Delta_{234},$$
$$1-c_3/1-c_2/1-c_1 = \Delta_{1234}/\Delta_{123}.$$

3. Since $3\{4\}2\{3\}2$ and $3\{3\}3\{4\}2$ are 3-symmetric and 6-symmetric (see §12·2 and §12·4), these apeirogons $p_1\{q\}r$ are

$$3\{6\}3 \quad \text{and} \quad 3\{4\}6.$$

§13·3

1.

```
1   1   1   1   1   1   1   1
  1   2   2   2   2   1   5   1
    1   3   3   3   1   4   4   1
      1   4   4   1   3   3   3   1
        1   5   1   2   2   2   2   1
          1   1   1   1   1   1   1   1
```

2. $$\frac{{}_0R}{l} = \sqrt{\frac{(-1,1)\,(0,4)}{(-1,4)\,(0,1)}} = \sqrt{\frac{\sqrt{3}\times\sqrt{3}}{1\times 1}} = \sqrt{3},$$

$$\frac{{}_1R}{l} = \sqrt{\frac{(-1,0)\,(1,4)}{(-1,4)\,(0,1)}} = \sqrt{\frac{1\times 2}{1\times 1}} = \sqrt{2},$$

$$\frac{{}_2R}{l} = \sqrt{\frac{(-1,0)\,(-1,1)\,(2,4)}{(-1,2)\,(-1,4)\,(0,1)}} = \sqrt{\frac{3}{2}}, \quad \frac{{}_3R}{l} = \sqrt{\frac{(-1,0)\,(-1,1)\,(3,4)}{(-1,3)\,(-1,4)\,(0,1)}} = 1.$$

3. A regular hexagon with all its diagonals and diameters.

4.

```
1   1   1   1   1   1   1   1
  3   1   3   1   3   1   3   ...
2   2   2   2   2   2   2   2
  1   3   1   3   1   3   1   ...
1   1   1   1   1   1   1   1
```

5. The given sequence of values of $\cos^2 \sigma_\nu$ yields

$$\sigma_1 = \sigma_2 = \tfrac{1}{2}\pi-\kappa, \quad \sigma_3 = \tfrac{1}{4}\pi, \quad \sigma_4 = \tfrac{1}{3}\pi, \quad \sigma_5 = \tfrac{1}{4}\pi.$$

By (13·32) with $n = 3$, we find $\phi = \psi = \frac{1}{4}\pi$, $\chi = \frac{1}{3}\pi$.

6.

```
 1   1   1   1   1            1   1   1   1   1
√2  √2  √2  √2  ...    or     2   1   2   1   ...
 1   1   1   1   1            1   1   1   1   1
```

§13·4

1. It becomes the symmetric group $\mathfrak{S}_n$, generated by the transpositions $(1_1 1_2), (1_2 1_3), \ldots, (1_{n-1} 1_n)$.

2. Since $2\{3\}2\ldots\{3\}2$ is the regular simplex α_n, $2[3]2\ldots[3]2$ is the symmetric group $\mathfrak{S}_{n+1}$.

3. Assuming (13·46) and defining $R = R_1 R_2 \ldots R_n$, we deduce

$$RR_1R^{-1} = R_1R_2R_1R_2{}^{-1}R_1{}^{-1} = R_2,$$
$$RR_2R^{-1} = R_1R_2R_3R_2R_3{}^{-1}R_2{}^{-1}R_1{}^{-1} = R_1R_3R_1{}^{-1} = R_3,$$

and so on. Thus $R_{\nu+1} = R^\nu R_1 R^{-\nu}$ $(\nu = 1, \ldots, n-1)$ and

$$R = R_1 \cdot RR_1R^{-1} \cdot R^2R_1R^{-2} \ldots R^{n-1}R_1R^{1-n} = (R_1R)^n R^{-n}.$$

Moreover, R_1 commutes with $R^\nu R_1 R^{-\nu}$ $(\nu > 1)$ and therefore also with $R^{-\nu}R_1R^\nu$.

Conversely, assuming (13·47) and defining $R_{\nu+1} = R^\nu R_1 R^{-\nu}$, we deduce $R = (R_1R)^nR^{-n} = R_1R_2\ldots R_n$. Since R^{n+1} is central, the values of ν for which R_1 commutes with $R^{-\nu}R_1R^\nu$ extend all the way from 2 to $n-1$; therefore R_μ commutes with R_ν whenever $\mu \leqslant \nu-2$. Finally,

$$R_2R_1R_2 = R_2R(R_3\ldots R_n)^{-1} = RR_1(R_3\ldots R_n)^{-1}$$
$$= R(R_3\ldots R_n)^{-1}R_1 = R_1R_2R_1,$$

and it follows easily that $R_{\mu+1}R_\mu R_{\mu+1} = R_\mu R_{\mu+1} R_\mu$ $(\mu \leqslant n-1)$.

§13·5

By (13·53), $m_\lambda = k_\lambda r - 1$, where k_λ is an integer. By (13·52), since $\epsilon^\lambda = 1$, the characteristic roots of $R^{h/r}$ are all equal:

$$\epsilon^{m_\lambda h/r} = \epsilon^{h\,k_\lambda - 1\cdot r)} = \epsilon^{-h/r}.$$

§13·6

1. Since these groups are real, the characteristic roots ϵ^{m_λ} occur in complex conjugate pairs:

$$m_1+m_n = m_2+m_{n-1} = \ldots = h \quad (n = 3 \text{ or } 4).$$

For [3, 5], $h = 10$,

$$m_1 = 1, \quad m_2 = 5, \quad m_3 = 9, \quad g = 2\times 6\times 10 = 120.$$

For [3, 4, 3], $h = 12$,

$$m_1 = 1, \quad m_2 = 5, \quad m_3 = 7, \quad m_4 = 11, \quad g = 2\times 6\times 8\times 12 = 1152.$$

For [3, 3, 5], $h = 30$,

$$m_1 = 1, \quad m_2 = 11, \quad m_3 = 19, \quad m_4 = 29, \quad g = 2\times 12\times 20\times 30 = 14400$$

(see Coxeter 1963, pp. 221, 226).

2. Since

$$\sin\left(\frac{m_\lambda\pi}{h}+\frac{\pi}{p}\right) = \pm\cos\frac{\pi}{p} \quad \text{and} \quad \frac{1}{h} = \frac{1}{p}+\frac{1}{q}-\frac{1}{2},$$

we have, for $p[q]p$ (that is, for 3[3]3, 3[4]3, 4[3]4, 3[5]3, or 5[3]5),

$$\frac{m_\lambda}{h}+\frac{1}{p} = \frac{1}{2}+\frac{1}{q} \quad \text{or} \quad \frac{3}{2}-\frac{1}{q}$$

$$= \frac{2}{q}-\frac{1}{h}+\frac{1}{p} \quad \text{or} \quad 1-\frac{1}{h}+\frac{1}{p}.$$

Thus $$m_1+1 = \frac{2h}{q}, \quad m_2+1 = h.$$

3. By (13·681), $\dfrac{m_\lambda}{h} = \dfrac{1}{6}+\dfrac{\lambda}{n+1} = \dfrac{\lambda+1}{n+1} - \dfrac{5-n}{6(n+1)}$. Since $5-n$ divides 6, it follows that

$$h = \frac{6(n+1)}{5-n}, \quad \frac{m_\lambda+1}{h} = \frac{\lambda+1}{n+1}, \quad r = \frac{h}{n+1} = \frac{6}{5-n},$$

and this value for r can be substituted in (13·69).

§13·7

1. By (13·691),

$$N_\mu = (n+1)!\left(\frac{6}{5-n}\right)^n \bigg/ (\mu+1)!\left(\frac{6}{5-\mu}\right)^\mu (n-\mu)!\left(\frac{6}{6-n+\mu}\right)^{n-\mu-1}.$$

2. In 3[3]3[3]3[3]3[3]3, the order of the subgroup generated by all the R_ν except $R_{\mu+1}$ is

$$(\mu+1)!\left(\frac{6}{5-\mu}\right)^\mu (5-\mu)!\left(\frac{6}{\mu+1}\right)^{4-\mu} = \frac{6^4(\mu+1)!\,(4-\mu)!}{(5-\mu)^{\mu-1}\,(\mu+1)^{4-\mu}}.$$

The desired expression is $6^4 5!$ divided by this number.

§13·8

1. Using Dickson's symbols 1, ..., 9 for the nine inflexions (Miller, Blichfeldt and Dickson 1916, p. 335) and comparing (12·34) with (12·38), we obtain

$$R_1 = (4\ 5\ 6)\ (7\ 9\ 8), \quad R_2 = (2\ 8\ 5)\ (3\ 6\ 9), \quad R_3 = (1\ 2\ 3)\ (4\ 6\ 5),$$

whence $$R = R_1R_2R_3 = (1\ 2\ 8\ 7)\ (3\ 5\ 9\ 4).$$

2. For R_1 we can quote (12·58); also

$$R = R_1R_2R_3R_4 = (011\ 302\ 301\ 201\ 330)\ (012\ 202\ 230\ 120\ 023)$$
$$\cdot(021\ 303\ 110\ 022\ 203)\ (031\ 210\ 032\ 310\ 103)\ (033\ 101\ 320\ 102\ 220).$$

§13·9

1. Since R_1 interchanges u_1^3 and u_2^3 while leaving u_3^3 unchanged, it transforms $\mathfrak{C}_9$ into $-\mathfrak{C}_9$.

2. Leopold Flatto (*American Journal of Mathematics* **92** (1970), 552–61) has observed that one basic set of invariants for a real group $[p, q, \ldots]$ consists of the n forms

$$\sum_{\lambda=1}^{N} (x_1^{(\lambda)}u_1+\ldots+x_n^{(\lambda)}u_n)^{m_\nu+1} \quad (\nu = 1, \ldots, n),$$

where the summation is over the N vertices $(x_1^{(\lambda)}, \ldots, x_n^{(\lambda)})$ $(\lambda = 1, \ldots, N)$ of the regular polytope $\{p, q, \ldots\}$. For instance, in the case of $\{3, \ldots, 3, 4\}$ we have $x_\mu^{(\lambda)} = \delta_\mu^\lambda$ and $m_\nu+1 = 2\nu$, so that the forms for $[3, \ldots, 3, 4]$ are

$$\sum_{\lambda=1}^{2n} u_\lambda^{2\nu} \quad (\nu = 1, \ldots, n).$$

Applying this procedure to the icosahedron whose vertices are cyclic permutations of $(0, \pm\tau^{\frac{1}{2}}, \pm\tau^{-\frac{1}{2}})$, one easily finds for [3, 5] the forms

$$\Sigma u_1^2 = u_1^2+u_2^2+u_3^2,$$
$$2\Sigma u_1^6+3\sqrt{5}\,\tau\Sigma u_1^4u_2^2+3\sqrt{5}\,\tau^{-1}\,\Sigma u_1^2u_2^4,$$
$$\sqrt{5}\,\Sigma u_1^{10}+9\tau^3\Sigma u_1^8u_2^2+42\tau\Sigma u_1^6u_2^4+42\tau^{-1}\Sigma u_1^4u_2^6+9\tau^{-3}\Sigma u_1^2u_2^8,$$

where $\Sigma u_1^4u_2^2$ means $u_1^4u_2^2+u_2^4u_3^2+u_3^4u_1^2$, and so on, the subscripts being cyclically permuted.

Using the permutations of $(\pm 1, \pm 1, 0, 0)$, one finds similarly, for [3, 4, 3],

$$\Sigma u_\lambda^2,$$
$$\Sigma u_\lambda^6+5\Sigma u_\lambda^4u_\mu^2,$$
$$3\Sigma u_\lambda^8+28\Sigma u_\lambda^6u_\mu^2+70\Sigma u_\lambda^4u_\mu^4,$$
$$\Sigma u_\lambda^{12}+22\Sigma u_\lambda^{10}u_\mu^2+165\Sigma u_\lambda^8u_\mu^4+308\Sigma u_\lambda^6u_\mu^6,$$

where $\Sigma u_\lambda^a u_\mu^b$ means a sum of six terms if $a = b$, twelve terms if $a \neq b$.

Bibliography

Friedrich Bachmann, 1959. *Aufbau der Geometrie aus dem Spiegelungsbegriff*. Springer, Berlin.

H. F. Baker, 1946. *A locus with 25920 linear self-transformations*. Cambridge University Press, London.

W. W. Rouse Ball, 1967. *Mathematical Recreations and Essays* (11th ed.). Macmillan, London; 12th ed., University of Toronto Press, 1974.

C. T. Benson and L. C. Grove, 1971. *Finite Reflection Groups*. Bogden and Quigley, Tarrytown-on-Hudson.

Garrett Birkhoff and Saunders MacLane, 1965. *A Survey of Modern Algebra* (3rd ed.). Collier–Macmillan, New York.

Wilhelm Blaschke, 1954. *Projektive Geometrie* (3rd ed.). Birkhäuser, Basel.

Nicolas Bourbaki, 1968. *Groupes et Algèbres de Lie*, Chapitres 4, 5, 6. Hermann, Paris.

William Burnside, 1911. *Theory of Groups of Finite Order* (2nd ed.). Cambridge University Press, London. (Also Dover, New York, 1955.)

G. K. Chesterton, 1912. *Manalive*. Nelson, London.

W. K. Clifford, 1882. *Mathematical Papers*. Macmillan, London.

H. S. M. Coxeter, 1963. *Regular Polytopes* (2nd ed). Collier–Macmillan, New York; 3rd ed., Dover, New York, 1973.

1965. *Non-Euclidean Geometry* (5th ed.). University of Toronto Press.

1968. *Twelve Geometric Essays*. Southern Illinois University Press, Carbondale.

1969. *Introduction to Geometry* (2nd ed.). Wiley, New York.

1970. *Twisted Honeycombs*. Regional Conference Series in Mathematics, Number 4, American Mathematical Society.

H. S. M. Coxeter and S. L. Greitzer, 1967. *Geometry Revisited*. Random House, New York.

H. S. M. Coxeter and W. O. J. Moser, 1972. *Generators and Relations for Discrete Groups* (3rd ed.). Springer, Berlin.

H. M. Cundy and A. P. Rollett, 1961. *Mathematical Models* (2nd ed.). Oxford University Press, London.

Patrick Du Val, 1964. *Homographies, Quaternions and Rotations*. Oxford University Press, London.

László Fejes Tóth, 1964. *Regular Figures*. Pergamon, New York.

Branko Grünbaum, 1967. *Convex Polytopes*. Interscience, New York.

R. C. Gunning, 1962. *Lectures on Modular Forms*. Princeton University Press.

Marshall Hall, 1959. *The Theory of Groups*. Collier–Macmillan, New York.

P. R. Halmos, 1958. *Finite-dimensional Vector Spaces* (2nd ed.). Van Nostrand, New York.

G. H. Hardy, 1967. *A Mathematician's Apology* (reprinted). Cambridge University Press, London.

G. H. Hardy and E. M. Wright, 1960. *Theory of Numbers* (4th ed.). Clarendon Press, Oxford.

Archibald Henderson, 1911. *The Twenty-seven Lines upon the Cubic Surface*. Cambridge University Press, London.

David Hilbert and S. Cohn-Vossen, 1952. *Geometry and the Imagination*. Chelsea, New York. Translation of *Anschauliche Geometrie*, Berlin, 1973.

E. W. Hobson, 1925. *A Treatise on Plane Trigonometry* (6th ed.). Cambridge University Press, London.

Johannes Kepler, 1611. *The Six-cornered Snowflake*. Oxford University Press, London, 1966.

C. J. Keyser, 1907. *Mathematics, Philosophy, Science and Art*. Columbia University Press, New York.

Felix Klein, 1913. *Lectures on the Icosahedron* (2nd ed.). Kegan Paul, London.

J. L. Lagrange, 1795. *Lectures on Elementary Mathematics* (2nd ed.). Open Court, Chicago, 1901. Translation of *Leçons élémentaires sur les mathématiques* in Vol. VII of *Oeuvres de Lagrange*, Gauthier-Villars, Paris, 1877.

Horace Lamb, 1920. *Higher Mechanics*. Cambridge University Press, London.

H. P. Manning, 1914. *Geometry of Four Dimensions*. Macmillan, New York.

G. A. Miller, H. F. Blichfeldt and L. E. Dickson, 1916. *Theory and Applications of Finite Groups*. Wiley, New York.

Ludwig Schläfli, 1950. *Gesammelte Mathematische Abhandlungen, I*. Birkhäuser, Basel.

1953. *Gesammelte Mathematische Abhandlungen, II*. Birkhäuser, Basel.

P. H. Schoute, 1902. *Mehrdimensionale Geometrie, I.* Göschen, Leipzig.
1905. *Mehrdimensionale Geometrie, II.* Göschen, Leipzig.

Beniamino Segre, 1942. *The Non-singular Cubic Surfaces.* Clarendon Press, Oxford.
1961. *Lectures on Modern Geometry.* Cremonese, Rome.

Herbert Seifert and William Threlfall, 1947. *Lehrbuch der Topologie.* Chelsea, New York.

V. I. Smirnoff, 1964. *A Course of Higher Mathematics*, vol. v. Pergamon, Oxford.

D. M. Y. Sommerville, 1929. *Geometry of n Dimensions.* Methuen, London.

S. P. Thompson, 1910. *Life of William Thomson, Baron Kelvin of Largs.* Macmillan, London.

Oswald Veblen and J. W. Young, 1910. *Projective Geometry, I.* Ginn, Boston, Mass.
1918. *Projective Geometry, II.* Ginn, Boston, Mass.

B. L. van der Waerden, 1950. *Modern Algebra, II.* Ungar, New York.
1961. *Science Awakening*, Oxford University Press, London.

Magnus J. Wenninger, 1971. *Polyhedron Models.* Cambridge University Press, London.

Hermann Weyl, 1952. *Symmetry.* Princeton University Press.

Index